Lecture Notes in Mechanical Engineering

Lecture Notes in Mechanical Engineering (LNME) publishes the latest developments in Mechanical Engineering—quickly, informally and with high quality. Original research or contributions reported in proceedings and post-proceedings represents the core of LNME. Volumes published in LNME embrace all aspects, subfields and new challenges of mechanical engineering.

To submit a proposal or request further information, please contact the Springer Editor of your location:

Europe, USA, Africa: Leontina Di Cecco at Leontina.dicecco@springer.com
China: Ella Zhang at ella.zhang@springer.com
India: Priya Vyas at priya.vyas@springer.com
Rest of Asia, Australia, New Zealand: Swati Meherishi at swati.meherishi@springer.com

Topics in the series include:

- Engineering Design
- Machinery and Machine Elements
- Mechanical Structures and Stress Analysis
- Automotive Engineering
- Engine Technology
- Aerospace Technology and Astronautics
- Nanotechnology and Microengineering
- Control, Robotics, Mechatronics
- MEMS
- Theoretical and Applied Mechanics
- Dynamical Systems, Control
- Fluid Mechanics
- Engineering Thermodynamics, Heat and Mass Transfer
- Manufacturing Engineering and Smart Manufacturing
- Precision Engineering, Instrumentation, Measurement
- Materials Engineering
- Tribology and Surface Technology

Indexed by SCOPUS, EI Compendex, and INSPEC.

All books published in the series are evaluated by Web of Science for the Conference Proceedings Citation Index (CPCI).

To submit a proposal for a monograph, please check our Springer Tracts in Mechanical Engineering at https://link.springer.com/bookseries/11693.

Faiz Ahmad · Taib Iskandar · Khairul Habib
Editors

ICREEM 2022

Proceedings of the International Conference on Renewable Energy and E-mobility

Springer

Editors
Faiz Ahmad
Department of Mechanical Engineering
Universiti Teknologi Petronas
Seri Iskandar, Malaysia

Taib Iskandar
Department of Mechanical Engineering
Universiti Teknologi Petronas
Seri Iskandar, Malaysia

Khairul Habib
Department of Mechanical Engineering
Universiti Teknologi Petronas
Seri Iskandar, Malaysia

ISSN 2195-4356 ISSN 2195-4364 (electronic)
Lecture Notes in Mechanical Engineering
ISBN 978-981-99-5945-7 ISBN 978-981-99-5946-4 (eBook)
https://doi.org/10.1007/978-981-99-5946-4

This Springer imprint is published by the registered company Springer Nature Singapore Pte Ltd.
The registered company address is: 152 Beach Road, #21-01/04 Gateway East, Singapore 189721, Singapore

Paper in this product is recyclable.

Contents

Materials and Automotive

About the Editors

Faiz Ahmad is Professor of Universiti Teknologi PETRONAS, Malaysia. His Ph.D. is from Brunel University, UK, and his B.Sc. in Metallurgy & Materials Science Engineering from Punjab University, Pakistan. He worked as a Principal Research Officer for the Pakistan government in his early career. After that, he joined the National University of Sciences and Technology, Islamabad. Since 2003, he has been with his present employer, and his major research secured $9 million in research grants. He published 335 articles, 03 books, and 11 patents. He is a Fellow of two technical societies and a Professional Member of several societies. He is a Member of three committees of ASTM on developing standards.

Taib Iskandar is currently an Associate Professor at Mechanical Engineering Department, Universiti Teknologi PETRONAS, Malaysia. Previously, he worked at Mechanical Engineering Technology Department, Yanbu Industrial College, and was Head of the Research Project Unit at Yanbu Research Center, both under the Royal Commission of Yanbu—Colleges and Institutes (RCYCI) Division, Kingdom of Saudi Arabia. Additionally, he is entrusted to lead RCYCI's Renewable Energy Center and is an Editorial Board Member for Yanbu Journal of Engineering and Science. Previously, he worked at the National University of Malaysia (UKM) as a Senior Lecturer in the Department of Mechanical and Materials Engineering between 1996 and 2015. He was also Founding Deputy Head and Associate Senior Fellow at the Centre for Automotive Research. He completed his Ph.D. at Cranfield University (2006), M.Eng. at Vanderbilt University (1999), and BSME at University of Arizona University (1996). His research interests lie in natural gas engines, direct fuel injection, and engine component design. His continuing interest in advanced engines includes TCDI, RCCI, and GCI engine. His secondary interest is solar thermal engineering, particularly in the development of a solar hot water system using polymer thermal absorbers. To date, he has completed three funded research projects with about MYR980,000 total funding from the government of Malaysia. In 2013, his SAR2 million project proposal to Saudi Arabia's NSTIP funding on turbocharged direct injection gas engine in collaboration with Clean Combustion Research Center at KAUST and Water & Energy Research Institute at KACST.

He has more than 70 technical publications in refereed journals, chapters, a book, and various conference proceedings. He is a Regular Reviewer for reputable journals in the energy area—Energy Conversion and Management, Fuel, Renewable and Sustainable Energy Review, Journal of Natural Gas Science and Engineering, Proceedings of IMechE—Journal of Automobile Engineering, Applied Thermal Engineering and many more. His teaching specialities are thermodynamics, power plant engineering, renewable energy systems, and internal combustion engines. In 2010, he received an "Excellent Educator" Award while in UKM. In the same year, he led a team of students to win the Gasoline Alternative Category awards in the Shell Eco-Marathon Asia 2010 with their LPG-converted engine with 479 km/liter.

Khairul Habib obtained his B.Sc. (Hons.) and M.Eng. degrees from Bangladesh University of Engineering and Technology and National University of Singapore in 2002 and 2005, respectively. He received his Ph.D. in 2009 from the Kyushu University, Japan. He worked as a Research Scientist in Solar Energy Research Institute of Singapore prior to joining the Mechanical Engineering Department of Universiti Teknologi PETRONAS in 2011 as a Senior Lecturer. His main research interests are thermally powered sorption systems, nanofluid enhanced heat transfer, zero energy building, and energy efficiency assessment. He has published more than 95 articles in peer reviewed journals and international conference proceedings. He is a registered Chartered Engineer from the Institute of Mechanical Engineers (IMechE), UK and a member of the American Society of Mechanical Engineers (ASME).

Renewable Energy

Application of Circuit Model for Early Fault Detection

Easter Joseph, Balbir Singh Mahinder Singh, and Dennis Ling Chuan Ching

Abstract A Matlab/Simulink-based simulation study of the PV string/array is conducted that focuses on the current parameters. The paper illustrates the variation of current parameters of the PV string and PV array according to two environmental conditions: solar radiation and temperature. The photocurrent of the PV string and current output of the PV array is highly sensitive to the amount of solar radiation while both saturation and reverse saturation currents escalate with the temperature. The percentage of current output reduction is categorized into three groups which correspond to the amount of solar radiation absorbed by the PV array. The simulation results are used as the performance indicator in developing a simple fault detection algorithm.

Keywords PV string · PV array · Simulink · Algorithm · Fault detection

1 Introduction

The development of renewable energy progressed significantly as an alternative to depleting fossil fuels. With the increasing population size and the global energy demand, solar particularly is the best solution to provide and meet the growing energy demand. Recent research has seen remarkable growth in solar photovoltaic (PV) installation for both grid-connected and remote electrification [1]. Off-grid PV is an effective way of supplying electricity in isolated areas since extending the

E. Joseph (✉) · B. S. M. Singh · D. L. C. Ching
Department of Fundamental and Applied Sciences (FASD), Universiti Teknologi PETRONAS, 32610 Bandar Seri Iskandar, Perak, Malaysia
e-mail: easterjoseph@gmail.com

B. S. M. Singh
e-mail: balbir@utp.edu.my

D. L. C. Ching
e-mail: dennis.ling@utp.edu.my

F. Ahmad et al. (eds.), *ICREEM 2022*, Lecture Notes in Mechanical Engineering,
https://doi.org/10.1007/978-981-99-5946-4_1

national grid to these locations involves extreme cost production and is technically difficult.

Off-grid PV also known as a standalone system electricity generating system (SEGS) is an independent system, which requires a proper monitoring approach, as well as a self-health check to fully exploit the usage of the system. Electricity generation is highly dependent on the performance of the PV system and therefore affects the investment in the system. The major factors that may constantly degrade the performance of PV modules are solar radiation, temperature, shading, dirt or soiling, and the modules' tilt angle and orientation [2–4]. The occurrence of faults caused by these factors could prompt a high amount of energy loss leading to critical financial loss as the faults are difficult to observe and encompass. For this reason, energy prediction in a standalone PV system is essential through developing an algorithm to facilitate the detection of possible faults [5]. Recent technology developments have improved the knowledge of energy prediction, especially for PV systems with new monitoring and diagnostic devices, improving rapidly, yet the employment of advanced surveillance systems in the small-scale SEGS is not practical and uneconomical [6]. Those existing monitoring systems that are available in the market are such as supervisory control and data acquisition (SCADA), DT-80 Data taker, seaward solar data logger, and Laboratory Virtual Instrumentation Engineering Workbench (LabVIEW) data acquisition software. The similarities of these monitoring systems are their complexity and adaptability to the gird connected-PV system which is not suitable to be applied in a small-scale standalone SEGS.

There are two main fault detection methods for PV fault diagnosis: a process history-based approach and a model-based approach. This paper focuses on the model-based approaches that compare the analytically computed outputs with measured values and prompt a signal as an alarm [7]. Tina et al. proposed a fault detection method by evaluating the absolute performance ratio error (APRE) with a threshold which afterward generates a diagnostic signal. The DC-AC power ratio as an indicator was used to detect the failure that occurred in the inverter [8]. Dhoke et al. suggested generating fault indicator signals called 'residual' to automatically detect and locate intra-string line-line faults in a large-scale PV system. The residual-based outlier detection is controlled by the pre-defined threshold, which means any anomaly outside the boundary of the threshold indicates a fault. Consequently, the fault location algorithm which uses regression-based expressions estimates the fault location within the string [9]. Other studies used an automatic monitoring and fault detection system based on power loss analysis where the researchers compared the monitored data with the simulation results and the processed data was used to generate a faulty signal [10]. Further to this, some studies were using fault detection methods based on module level or array level. Each PV module or PV array was installed with local sensors to measure either temperature, irradiance, maximum power point (MPPT), or energy production. The data obtained from the measured values were then compared with the estimated values to detect faults [11, 12]. Despite these research contributions, the application of monitoring and diagnostic systems in small PV plants still needs to be investigated in depth and most of the studies were centered on the on-grid and large-scale PV plants.

The objective of this research paper is to develop a simple algorithm for fault detection based on the photocurrent value at the PV string level and the overall current output of the PV array. This research is simulated using Matlab/Simulink.

2 Methodology

In this simulation, two commercial PV modules were connected in series to form a string where each module has 36 cells. Four strings then were connected in parallel to form an array as shown in Fig. 1. The equivalent circuit of a PV cell obtained in [13] is comprised of a photo-current source (I_{ph}) in parallel with a single diode (D), a shunt resistor (R_{sh}) and a series resistor (R_s).

The photovoltaic panel can be modeled mathematically as given in Eq. (1)–(5) [13].

Module photocurrent, I_{ph}:

$$I_{ph} = [I_{sc} + k_i(T - 298)]\frac{G}{1000} \tag{1}$$

The I_{ph} is linearly dependent on the solar irradiation, G, and temperature, T. Module diode reverses saturation current, I_{rs}:

$$I_{rs} = \frac{I}{\left[e^{\left(\frac{qV_{oc}}{nN_sKT}\right)-1}\right]} \tag{2}$$

Module saturation current, I_0:

Fig. 1 An array made up of 4 strings

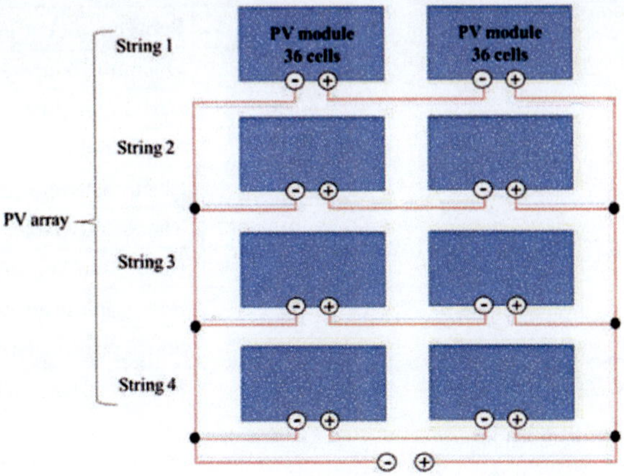

$$I_0 = I_{rs}\left(\frac{T}{T_n}\right)^3 exp\left[\frac{qE_g\left(\frac{1}{T_n} - \frac{1}{T}\right)}{nK}\right] \quad (3)$$

Module I_0 varies with cell temperature as shown in Table 3.
Module shunt resistance, I_{sh}:

$$I_{sh} = \left(\frac{V + IR_s}{R_{sh}}\right) \quad (4)$$

R_{sh} is the equivalent parallel resistance that is caused by the current leakages, tunnel effect, breakdown by microplasmas, leaks along surface channels, etc. Normally, the value of R_{sh} is generally high meanwhile R_s is very small. The V_{oc} and fill factor will reduce when the R_{sh} is particularly small.

Module Current output, I:

$$I = I_{ph} - I_D - I_{sh}$$
$$I = N_p I_{ph} - N_p I_0\left[exp^{\left(\frac{q(V+R_S)}{nKN_sT}\right)} - 1\right] - I_{sh} \quad (5)$$

The descriptions of each mathematical symbol are depicted in Table 1 while the datasheets of the reference PV module for this simulation are given in Table 2.

Using all the mathematical equations mentioned above, five (5) subsystems were created to calculate I_{ph}, I_0, I_{sh}, I_{rs}, and I. All the subsystems are interconnected to produce a PV array model as illustrated in Fig. 2. This paper focuses on three stages to develop a simple fault detection algorithm as displayed in Table 3.

Table 1 Details of mathematical symbols

Symbol	Name	Value
ki	Short circuit current of a cell	0.0032
T	Operating temperature (K)	T
Tn	Nominal temperature (K)	298
G	Solar irradiance (W/m2)	1000
q	Electron charge (C)	1.6×10^{-19}
n	The ideality factor of the diode	1.3
K	Boltzmann's constant (J/K)	1.38×10^{-23}
Eg	Band gap of semiconductor (eV)	1.1
Rs	Series resistance (Ω)	0.1
Rsh	Shunt resistance (Ω)	500

Table 2 Electrical characteristics data of reference PV module

Symbol	Name	Value
I_{sc}	Short circuit current (A)	8.21
V_{oc}	Open circuit voltage (V)	43.92
V_{max}	Voltage at max power (V)	39.58
I_{max}	Current at max power (A)	7.58
P_{max}	Rated power (W)	300
N_s	Number of cells connected in series	72 cells
N_p	Number of PV modules connected in parallel	4

Table 3 Three stages of developing a simple fault detection algorithm

Stage	Descriptions
(1) Photocurrent at the string level	Variation of • Solar radiation: 1000, 700, 500, 200, 100 W/m^2 • Temperature: 25, 30, 40 °C
(2) Saturation and reverse saturation current at the string level	• Temperature varied at 25, 30, 40, 50 & 90 °C • Solar radiation fixed at 1000 W/m^2
(3) Current output at the array level	Variation of • Solar radiation: Increment of 100 W/m^2 until 1000 W/m^2 • Temperature: 25, 30, 40 °C • Percentage current reduction • Current from the datasheet: 32.84 A

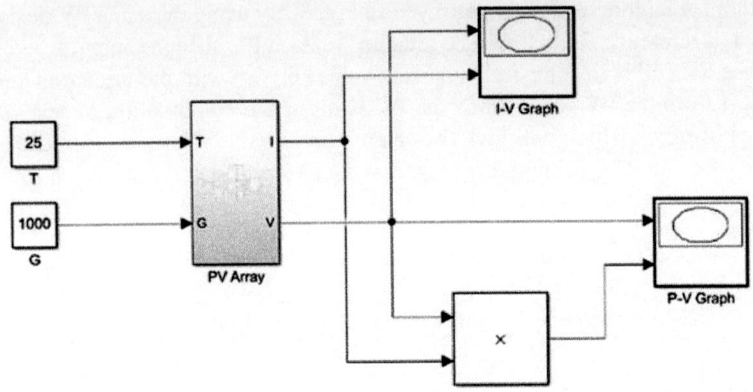

Fig. 2 PV array model

3 Results and Discussion

In the first stage, solar radiation and temperature were varied to observe their effect on I_{ph} both at the string level and the overall I of the PV array. Table 4 depicts the values of I_{ph} that are decreasing as the solar radiation decreases for all temperature levels. On the other hand, the temperature rise does not have much effect on the I_{ph}. Similar results were obtained for the I. However, the values of I were slightly dropped as the temperature approaches 40 °C for all solar radiation levels.

To reveal the reason for the decrement of the I, the second stage addresses the values of I_{rs} and I_0. Since these two parameters are temperature dependent, the temperatures were varied as exhibited in Table 5 while maintaining the solar radiation level at 1000 W/m^2. The table result exposes that both current parameters gradually increased with the temperature which mainly contributes to the high internal carrier recombination in the PV cell [14]. Fundamentally, I rely on the values of I_{ph}, I_0, and I_{rs} as given in (5). The third stage highlights the percentage of I reduction for different solar radiation and temperature values as indicated in Table 6. Despite the working conditions at the STC values, there is about a 1.38% of I reduction in the PV array as a result of heat loss and internal recombination. Based on [15], the data table can be categorized into three groups according to the value of percentage I reduction. The first group is for the percentage I reduction that is less than 10% which reflects about 900–1000 W/m^2 solar radiation. The second group is when the percentage I reduction is approximately 10% to 40% which refers to around 500–900 W/m^2 Solar radiation. The third group corresponds to the percentage I reduction of more than 40% that falls under less than 500 W/m^2 solar radiation.

Based on this framework, crucial information can be extracted to develop a simple algorithm for fault detection as portrayed in Fig. 3. By using the same PV design as in Fig. 1, a reference PV cell will be attached to each PV string. The purpose of the reference PV cell is to compare its current value ($I_{ref.cell}$) with the measured current ($I_{measured}$) from the PV string since the PV string is the combination of several PV modules in series which will give the same current value. If the measured current is lower than the $I_{ref.cell}$, the percentage I reduction will determine which group it

Table 4 Values of photocurrent and output current at different solar radiation and temperature levels

No	Solar radiation (W/m^2)	String (Series)			Array (Parallel)		
		I_{ph} (A)			*I (A)*		
		25 °C	*30 °C*	*40 °C*	*25 °C*	*30 °C*	*40 °C*
1	1000	8.21	8.37	8.69	32.39	32.65	32.04
2	700	5.75	5.86	6.08	22.66	22.86	22.45
3	500	4.11	4.19	4.35	16.16	16.29	15.92
4	200	2.05	2.09	2.17	8.01	8.03	7.62
5	100	0.82	0.84	0.87	3.11	3.06	2.59

Table 5 Values of diode reverse saturation current and saturation at different temperature values

No	Solar radiation (1000 W/m^2)		
	Temperature (°C)	I_{rs} *(A)*	I_0 *(A)*
1	25	9.58×10^{-8}	9.68×10^{-8}
2	30	1.30×10^{-7}	2.37×10^{-7}
3	40	2.32×10^{-7}	1.30×10^{-6}
4	50	3.97×10^{-7}	6.47×10^{-6}
5	90	2.54×10^{-6}	1.67×10^{-3}

Table 6 Percentage of current output reduction at different solar radiation and temperature levels

No	Solar radiation (W/m^2)	Array (Parallel)					
		I (A)					
		25 °C	*% I reduction*	*30 °C*	*% I reduction*	*40 °C*	*% I reduction*
1	1000	32.39	1.38	32.65	0.58	32.04	2.44
2	900	29.15	11.24	29.4	10.48	28.88	12.06
3	800	25.91	21.10	26.13	20.43	25.68	21.80
4	700	22.66	31.00	22.86	30.00	22.45	31.64
5	600	19.41	40.90	19.58	40.38	19.20	41.53
6	500	16.16	50.79	16.29	50.40	15.92	51.52
7	400	12.90	60.72	12.99	60.44	12.61	61.60
8	300	9.64	70.64	9.68	70.52	9.29	71.71
9	200	8.01	75.64	8.03	75.55	7.62	76.80
10	100	3.11	90.53	3.06	90.68	2.59	92.11

belongs to. As reported in [16], the first group or less than 10% I reduction does not necessarily indicate a fault. The possible factors could be generally clouds and rain. Consecutively, the feasible faults for the second group are short circuit faults, shading, and soiling which have a limit of current reduction of about 40%. Shading and soiling occur when the generated photocurrents of the individual cells or modules in series are much lower than the normal cell, resulting in the shaded modules becoming revered biased and allowing the current to flow in the opposite direction. In the case of partial shading, a large fraction of power generated by the normal modules could be dissipated due to high resistance in the diode.

Meanwhile, the major fault in group three is predominantly coming from the hotspot phenomena that can damage the PV system. Hotspot or overheating is a condition when the difference in illumination level between two cells or modules is extremely high that could plummet the energy production of the PV system. Therefore, the utmost concern to be given priority is for groups two and three as they could accelerate the deterioration of the cells and even halt the entire PV system if no rectification steps are being taken [17]. Accordingly, the next step to be considered is to locate the fault at the PV string level. In this step, the $I_{ref.\ cell}$ will be compared again

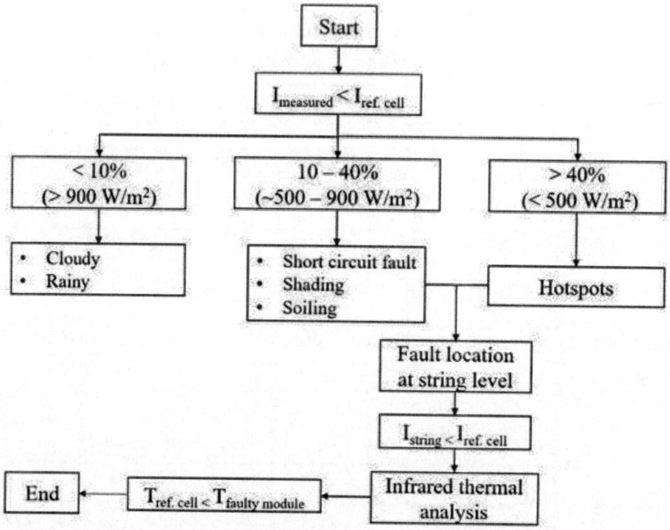

Fig. 3 General algorithm for early fault detection

with the $I_{measured}$ from the faulty PV string. In order to determine which PV module is faulty or not functioning in optimum conditions, the infrared thermal analysis will be used to justify the temperature of the faulty PV module with the reference cell. Subsequently, countermeasures can be taken accordingly once the fault module has been identified.

4 Conclusion

The objective of this study is to develop a simple algorithm based on the photocurrent and current output values obtained from the MATLAB/Simulink simulations. Subsystem blocks were developed to design a PV array model with simple and practical icons and dialogs to produce a meaningful outcome. Both photocurrent and current output are highly sensitive to the amount of solar radiation while saturation and reverse saturation currents escalate with the temperature. The percentage of current output reduction can be categorized into three groups that also reflect the amount of solar radiation absorbed by the PV array. A simple fault detection algorithm is developed based on the categories and their possible faults.

Acknowledgements This research work is supported by the Yayasan Universiti Teknologi PETRONAS (YUTP) (015LC0-294).

References

1. Zhang M, Zhang Q, Zhou D, Wang L (2021) Punishment or reward? Strategies of stakeholders in the quality of photovoltaic plants based on evolutionary game analysis in China. Energy 220:119754
2. Hussin MZ, Hamid MHA, Zain ZM, Ab Rahman R (2010) An evaluation data of solar irradiation and dry bulb temperature at Subang under Malaysian climate, 2010
3. Kaplanis S, Kaplani E (2011) Energy performance and degradation over 20 years performance of BP c-Si PV modules. Simul Model Pract Theory 19:1201–1211, 2011/04/01
4. Maghami M, Hizam H, Gomes C, Hajighorbani S, Rezaei N (2015) Evaluation of the 2013 Southeast Asian Haze on Solar Generation Performance. PLoS One 10:e0135118
5. Rahman MM, Selvaraj J, Rahim NA, Hasanuzzaman M (2018) Global modern monitoring systems for PV based power generation: a review. Renew Sustain Energy Rev 82:4142–4158
6. Drews A, de Keizer AC, Beyer HG, Lorenz E, Betcke J, van Sark W et al (2007) Monitoring and remote failure detection of grid-connected PV systems based on satellite observations 81
7. Harrou F, Nounou MN, Nounou HN, Madakyaru M (2015) PLS-based EWMA fault detection strategy for process monitoring. J Loss Prev Process Ind 36:108–119
8. Chine W, Mellit A, Pavan AM, Kalogirou SA (2014) Fault detection method for grid-connected photovoltaic plants. Renew Energy 66:99–110, 2014/06/01
9. Dhoke A, Sharma R, Saha T (2019) An approach for fault detection and location in solar PV systems. Solar Energy 194:197–208 11/12/2019
10. Davarifar M, Rabhi A, El-Hajjaji A, Dahmane M (2013) Real-time model base fault diagnosis of PV panels using statistical signal processing. Int Conf Renew Energy Res Appl (ICRERA) 2013:599–604
11. Ali MH, Rabhi A, Hajjaji AE, Tina GM (2017) Real time fault detection in photovoltaic systems. Energy Procedia 111:914–923
12. Sun X, Chavali RVK, Alam MA (2018) Real-time monitoring and diagnosis of photovoltaic system degradation only using maximum power point—the Suns-Vmp method. Prog Photovoltaics Res Appl 27:55–66
13. Ahmad Hamdi RT (2017) Solar cell system simulation using Matlab-Simulink. Kurd J Appl Res 2:45–51
14. Yaprak D, Spielberg ET et al, A roadmap to uranium ionic liquids: anti-crystal engineering. Chem—A Eur J 6482–6493
15. Xu X, Wang H, Xu X, Zuo Y (2011) Method for diagnosing photovoltaic array fault in solar photovoltaic system. Asia-Pac Power Energy Eng Conf 2011:1–5
16. Pandiarajan N, Ramaprabha R, Muthu R (2012) Application of circuit model for photovoltaic energy conversion system. Int J Photoenergy 2012:1–14
17. Ramabadran R, Badrilal M (2009) MATLAB based modelling and performance study of series connected SPVA under partial shaded conditions. J Sustain Dev 2:10/20

A Theoretical Study on the Properties of Submicron b-Si Nanostructures for Improved Absorption in b-Si Thermophotovoltaic Cells

Jasman Y. H. Chai, Basil T. Wong, and Saulius Juodkazis

Abstract Black silicon (b-Si) is a nanostructured surface modification with excellent optical properties and is an attractive candidate for optoelectronic applications. In this work, a preliminary simulation study is done to determine the optical properties of b-Si layers on silicon photovoltaic cells for thermophotovoltaic applications. Finite-difference time-domain method was used to simulate the optical responses of b-Si photovoltaic cells. Black silicon nanostructures of about 50 to 650 nm in diameter and 200 to 1800 nm in height were studied. The optical properties improved with decreasing nanostructure diameter and increasing height. The light trapping properties in these nanostructures were discussed and were attributed to the refractive index gradients of the b-Si layer.

Keywords Black silicon · Radiation · Optical simulation · Photovoltaics · Nanotechnology

1 Introduction

Thermophotovoltaic (TPV) systems convert heat energy, often from waste heat sources, to useful electrical energy. In silicon TPV systems, the thermal emitters typically need to be heated to extremely high temperatures, leading to low popularity among researchers. Low band gap materials such as germanium (0.67 eV) are attractive material choices for the photovoltaic cell, but silicon (1.12 eV) has the

J. Y. H. Chai · B. T. Wong (✉)
Department of Mechanical Engineering, School of Engineering, Faculty of Engineering, Computing, and Science, Swinburne University of Technology Sarawak Campus, Kuching, Malaysia
e-mail: twong@swinburne.edu.my

S. Juodkazis
Optical Sciences Center and ARC Surface Engineering and Advanced Materials (SEAM) Center, School of Science, Swinburne University of Technology, Hawthorn, Australia
e-mail: sjuodkazis@swin.edu.au

F. Ahmad et al. (eds.), *ICREEM 2022*, Lecture Notes in Mechanical Engineering,
https://doi.org/10.1007/978-981-99-5946-4_2

advantage of being low-cost and the manufacturing technologies for silicon photovoltaic cells are mature. Furthermore, in some applications, silicon TPV systems with 1–2% efficiencies are sufficient, such as for domestic water heating [2].

Black silicon (b-Si) is a type of surface modification of silicon surfaces. It has a textured surface with silicon nanostructures which can assist light trapping and may potentially improve the efficiency of photovoltaic cells or the performance of photoelectric devices. The textured surface can be produced by various etching methods [3] and is especially useful for photovoltaic applications due to its exceptional light trapping properties and low production cost [4]. In this study, a simulation model was developed to determine theoretically the optimum nanostructure dimensions, and the light trapping properties of b-Si nanostructures are discussed in detail.

In several studies, the b-Si layer was modelled using an effective medium approximation (EMA) or similar techniques [5–9]. In essence, these methods use the volume fraction of silicon in the b-Si layer to calculate an effective refractive index for the layer, and apply ray tracing, transfer matrix method, or finite-difference time-domain (FDTD) simulations to simulate the optical responses of the surfaces. However, the approximation contributes to losses in accuracy especially when fine tuning the properties of the b-Si surface at the nanoscale [10]. In other studies, attempts were made to model the nanostructures using 3D FDTD [8, 11–14], finite element method (FEM) [15, 16], and rigorous coupled wave analysis (RCWA) [17, 18]. Majority of these studies used a rough surface model to represent the random sized nanostructures found in some b-Si surfaces. Utilising a rough surface model can be time and memory consuming, as it often requires modelling of a large number of nanostructures to replicate the b-Si surface. Another approach is to assume the nanostructures have an average size and model just one nanostructure with a periodic cell. This was done by Wang and Leu [14] and they have studied the impact of height for cylindrical and truncated cone nanostructures, but the effects of nanostructure diameter were not explored. In a more comprehensive study by Nguyen et al. [15], they studied the impacts of heights and diameters of conical nanostructures on the reflectance of the b-Si surface. The simulated dimensions are >1 μm in both heights and diameters. Based on the analysis of high-performing b-Si photovoltaic cells in [3], the dimensions of typical b-Si nanostructures are in the submicron range. Hence, in this study, a theoretical study is intended to optimise the dimensions of b-Si nanostructures for improved optical absorption for thermophotovoltaic applications.

2 Methodology

In the submicron length scale, macroscale optical analysis methods like ray tracing are inappropriate as they do not account for the diffraction effects of light in the submicron wavelength range. Hence, in this study, an in-house developed FDTD model [10] was used to simulate the b-Si nanostructures as it accounts for diffraction effects at this length scale. In the FDTD method, the update equations for the propagation of electric and magnetic fields ($\vec{E}$ and $\vec{H}$) in a staggered Yee grid [19] are given by:

$$-\frac{1}{\mu}\vec{\nabla} \times \vec{E} = \frac{\vec{H}\left(t + \frac{\Delta t}{2}\right) - \vec{H}\left(t - \frac{\Delta t}{2}\right)}{\Delta t} \tag{1}$$

$$\frac{1}{\varepsilon}\vec{\nabla} \times \vec{H} = \frac{\vec{E}(t + \Delta t) - \vec{E}(t)}{\Delta t} \tag{2}$$

where μ is the permeability and ε is the permittivity.

These update equations describe the propagation of the electromagnetic waves in a simulation domain with specified material regions. A full description of the FDTD method can be found elsewhere [20, 21]. An illustration of the simulation domain is shown in Fig. 1. Periodic boundary conditions are used in the horizontal directions. On the top and bottom boundaries, the perfectly matched layer (PML) boundary condition is used. The complex refractive index of silicon was obtained from [22]. To obtain the output data, the steady-state field values were obtained by applying discrete Fourier transform onto the time-domain field values. Power flowing through the reflection and transmission planes, and the normalised reflectance (R) and transmittance (T) were calculated using the equations:

$$P_{z,R/T}(\lambda, r) = \frac{1}{2}\mathrm{Re}\big(E_{x,R/T}(\lambda, r)H^{*}_{y,R/T}(\lambda, r) - E_{y,R/T}(\lambda, r)H^{*}_{x,R/T}(\lambda, r)\big) \cdot A(r) \tag{3}$$

$$R/T(\lambda) = \frac{\sum_r P_{z,R/T}(\lambda, r)}{P_{src}(\lambda)} \tag{4}$$

where P is the power flowing through the area A, λ is wavelength, and r is position. The power absorbed can be calculated from the steady-state electric field using the equation:

$$P_A(\lambda, r) = -\frac{1}{2}\left|\vec{E}(\lambda, r)\right|^2 \cdot \sigma(\lambda, r) \cdot V_{cell} \tag{5}$$

and the absorptance can be calculated using Equation (4) of a similar form. Here, σ is the conductivity.

The range of dimensions of the simulated b-Si nanostructures is selected based on the analysis from [3] on the typical b-Si dimensions producible using current technologies. Diameters range from 50–650 nm, and heights range from 200–1800 nm. Nanostructures were modelled as cones and paraboloids and the optical responses were compared for both. Simulation models in the literature assumed the shape of nanostructures to be cylindrical or conical, however, the actual shape of some b-Si nanostructures resembles a paraboloid as well [1]. Figure 2 shows a side-by-side comparison of the paraboloid model used in this study and an SEM image of a real b-Si surface. The thermophotovoltaic spectrum used was from a Yb_2O_3 emitter at

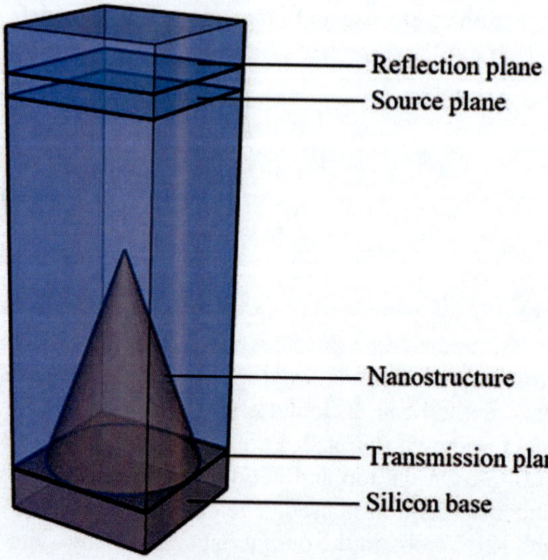

Fig. 1 Illustration of the simulation domain showing a b-Si nanostructure

1735 K [23, 24], which is the typical selective emitter used for silicon thermophotovoltaic applications. The spectrum is shown in Fig. 3. For a more intuitive comparison, the integrated reflectance, transmittance, or absorptance (RTA) with respect to the source spectrum was calculated using the proposed equation:

$$\text{Integrated RTA} = \frac{\int_{300nm}^{1100nm} RTA(\lambda) P_{in}(\lambda) d\lambda}{\int_{300nm}^{1100nm} P_{in}(\lambda) d\lambda} \tag{6}$$

3 Results and Discussion

3.1 Impact of Nanostructure Diameter and Height

For each dimension simulated, a set of RTA curves can be obtained, as shown in Fig. 4. In the figure, the simulated nanostructure is a paraboloid shaped nanostructure with a diameter of 200 nm, height of 800 nm, and period of 200 nm, based on the average dimensions of nanostructures in the work of Savin et al. [1]. The simulation results are in agreement with experimental measurements from their work. To simplify the analysis, the integrated reflectance with respect to the source spectrum was calculated. For the configuration in Fig. 4, the RTA curve can be condensed into integrated RTA values as shown in Table 1.

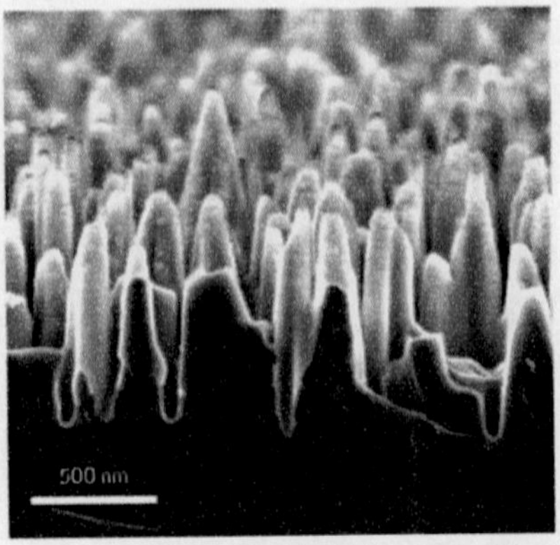

(a)

(b)

Fig. 2 **a** SEM image of a b-Si surface (reprinted by permission from Springer Nature: [1]), and **b** illustration of a paraboloidal nanostructure model

Fig. 3 Spectral power emitted by Yb_2O_3 emitter at 1735 K

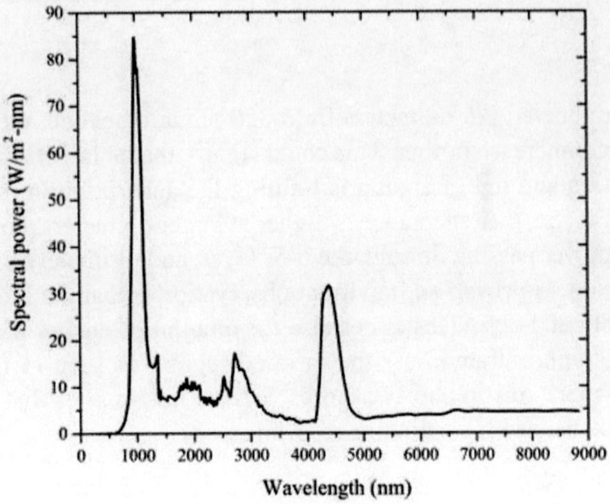

In Fig. 5, the integrated reflectance and absorptance with respect to Yb_2O_3 emission spectrum are shown for different diameters and heights for the b-Si nanostructures. From the figure, the reflectance curve shows that for every diameter size, the integrated reflectance decreases with increasing nanostructure height, and beyond a height of about 800–1000 nm, the reflectance is near constant. When observing the absorptance curve, the integrated absorptance increases linearly with height for every diameter size. This shows that the amount of absorption depends on the height and amount of volume available for absorption, since more height indicates a large volume of material. Increasing diameter also increases the volume of the nanostructure, but the integrated absorptance does not increase linearly with increasing

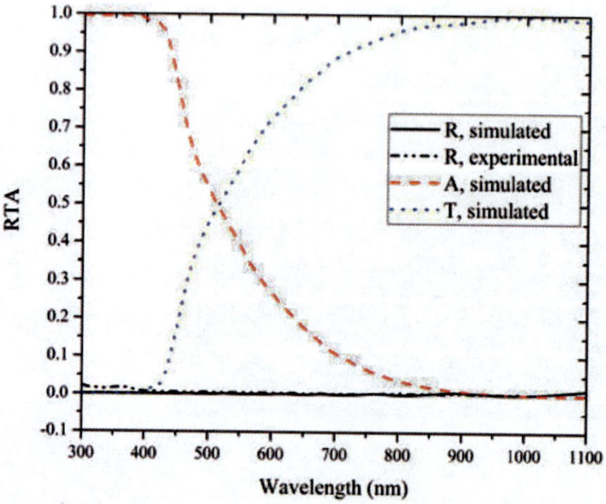

Fig. 4 RTA curve for a b-Si surface. The experimental R was included for comparison

Table 1 Integrated RTA for the configuration in Fig. 4

Integrated R	Integrated T	Integrated A
0.0024%	99.3%	0.61%

diameter. For diameters from 350 nm and beyond, the integrated absorptance does not increase further. This could signify that at larger diameters, the amount of reflection and light trapping is limiting the absorption in the nanostructures. As seen in Fig. 5a, the reflectance is higher at larger diameters, hence decreasing the amount of power passing through the b-Si layer, and ultimately limiting the amount of absorption. However, an important observation is that the integrated absorptance is low at about 1–2%. This is because the amount of energy present in the Yb_2O_3 spectrum is concentrated near the infrared region, as seen in the peak near 800–1100 nm, where silicon poorly absorbs. Hence, this implies that a thick silicon substrate will be needed for optimum absorption.

3.2 *Light Trapping Properties in B-Si Nanostructures*

To discuss more on the light trapping properties in b-Si nanostructures, different nanostructure shapes are modelled and compared with a flat silicon surface. The wavelength-dependent reflectance is shown in Fig. 6. All nanostructures were modelled using the same dimensions: 350 nm diameter, 600 nm height, and 350 nm period. In comparison, while the cylindrical nanostructures reduced the reflectance, it is not as significant as the conical or paraboloidal shaped nanostructure. This can be attributed to the refractive index gradients that the shapes contribute to [3, 25]. An illustration of the refractive index gradients of the different nanostructures is shown in Fig. 7. These refractive index profiles are estimated based on the volume fraction

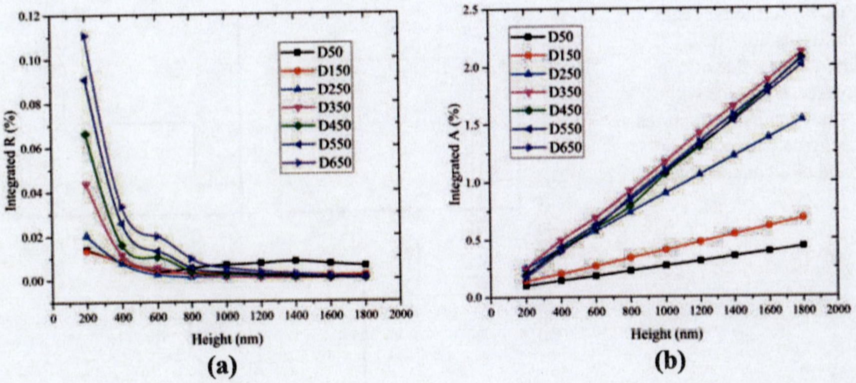

Fig. 5 Integrated **a** reflectance and **b** absorptance for different b-Si nanostructure dimensions. D50 means a diameter of 50 nm and so on

of each material, as described by the Volume Average Theory [3]. A sudden jump in refractive index contributes to Fresnel reflection, hence we observe a large amount of reflection in the flat silicon and surface with cylinder nanostructures. The refractive index profiles for these two surfaces show sudden, steep changes in refractive index across the interface, hence contributing to some Fresnel reflection. In contrast, the conical and paraboloidal nanostructures have a tapered shape, and hence the refractive index increases gradually towards the silicon substrate. Hence, for efficient absorption in thermophotovoltaic cell, it is important to have nanostructures with refractive index gradients to minimise reflection losses and improve the light trapping in the cell.

Fig. 6 Reflectance of b-Si nanostructures modelled using different shapes, compared to a flat Si surface

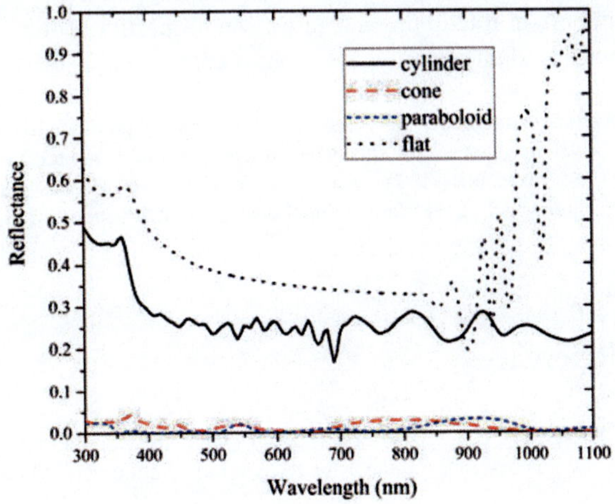

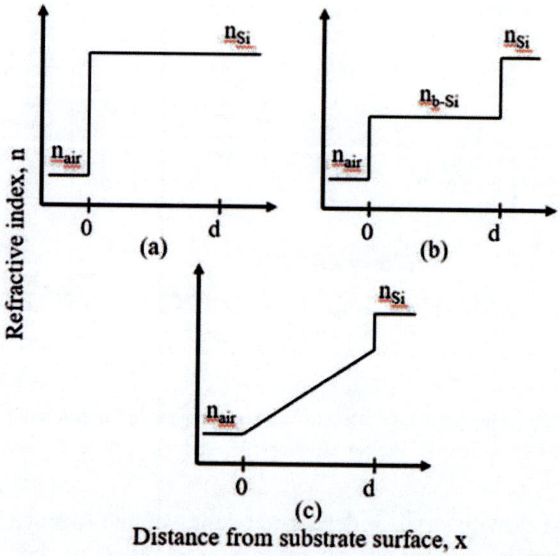

Fig. 7 Refractive index profiles at air-silicon interfaces: **a** flat silicon surface, **b** surface with cylindrical nanostructures, **c** surface with cone/paraboloid nanostructures

4 Conclusion

To sum up, a study on the impact of nanostructure size on the optical properties of a black silicon surface was conducted for silicon thermophotovoltaic applications. The reflectance for a b-Si surface is typically lower for longer nanostructures or deeper etching, and nanostructures with smaller diameters. The surface is able to achieve an integrated reflectance of as low as 0.01%. However, an important point to note is that it is more difficult to passivate the surfaces with decreasing diameters and increasing nanostructure heights. Absorption increases with height and diameter, but is typically low, and majority of the absorption in a b-Si cell for thermophotovoltaic sources will be in the substrate. The shape of the b-Si nanostructures also plays an important role in the anti-reflection properties of the surface. Nanostructure shapes with gradual refractive index gradients are desirable for a low-reflectance surface.

Acknowledgements The authors wish to express gratitude to the Ministry of Education Malaysia (MoE) for sponsoring this work under the Fundamental Research Grant Scheme (FRGS/1/2018/TK07/SWIN/02/1). In addition, Jasman Y. H. Chai is grateful to Swinburne University of Technology Sarawak Campus, Malaysia for partial support of his PhD scholarship.

References

1. Savin H, et al (2015) Black silicon solar cells with interdigitated back-contacts achieve 22.1% efficiency. Nat Nanotechnol 10:624. https://doi.org/10.1038/nnano.2015.89

2. Bitnar B (2003) Silicon, germanium and silicon/germanium photocells for thermophotovoltaics applications. Semicond Sci Technol 18(5):S221–S227. https://doi.org/10.1088/0268-1242/18/5/312
3. Chai JYH, Wong BT, Juodkazis S (2020) Black-silicon-assisted photovoltaic cells for better conversion efficiencies: a review on recent research and development efforts. Mater Today Energy 18:100539. https://doi.org/10.1016/j.mtener.2020.100539
4. Liu X, Coxon P, Peters IM, Hoex B, Cole JM, Fray D (2014) Black silicon: fabrication methods, properties and solar energy applications. Energy Environ Sci 7. https://doi.org/10.1039/c4ee01152j
5. Elsayed AA, Sabry Y, Marty F, Bourouina T, Khalil D (2018) Optical modeling of black silicon using an effective medium/multi-layer approach. Optics Express 26:13443. https://doi.org/10.1364/OE.26.013443
6. Lin C, Povinelli ML (2009) Optical absorption enhancement in silicon nanowire arrays with a large lattice constant for photovoltaic applications. Optics Express 17(22):19371–19381. https://doi.org/10.1364/OE.17.019371
7. Patchett S, Khorasaninejad M, Nixon O, Saini S (2013) Effective index approximation for ordered silicon nanowire arrays. J Opt Soc Am B Opt Phys 30:306. https://doi.org/10.1364/JOSAB.30.000306
8. Rahman T, Boden SA (2017) Optical modeling of black silicon for solar cells using effective index techniques. IEEE J Photovolt 1–7. https://doi.org/10.1109/JPHOTOV.2017.2748900
9. Ravindra N (2015) Modelling of optical properties of black silicon/crystalline silicon. J Sci Ind Metrol 1:1–7
10. Chai JYH, Wong BT, Juodkazis S (2022) Comparison between one-dimensional and three-dimensional optical modelling of ordered black silicon nanostructures using effective medium approach. In: Presented at the 10th international conference on nano and materials science 2022, Singapore, 12–14 February
11. Ma S, et al (2018) A theoretical study on the optical properties of black silicon. AIP Adv 8:035010. https://doi.org/10.1063/1.5018642
12. Rahman T, Bonilla RS, Nawabjan A, Wilshaw PR, Boden SA (2017) Passivation of all-angle black surfaces for silicon solar cells. Sol Energy Mater Sol Cells 160:444–453. https://doi.org/10.1016/j.solmat.2016.10.044
13. Steglich M, Käsebier T, Zilk M, Pertsch T, Kley E-B, Tünnermann A (2014) The structural and optical properties of black silicon by inductively coupled plasma reactive ion etching. J Appl Phys 116(17):173503. https://doi.org/10.1063/1.4900996
14. Wang B, Leu PW (2012) Enhanced absorption in silicon nanocone arrays for photovoltaics. Nanotechnology 23(19):194003. https://doi.org/10.1088/0957-4484/23/19/194003
15. Nguyen K, et al (2012) Study of black silicon obtained by cryogenic plasma etching: approach to achieve the hot spot of a thermoelectric energy harvester. Microsyst Technol 18:1807–1814. https://doi.org/10.1007/s00542-012-1486-0
16. Tyson J, Rahman T, Boden SA (2019) Optical simulation of black silicon surfaces using geometric randomisation and unit-cell based averaging. In: Presented at the 15th photovoltaic science, application and technology conference, University of Warwick, UK
17. Jürgen Bett A, et al. (2016) Wave optical simulation of the light trapping properties of black silicon surface textures. Optics Express 24:A434–A445. https://doi.org/10.1364/OE.24.00A434
18. Tucher N, Gebrewold HT, Blasi B (2018) Field stitching approach for the wave optical modeling of black silicon structures. Opt Express 26(22):A937–A945. https://doi.org/10.1364/OE.26.00A937
19. Kane Y (1966) Numerical solution of initial boundary value problems involving maxwell's equations in isotropic media. IEEE Trans Antennas Propag 14(3):302–307. https://doi.org/10.1109/TAP.1966.1138693
20. Sullivan DM (2013) Electromagnetic simulation using the FDTD method, 2nd edn. IEEE Microwave Theory Tech Soc. https://doi.org/10.1002/9781118646700.ch1.

21. Rumpf R (2020) Electromagnetic analysis using finite-difference time-domain. https://emposs ible.net/academics/emp5304/. Accessed 10 March 2020
22. Green MA, Keevers MJ (1995) Optical properties of intrinsic silicon at 300 K. Prog Photovolt: Res Appl. 3(3):189–192. https://doi.org/10.1002/pip.4670030303
23. Bitnar B, Durisch W, Holzner R (2013) Thermophotovoltaics on the move to applications. Appl Energy 105:430–438. https://doi.org/10.1016/j.apenergy.2012.12.067
24. Bitnar B, Durisch W, Mayor JC, Sigg H, Tschudi HR (2022) Characterisation of rare earth selective emitters for thermophotovoltaic applications. Sol Energy Mater Sol Cells 73(3), 221–234. https://doi.org/10.1016/S0927-0248(01)00127-1
25. Striemer CC, Fauchet PM (2002) Dynamic etching of silicon for broadband antireflection applications. Appl Phys Lett 81(16):2980–2982. https://doi.org/10.1063/1.1514832

A Theoretical Study on the Efficiencies of Black Silicon Photovoltaic Cells in Thermophotovoltaic Applications

Jasman Y. H. Chai, Basil T. Wong, and Saulius Juodkazis

Abstract Silicon photovoltaic cells have been widely used in harvesting solar energy, and research efforts have driven significant improvements in the efficiencies of the photovoltaic cells. However, research on the application of silicon photovoltaic cells in thermophotovoltaic systems is significantly underdeveloped. In this study, a simulation study was conducted to study the performance of a black silicon photovoltaic cell in thermophotovoltaic applications. The photovoltaic cell was paired with two emitters made of different materials: a 1735 K Yb_2O_3 emitter, and a 1500 K Ta PhC emitter. The results of the simulation showed that when paired with the Yb_2O_3 emitter and the Ta PhC emitter, the cell efficiency was 1.7% and 0.76% respectively. The low efficiency in both was attributed to the limited amount of energy that is available above the band gap of silicon. The simulation results indicate that the Yb_2O_3 yielded better efficiency, as the selectivity of the emitter was better compared to the Ta PhC emitter.

Keywords Radiation · Simulation · Nanotechnology · Photovoltaics · Waste heat

1 Introduction

Silicon thermophotovoltaics is an application of silicon photovoltaic cells for the harvesting of thermal energy. In thermophotovoltaic (TPV) applications, instead of solar energy, photovoltaic cells collect energy from thermal sources. One of the most important sources of thermal energy is from waste heat produced from primary

J. Y. H. Chai · B. T. Wong (✉)
Department of Mechanical Engineering, Faculty of Engineering, Computing, and Science, School of Engineering, Swinburne University of Technology, Sarawak Campus, Kuching, Malaysia
e-mail: twong@swinburne.edu.my

S. Juodkazis
School of Science, Optical Sciences Center and ARC Surface Engineering and Advanced Materials (SEAM) Center, Swinburne University of Technology, Hawthorn, Australia
e-mail: sjuodkazis@swin.edu.au

F. Ahmad et al. (eds.), *ICREEM 2022*, Lecture Notes in Mechanical Engineering,
https://doi.org/10.1007/978-981-99-5946-4_3

energy consumption. Based on a review done by Forman et al. [1] in 2016, 72% of the global primary energy consumption or about 341.4 PW of energy is wasted annually. In thermophotovoltaics, low band gap materials such as germanium are desirable as they can absorb low-energy photons better [2–5]. However, in certain applications such as for residential water heating, efficiencies of 1–2% are sufficient [6], which are typical for silicon TPV systems, and silicon has the advantage of being cheap and having mature manufacturing technologies.

Typical TPV systems yield lower conversion efficiencies compared to photovoltaic cells under solar illumination. For instance, a SunPower solar cell achieved a solar efficiency of 20.5% but a TPV system using the same cell yielded a 0.76% efficiency [7]. Limited simulation or experimental studies on silicon thermophotovoltaics are available, and the most extensive ones have been done by Bitnar et al. [3], Qiu and Hayden [8], and Yeng et al. [9]. Materials used for thermal emitters in these studies are dominantly Yb_2O_3 due to the high selective emission in the absorbing wavelength range of silicon. In the work of Bitnar et al., they demonstrated a TPV system with system efficiency of 2% using a Yb_2O_3 emitter at 1735 K and a 19%-solar efficient silicon photovoltaic cell. Similarly, Qiu and Hayden used a Yb_2O_3 emitter to obtain a TPV system with 0.2 W/cm^2 output power density. In a more recent study, Yeng et al. studied tantalum photonic crystals (Ta PhC) based thermal emitters for silicon thermophotovoltaics. In their work, they achieved a TPV efficiency of 1.18% using Ta PhC emitters at 1380 K.

Black silicon (b-Si) is an emerging textured silicon surface with exceptional light trapping properties, and thorough reviews can be found in the literature [10, 11]. An illustration of b-Si nanostructures is shown in Fig. 1. The surface consists of etched nanostructures which help transmit light efficiently through the surface, and hence are desirable for photovoltaic applications. However, to the best of our knowledge, b-Si photovoltaic cells have been studied only under solar illumination in previous studies. Efficiencies of b-Si photovoltaic cells in silicon thermophotovoltaic applications have yet to be studied. Hence, in this study, a simulation model was developed to determine the efficiencies of b-Si photovoltaic cells under different thermophotovoltaic sources.

2 Methodology

The simulations are divided into two parts: the optical simulation and electrical simulation. For the optical simulation, the FDTD method was used and for the electrical simulation, the semiconductor equations were solved for the electrical performance of the b-Si cells under thermophotovoltaic sources. An illustration of the simulated cell is shown in Fig. 2.

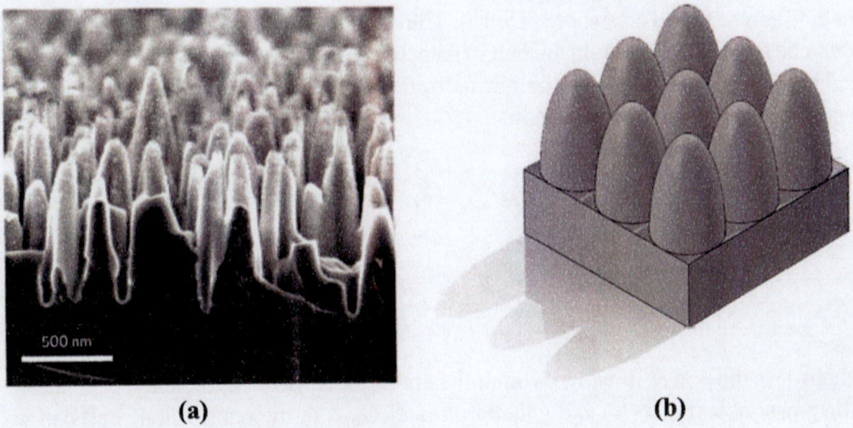

Fig. 1 **a** SEM image of a b-Si surface (reprinted by permission from Springer Nature: [12]), and **b** illustration of a nanostructure model

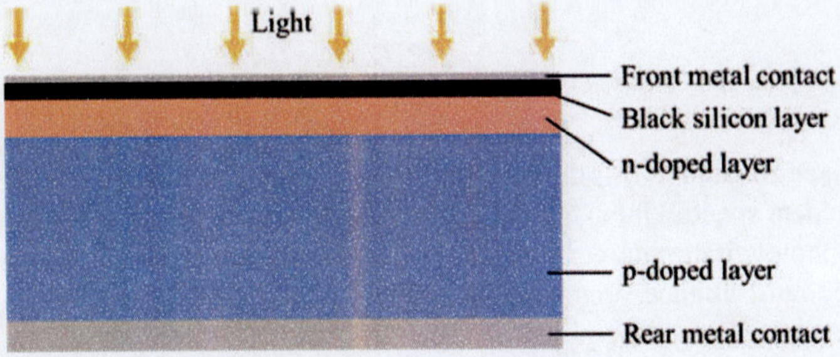

Fig. 2 Illustration of the simulated cell

2.1 *Optical Model*

An in-house developed FDTD model was used to simulate the b-Si photovoltaic cells. In the FDTD method, the update equations for the propagation of electric and magnetic fields ($\vec{E}$ and $\vec{H}$) in a staggered Yee grid [13] are given by [14]:

$$-\frac{1}{\mu}\vec{\nabla} \times \vec{E} = \frac{\vec{H}\left(t + \frac{\Delta t}{2}\right) - \vec{H}\left(t - \frac{\Delta t}{2}\right)}{\Delta t} \tag{1}$$

$$\frac{1}{\varepsilon}\vec{\nabla} \times \vec{H} = \frac{\vec{E}\,(t + \Delta t) - \vec{E}\,(t)}{\Delta t} \tag{2}$$

where μ is the permeability and ε is the permittivity.

These update equations describe the propagation of the electromagnetic waves in a simulation domain with specified material regions. A full description of the FDTD method can be found elsewhere [15, 16]. The complex refractive index of silicon was obtained from [17]. To obtain the output data, the power flowing through the reflection and transmission planes, and the normalized reflectance (R) and transmittance (T) were calculated using the equations:

$$P_{z,\mathrm{R/T}}(\lambda, r) = \frac{1}{2}\mathrm{Re}\big(E_{x,\mathrm{R/T}}(\lambda, r)H^{*}_{y,\mathrm{R/T}}(\lambda, r) - E_{y,\mathrm{R/T}}(\lambda, r)H^{*}_{x,\mathrm{R/T}}(\lambda, r)\big) \cdot A(r) \tag{3}$$

$$R/T(\lambda) = \frac{\sum_r P_{z,\mathrm{R/T}}(\lambda, r)}{P_{\mathrm{src}}(\lambda)} \tag{4}$$

where P is the power flowing through the area A, λ is wavelength, and r is position. The power absorbed (P_{A}) and generation rate (G_{opt}) in the photovoltaic cell can be calculated from the steady-state electric field using the equations:

$$P_{\mathrm{A}}(\lambda, r) = -\frac{1}{2}\left|\vec{E}(\lambda, r)\right|^{2} \cdot \sigma(\lambda, r) \tag{5}$$

$$G_{\mathrm{opt}}(r, f) = \frac{\int P_{\mathrm{A}}(\lambda, r) \cdot I_{\lambda} d\lambda}{hf} \tag{6}$$

In the equations above, σ is the conductivity, h is Planck's constant, f is frequency, and the incident spectral intensity, I_{λ} is obtained from the spectral irradiance curves of thermophotovoltaic sources, as shown in the results section. The integrated reflectance, transmittance, or absorptance (RTA) with respect to the source spectrum can be calculated using the equation:

$$\text{Integrated RTA} = \frac{\int_{300\mathrm{nm}}^{1100\ \mathrm{nm}} \mathrm{RTA}(\lambda) \cdot P_{\mathrm{in}}(\lambda) d\lambda}{\int_{300\mathrm{nm}}^{1100\mathrm{nm}} P_{\mathrm{in}}(\lambda) d\lambda} \tag{7}$$

2.2 *Electrical Model*

To simulate the electrical performance of the b-Si photovoltaic cell under the thermophotovoltaic sources, the semiconductor equations are used. The semiconductor equations consist of the following set of equations, including the Poisson equation, carrier continuity equations, and drift–diffusion current equations [18, 19]. To couple the optical and electrical models, the generation rate is obtained from the optical simulation.

$$\nabla \cdot \nabla V = -\frac{q}{\varepsilon}(p - n + N_{\mathrm{D}} - N_{\mathrm{A}}) \tag{8}$$

$$\nabla \cdot \vec{J}_n - q\frac{\partial n}{\partial t} = q(R_{\text{net}} - G_{\text{opt}}) \tag{9}$$

$$\nabla \cdot \vec{J}_p + q\frac{\partial p}{\partial t} = -q(R_{\text{net}} - G_{\text{opt}}) \tag{10}$$

$$\vec{J}_n = qn\mu_n \vec{E} + qD_n \nabla n \tag{11}$$

$$\vec{J}_p = qp\mu_p \vec{E} - qD_p \nabla p \tag{12}$$

In the equations above, V is the potential, n/p is the electron/hole concentration, $N_{\text{D/A}}$ is the donor/acceptor concentration, R_{net} is the net recombination rate, $D_{n/p}$ is the diffusion coefficient for electrons/holes, and $\vec{J}_{n/p}$ is the electron/hole current density.

The solution method used here is similar to the one described in [20]. To solve these equations, the finite difference scheme is used to discretise the equations. The Slotboom formulation and Gummel's iterative scheme were applied to solve the equations. The Slotboom variables used are defined below:

$$\Phi_n = n_i e^{-\overline{\phi}_n} \tag{13}$$

$$\Phi_p = n_i e^{\overline{\phi}_p} \tag{14}$$

$$\overline{\phi}_n = \frac{q\phi_n}{k_B T} \tag{15}$$

$$\overline{\phi}_p = \frac{q\phi_p}{k_B T} \tag{16}$$

where T is the temperature, $\overline{\phi}_n$ and $\overline{\phi}_p$ are the normalized quasi-Fermi levels and are defined in (15) and (16). The electron and hole concentrations can be re-written in terms of the quasi-Fermi levels as:

$$n = n_i e^{\frac{q}{k_B T}(V - \phi_n)} \tag{17}$$

$$p = n_i e^{\frac{q}{k_B T}(\phi_p - V)} \tag{18}$$

The intrinsic carrier concentration, n_i can be calculated from the density of states near the conduction and valence bands, N_C and N_V [21]:

$$n_i = \sqrt{N_C N_V} e^{-\frac{E_g}{2k_B T}} \tag{19}$$

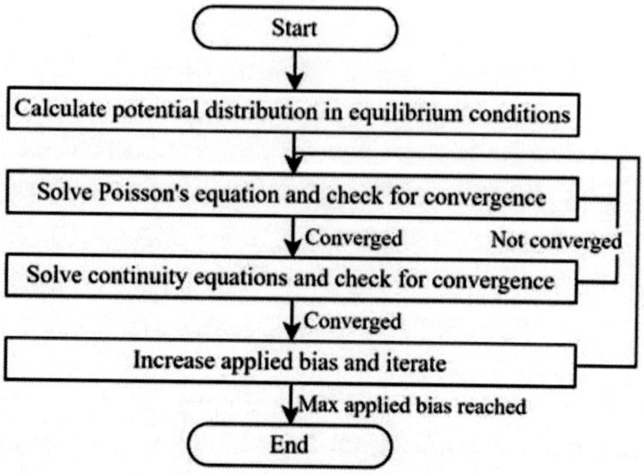

Fig. 3 Flowchart showing the iterative flow in Gummel's method

A direct solution to the set of semiconductor equations is difficult due to the complex coupling of the equations. Gummel's method and Newton's method are examples of iterative methods that are typically used to solve these equations. In this study, the Gummel's method is used. A flowchart of the method is shown in Fig. 3. The details in Gummel's method will not be discussed here as it is rather lengthy, hence the reader is referred to [20] for further information. Finally, the total current output (J_{total}) can be calculated and a J-V characteristic curve can be plotted. The total current output and efficiency of the device can be calculated as:

$$J_{\text{total, out}} = J_{n,\text{out}} + J_{p,\text{out}} \tag{20}$$

$$\text{Efficiency} = \frac{P_{\text{MPP}}}{P_{\text{incident}}} = \frac{J_{\text{MPP}} V_{\text{MPP}}}{P_{\text{incident}}} \tag{21}$$

where MPP denotes the maximum power point.

3 Results and Discussion

3.1 TPV Sources and b-Si Photovoltaic Cell

For silicon, the thermophotovoltaic sources of interest typically involve emitters made of Yb_2O_3 or Ta PhC. For this study, the emission spectrums from these emitters are used in the simulations. Figure 4a shows the emission spectrum for a 1735 K Yb_2O_3 emitter [2, 3], and Fig. 4b shows the emission spectrum for a 1500 K Ta PhC emitter [9]. The temperatures and spectra here are chosen based on the availability of data in the literature.

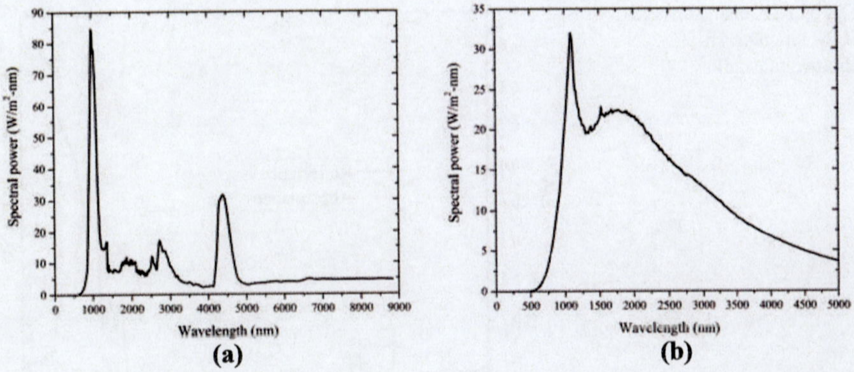

Fig. 4 Spectral power emitted by **a** Yb_2O_3 emitter at 1735 K and **b** Ta PhC emitter at 1500 K

Table 1 Specifications of b-Si photovoltaic cell used for the study

B-Si nanostructure shape	Paraboloid
B-Si nanostructure height	600 nm
B-Si nanostructure diameter	350 nm
B-Si nanostructure period	350 nm
Cell thickness (including b-Si layer)	50.6 μm
n-layer thickness	5.6 μm
n-layer dopant concentration	10^{19} cm^{-3}
p-layer thickness	45 μm
p-layer dopant concentration	10^{17} cm^{-3}
Rear metal layer (Al) thickness	300 nm

For b-Si photovoltaic cells, the dimensions of the nanostructures may affect the performance of the cell. However, the impacts of the sizes of the nanostructures are not the main subject of the study, and hence are discussed in a separate paper [22]. For this study, the specifications of the b-Si photovoltaic cell are listed in Table 1.

Figure 5 shows the RTA characteristics of the b-Si photovoltaic cell. Due to the presence of the rear metal contact layer, all the light incident on the metal layer is reflected off in the simulated wavelength range, hence the transmittance across the range is zero.

3.2 *Optical and Electrical Performance of the b-Si Cell Under Different TPV Sources*

The integrated reflectance and absorptance of the cells are calculated and shown in Table 2. From the optical results, it shows that about 47.4% of energy in the Yb_2O_3

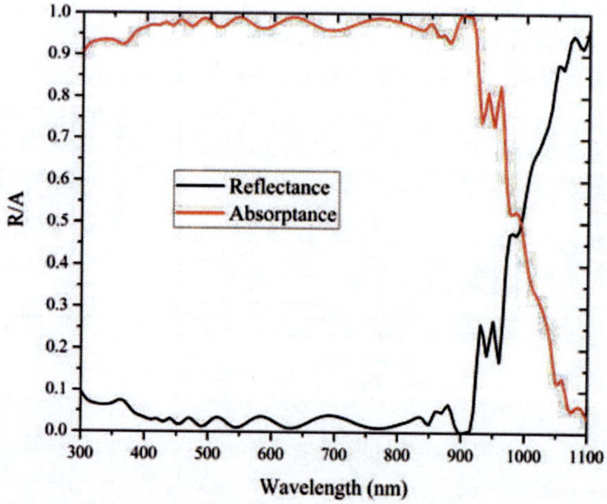

Fig. 5 RTA characteristics of the simulated b-Si photovoltaic cell

spectrum is absorbed, while 49.0% of energy in the Ta PhC spectrum is absorbed. The slight difference is due to the energy distribution of the emitted spectrum. This implies that the Ta PhC spectrum is slightly skewed to the left compared to Yb_2O_3 in the absorbing wavelength region. From the RTA curve of the photovoltaic cell, energies beyond 1000 nm are poorly absorbed by the cell. Hence, ideally, the incident spectrum should consist of energies with shorter wavelengths.

The electrical performances of the b-Si cell under the two TPV sources are shown below. Figure 6 shows the J-V curves for the cell under the illumination of the Yb_2O_3 and Ta PhC spectra. The parameters describing the performances of these cells are also extracted from the figures and recorded in Table 3. In comparison, the open-circuit voltage and fill factor are similar for both illumination sources. From the table, the notable difference is that for the Yb_2O_3 spectrum, there is a significantly higher short-circuit current, maximum power output, and also efficiency, when compared to the Ta PhC spectrum. The higher current and power output are expected as the Yb_2O_3 emitter is at a higher temperature, hence there would be a higher power emission from the emitter. This also indicates that in the absorbing wavelength region, the Yb_2O_3 emitter emits a much higher amount of energy compared to wavelengths above 1100 nm. The cell under the Yb_2O_3 spectrum has a higher efficiency, further supporting that a higher fraction of emitted power has wavelengths below 1100 nm than the Ta PhC spectrum. From the visual comparison in Fig. 4, it can also be seen that this is true. The Ta PhC spectrum has a significant fraction of energy above 1100 nm compared to the Yb_2O_3 spectrum.

Table 2 Integrated R and A for different TPV sources

Thermal emitter	Yb_2O_3 @ 1735 K	Ta PhC @ 1500 K
Integrated R (%)	52.6	51.0
Integrated A (%)	47.4	49.0

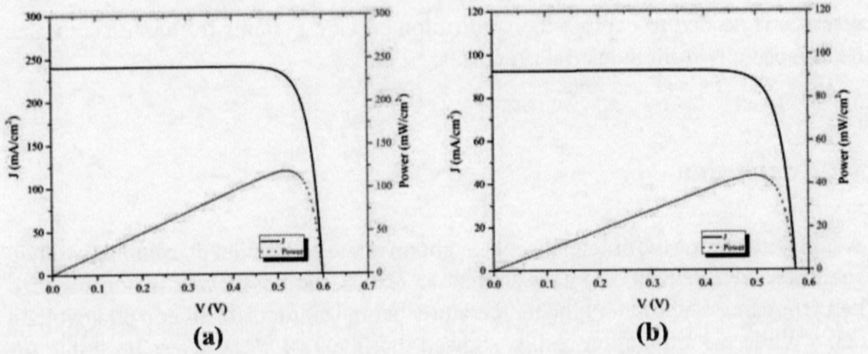

Fig. 6 J-V curve for b-Si photovoltaic cell under the illumination spectrum of **a** 1735 K Yb_2O_3 emitter and **b** 1500 K Ta PhC emitter

Table 3 Electrical performance metrics for the cell under different TPV sources

Thermal emitter	Yb_2O_3 @ 1735 K	Ta PhC @ 1500 K
Short-circuit current, J_{SC} (mA/cm^{-2})	240.7	92.7
Open-circuit voltage, V_{OC} (V)	0.599	0.574
Maximum power output (mW/cm^{-2})	119.1	43.7
Fill factor (%)	82.6	82.0
Efficiency (%)	1.7	0.76

From the discussion above, the Yb_2O_3 spectrum may be a better match for the b-Si photovoltaic cell with the specified configuration compared to the Ta PhC spectrum. From the distribution of the emission spectrum, the Yb_2O_3 emitter seems to be more selective compared to the other, as the fraction of area under the curve (equivalent to the integrated power) below 1100 nm is more significant compared to the Ta PhC spectrum. In the Ta PhC spectrum, a large fraction of integrated power lies on wavelengths above 1100 nm, which are not converted to useful electrical energy in silicon. However, the temperature difference between the two emitters is to be noted. The Ta PhC emitter may exhibit a different emission spectrum distribution at a higher temperature; however, this needs to be backed up with further research.

In addition, it is also observed that the conversion efficiencies are rather low compared to solar efficiencies. Typical solar cells operate at about 12–17% efficiency or higher. The discrepancy here is due to the temperature of the emitter, which affects the distribution of the emission spectrum. The sun is typically approximated as a blackbody at about 5778 K, which has a significant fraction of energy below 1100 nm. Hence, the energies for TPV sources are less efficiently absorbed compared to solar energy. However, the low conversion efficiency may still be useful—TPV systems with 1–2% efficiency may be sufficient for certain applications such as residential water heating. Besides, harvesting 1% of waste heat from high temperature industrial processes could amount to a significant portion of recycled energy. Hence, more research is needed to explore the application of TPV systems for harvesting waste heat, especially from industrial processes.

4 Conclusion

A study to examine the efficiency of b-Si photovoltaic cells under thermophotovoltaic sources was carried out. From the simulation results, the Yb_2O_3 was demonstrated to be a promising material for high temperature thermal emitters for silicon photovoltaic cells. While the simulation results showed that the Ta PhC was less desirable for silicon thermophotovoltaics, more research will be needed for the emitter at higher temperatures. One significant challenge in silicon TPV is the availability of materials that are able to withstand high temperatures while being able to selectively emit radiation at wavelengths below 1100 nm. Hence, more research will be needed to fine tune the properties of the emitters for better selective emission.

Acknowledgements The authors wish to express gratitude to the Ministry of Education Malaysia (MoE) for sponsoring this work under the Fundamental Research Grant Scheme (FRGS/1/2018/TK07/SWIN/02/1). In addition, Jasman Y. H. Chai is grateful to Swinburne University of Technology Sarawak Campus, Malaysia for partial support of his Ph.D. scholarship.

References

1. Forman C, Muritala IK, Pardemann R, Meyer B (2016) Estimating the global waste heat potential. Renew Sustain Energy Rev 57:1568–1579. https://doi.org/10.1016/j.rser.2015.12.192
2. Bitnar B, Durisch W, Holzner R (2013) Thermophotovoltaics on the move to applications. Appl Energy 105:430–438. https://doi.org/10.1016/j.apenergy.2012.12.067
3. Bitnar B, Durisch W, Mayor JC, Sigg H, Tschudi HR (2002) Characterisation of rare earth selective emitters for thermophotovoltaic applications. Solar Energy Mater Solar Cells 73(3):221–234. https://doi.org/10.1016/S0927-0248(01)00127-1
4. Burger T, Sempere C, Roy-Layinde B, Lenert A (2020) Present efficiencies and future opportunities in thermophotovoltaics. Joule 4(8):1660–1680. https://doi.org/10.1016/j.joule.2020.06.021
5. Daneshvar H, Prinja R, Kherani NP (2015) Thermophotovoltaics: fundamentals, challenges and prospects. Appl Energy 159:560–575. https://doi.org/10.1016/j.apenergy.2015.08.064
6. Bitnar B (2003) Silicon, germanium and silicon/germanium photocells for thermophotovoltaics applications. Semicond Sci Technol 18(5):S221–S227. https://doi.org/10.1088/0268-1242/18/5/312
7. Kushch AS, Skinner SM, Brennan R, Sarmiento PA (1997) Development of a cogenerating thermophotovoltaic powered combination hot water heater/hydronic boiler. AIP Conf Proc 401(1):373–386. https://doi.org/10.1063/1.53304
8. Qiu K, Hayden ACS (2006) Development of a silicon concentrator solar cell based TPV power system. Energy Convers Manag 47(4):365–376. https://doi.org/10.1016/j.enconman.2005.04.008

9. Yeng YX et al (2015) Photonic crystal enhanced silicon cell based thermophotovoltaic systems. Opt Express 23(3):A157–A168. https://doi.org/10.1364/OE.23.00A157
10. Chai JYH, Wong BT, Juodkazis S (2020) Black-silicon-assisted photovoltaic cells for better conversion efficiencies: a review on recent research and development efforts. Mater Today Energy 18:100539. https://doi.org/10.1016/j.mtener.2020.100539
11. Liu X, Coxon P, Peters IM, Hoex B, Cole JM, Fray D (2014) Black silicon: fabrication methods, properties and solar energy applications. Energy Environ Sci 7. https://doi.org/10.1039/c4ee01152j
12. Savin H et al (2015) Black silicon solar cells with interdigitated back-contacts achieve 22.1% efficiency. Nat Nanotechnol 10:624. https://doi.org/10.1038/nnano.2015.89
13. Kane Y (1966) Numerical solution of initial boundary value problems involving Maxwell's equations in isotropic media. IEEE Trans Antennas Propag 14(3):302–307. https://doi.org/10.1109/TAP.1966.1138693
14. Chai JYH, Wong BT, Juodkazis S (2022) Comparison between one-dimensional and three-dimensional optical modelling of ordered black silicon nanostructures using effective medium approach. Presented at the 10th international conference on nano and materials science 2022, Singapore, 12–14 Feb 2022
15. Sullivan DM (2013) Electromagnetic simulation using the FDTD method, 2nd edn. IEEE Microwave Theory Tech Soc. https://doi.org/10.1002/9781118646700.ch1
16. Rumpf R (2020) Electromagnetic analysis using finite-difference time-domain. https://empossible.net/academics/emp5304/. Accessed 10 March 2020
17. Green MA, Keevers MJ (1995) Optical properties of intrinsic silicon at 300 K. Prog Photovolt: Res Appl 3(3):189–192. https://doi.org/10.1002/pip.4670030303
18. Entner R (2012) Modeling and simulation of negative bias temperature instability. AV Akademikerverlag, Austria
19. Wuerfel P (2005) Physics of solar cells: from principles to new concepts. Wiley-VCH, p 198. https://doi.org/10.1002/9783527618545
20. Vasileska D, Goodnick SM, Klimeck G (2010) Computational electronics: semiclassical and quantum device modeling and simulation, 1st edn. CRC Press, Boca Raton, p 782
21. (2016) Atlas User's Manual: Silvaco Inc. Accessed 15 May 2022
22. Chai JYH, Wong BT, Juodkazis S (2022) A theoretical study on the properties of submicron b-Si nanostructures for improved absorption in b-Si thermophotovoltaic cells

Thermal Performance Evaluation of Heat Sink with Pin Fin, Metal Foam and Dielectric Coolant

Kim Leong Liaw, Amir Farid Haziq bin Rosle, Religiana Hendarti, and Jundika Candra Kurnia

Abstract Heat sink is an essential part in the cooling system for electronic operation specifically Central Processing Unit. Various enhancement methods which include metal foam have been studied to improve cooling performance. This study aims to analyze the forced convection of 10 PPI cooper metal foam with various shapes of pin fin (square, triangular, circular, and no-pin) heat sink under dielectric coolant in a range of Reynolds number (Re = 100, 500, 1000, 1500) at heat flux, q, of 100 kW/m^2. In addition, partial filled metal foam (PFMF) design is implemented to optimize the overall performance of the heat sink as compared to fully filled metal foam (FFMF) design. This numerical study is conducted by developing a 3D model in ANSYS, where accuracy is confirmed through grid independence test and validation. Result shows that the triangular pins model is having highest performance in FFMF while square pins model is the best in PFMF after considering both heat transfer and pressure drop. The reduction of metal foam decreases the friction factor, while drastically decreasing the heat transfer rate of triangular pins and no-pin models. This study serves as a new heat sink design by combining pin and metal foam for augmentation of heat transfer performance in electronic cooling system.

K. L. Liaw · A. F. H. Rosle
Mechanical Engineering Department, Universiti Teknologi PETRONAS, Perak Darul Ridzuan, Malaysia
e-mail: kim.liaw@mail.mcgill.ca

A. F. H. Rosle
e-mail: amir_16004572@utp.edu.my

K. L. Liaw
Department of Mining and Materials Engineering, McGill University, Montreal, Quebec, Canada

R. Hendarti
Architecture Department, Faculty of Engineering, Bina Nusantara University, Jakarta, Indonesia
e-mail: rhendarti@binus.edu

J. C. Kurnia (✉)
Department of Mechanical Engineering, Faculty of Engineering and Science, Curtin University, Sarawak, Malaysia
e-mail: jundika.kurnia@curtin.edu.my

F. Ahmad et al. (eds.), *ICREEM 2022*, Lecture Notes in Mechanical Engineering,
https://doi.org/10.1007/978-981-99-5946-4_4

Keywords Pore-scale metal foam · Pin fin · Electronic cooling · Laminar flow · Numerical simulations

1 Introduction

Advancement in technology has triggered the development of smaller and lighter microprocessors with significantly better performance [1]. The latent issue with small but mighty processors is the continuous increase of heat generation as it develops [2]. Allowing it to operate frequently at high temperatures will not only deteriorate its performance but may also lead to permanent degradation of the unit itself. Therefore, effective thermal management is essential to maintain optimum operating temperature and maximum performance [3]. The commonly adopted thermal management for electronic equipment is based on conduction cooling, natural convection and radiation cooling, forced-air cooling, liquid cooling, and immersion cooling [4]. Most of these require heat sink, a device that absorbs heat generated by electronic components or chips and dissipates it away to the surroundings. Typically, heat sink is equipped with an extended surface (pin or fin) to achieve high cooling efficiency [5]. Given its importance, numerous research have been conducted to increase the cooling capability and performance of heat sink including the application of metal foam as fin [6].

Metal foam has been utilized in various sectors namely aeronautic, automotive and construction industries due to its lightweight but excellent mechanical properties similar to its solid counterpart. Recently, metal foam has attracted thermal community attention due to its high thermal conductivity and large heat transfer surface area-to-volume ratio [7, 8]. For heat transfer application, an open-cell foam is more suitable for heat transfer because it allows fluid flow through. There are various open-cell metal foams available. However, the commonly adopted configurations are Kelvin cell and Weaire-Phelan, which are having similar impact on pressure drop and heat transfer [9]. Nevertheless, the fact that metal foam increases the pressure drop across foam as its structure obstructs the flow continuity is a hindrance in achieving higher efficiency heat transfer performance. Thus, optimal performance achieving a balance between the heat transfer and the pressure drop need to be found out [10].

A numerical study has been conducted by Contento et al. [11] on radiation heat transfer for Kelvin cell metal foam adopting Monte Carlo method, with four varying pore density (Pore Per Inch, PPI) and porosity (void-to-solid ratio) cells. Their result concluded that metal foam with high pore density have better radioactive conductivity with heat supplied 350–750 K because of higher surface area for heat transfer. Porosity and pore density provided significant differences in the hydraulic and thermal performance of metal foam [12]. Another study conducted by Morkos et al. [13] explored the effects of metal foam porosity, pore size and cell ligament towards flow efficiency. Seven blocks of metal foam with varied pore sizes (5 and 10 PPI) and porosities ($\varepsilon = 0.875$–0.952) undergone wind tunnel test with velocities ranging from 0 to 7.5 m/s. Their findings recorded that metal foam with larger porosity resulted in high downstream wind velocity due to more airflow managed to pass through. In relation to pore size, high pore size caused more obstruction to airflow due to many ligament structures. Sun et al. [14] investigated the possible structure of kelvin cell in maximizing the performance of the heat transfer, including the control of throat area and ellipse of kelvin cell. They found that altering the throat area was not feasible in enhancing the heat transfer performance. However, changing kelvin cell's structure elliptically on the axial and radial direction can decrease pressure drop with marginal decrease on heat transfer coefficient. Thus, the overall heat transfer performance was improved.

Various studies researched on heat transfer performance enhancement for heat sink equipped with metal foam under forced and natural convection. Among them, forced fluid convection has a better capability to dissipate absorbed thermal energy from the metal foam heat sink. Zhong et al. [15] applied forced air through metal foam attached to copper heat sink with a velocity of 1.33 m/s and focused on heat transfer coefficient, h value. Result of the study showed an increase of heat transfer coefficient relative to airflows velocity. Similar performance was also achieved by Almonti et al. [16], through experimental research on metal foam model. The model was developed by indirect additive manufacturing which was a combination of 3D printing and aluminum metal casting. They shaped metal foam into various pore sizes (4, 6, 8 and 10 PPI). They reported that the increase of inlet speed, pore density and ligament thickness all resulted in an increase of heat transfer coefficient. Heyhat et al. [17] experimentally investigated the use of metal foam in solar collectors. Their result was encouraging with maximum thermal efficiency increment of 79.3% and temperature increment of 16.6 °C when metal foam is embedded, despite the friction coefficient is increased up to 80 times.

Effective thermal performance of metal foam encourages the application of pins and fins in heat sink. Feng et al. [6] studied the natural convection in copper metal foam heat sinks. The experimental study used copper foam with 5 PPI and 0.91 porosity with different fin heights (range from 10 to 80 mm) and the number of strips. Metal foam fins were positioned with various gap differences (ranging from 2.86 to 20 mm) on the heat sink. The model was subjected to heat inputs between 0.8–24 W and placed in an enclosed space with ambient temperature. They found that the metal foam heat sink with a gap in between the fin exhibits greater performance compared to none as the gap provides better heat transfer from metal foam fins into air. The spaces provided good air circulation between the fins, which encouraged heat circulation for natural convection scenarios. Alfellag et al. [18] conducted research on a comparison between metal foam pin fin (MFPF) and solid pin fin involving Nusselt number over Reynolds number (*Re* 4000–*Re* 12,000). The Nusselt number increased linearly as the velocity increases. The authors also recorded a significant improvement in the performance of MFPF as achieved higher *Nu* compared to solid pin fin. This showed that good circulation in metal foam heat sink of fluid provides better heat transfer performance. Experimental study done by Li et al. [19] to study the thermal efficiency and properties of aluminum metal foam as heat sink with and without pin fins. The aluminum foams were having pore density of 10 PPI and 20 PPI, thermal conductivity of 218 W/m K, and porosity of 88%. Their result showed linear increase in *Nu* and *h* values relative to the increment of water velocity. The insertion of solid pin fins increased the flow obstruction and resistances. However, application of pin fins into the aluminum metal foam improved heat transfer performance of the heat sink as it achieved better *Nu* compared to no-pin fins metal foam on the same parameter. Enhancement of the metal foam was primarily due to the effective heat conductivity provided by the presence of pin fins. This study proved the application of pin fin combination with metal foam is beneficial for heat sink efficiency and performance.

Coolant is an important component in heat exchangers. Utilizing dielectric coolant as the working fluid is in trend nowadays for electronic packaging as it does not conduct electricity while have a high thermal capacity [20]. This can protect the electronic equipment from short circuit when leaks happen [21]. Such unique features encourage various studies on the heat performance of dielectric coolant. Patil et al. [22] investigated the cooling performance of dielectric fluid immersion cooling (DFIC) for lithium-ion (Li-ion) battery. Li-ion batteries are widely used for electric vehicles as an energy storage. The downside if it is having high heat generation during its operation thus requires efficient cooling. They decided to apply DFIC into their numerical study and record the heat transfer performance. The numerical simulation focused on Li-ion pouch cell and 50 V battery pack in ANSYS Fluent with the DFIC cooling system design. Their result showed that heat transfer occurred mainly due to single phase convection between the dielectric fluid along the battery's outer surface. The cooling system succeeded to reduce the temperature to 31.3 °C with flowing DFIC from initial battery temperature of 58.2 °C.

Another study related to dielectric coolant application was done by Alhusseny et al. [23]. The authors used graphitic foam heat sink along dielectric liquid to investigate the heat dissipation efficiency under extreme heat flux up to 100 W/cm^2. Graphitic foam was created based on graphene materials and compared to metal foam, graphitic foam performed superior heat transfer performance corresponding to its thermal conductivity up to 1900 W/m K. The foam was designed with properties of pore size range between 400–800 μm. As heat sinks, the foams were arranged in parallel and perpendicular to the coolant inlet flow. Dielectric liquid was supplied continuously onto the heated graphitic foam. Gathered data from the study managed to prove that effective cooling was achieved by graphitic foam with dielectric liquid based on the increase of *Nu* relative to the increment of *Re*. Besides, numerical simulation application of dielectric liquid as a cooling medium for servers in data center was researched by Chhetri et al. [24]. The aim of the study was to determine the cooling capability of dielectric liquid in handling high temperature servers. Conventionally, server operate using air as a cooling medium. Their study applied EC-110 dielectric liquid onto processing units powered by rate between 65–150 W. Positive results were recorded through the numerical analysis as the dielectric liquid managed to minimize the temperature of the processing unit from 130 °C to below 85 °C.

To the best of our knowledge, no research on the single-phase cooling performance of dielectric coolant in heat sink with metal foam and pin fin has been reported. This could hinder the potential application of this promising configuration which could solve the problem of microprocessor overheating. Thus, this study was conducted with the objective to investigate the performance of heat transfer using combined various shaped pin fin metal foam heat sink with dielectric coolant. The effect of partial filled metal foam (PFMF) will be examined as well. To achieve high fidelity data, pore-scale modeling approach was adopted in this study as it offers an accurate prediction of transport phenomena occurring within the investigated cooling system. The effect of Reynolds number, pin design, and partial fill of metal foam will be studied and discussed in terms of friction factor (f), Nusselt number (Nu), and Performance evaluation criterion (PEC). These shall offer insight on the application of pinned metal foam in electronic cooling with dielectric fluid within laminar region.

2 Mathematical Formulation

A three-dimensional computational fluid dynamics (CFD) model is created with conservation principle of mass, momentum and energy, which are defined as below respectively.

$$\nabla \cdot \rho \boldsymbol{u} = 0 \tag{1}$$

$$\nabla \cdot \left(\rho_f \boldsymbol{u}\boldsymbol{u}\right) = -\nabla \cdot p\boldsymbol{I} + \nabla \cdot \left[\mu\left(\nabla u + (\nabla u)^T\right)\right] + \rho_f \boldsymbol{g} \tag{2}$$

$$\rho_f c_{p,f} \boldsymbol{u} \cdot \nabla T = \nabla \cdot \left(k_f \nabla T\right) \tag{3}$$

The conservation of energy for solid is

$$k_s \nabla^2 T = 0 \tag{4}$$

The geometry is developed based on the experimental test section by Li et al. [19] with pore-scale kelvin cell metal foam, as illustrated in Fig. 1 with arrangement of fully filled metal foam (FFMF) in (c) and partially filled metal foam (PFMF) in (d).

There are variations of pin geometry design of square, triangular, circular and no-pin. Furthermore, the parameters of the geometry and the operating conditions are listed in Table 1.

The working fluid used in this study is a dielectric hydrofluoroether-based (HFE) Novec fluids according to 3 M™ product, Novec™ 7500 Engineering Fluid [25]. The thermophysical properties of the working fluid are referring to the product information handbook [25], with piecewise linear approximation in terms of temperature since the properties are almost linearly proportional to the temperature changes. The equations of Reynolds number (Re), friction factor (f), Nusselt number (Nu), and performance evaluation criterion (PEC), are listed below to evaluate the heat transfer performance in this study.

$$Re = \rho_f u D_h / \mu \tag{5}$$

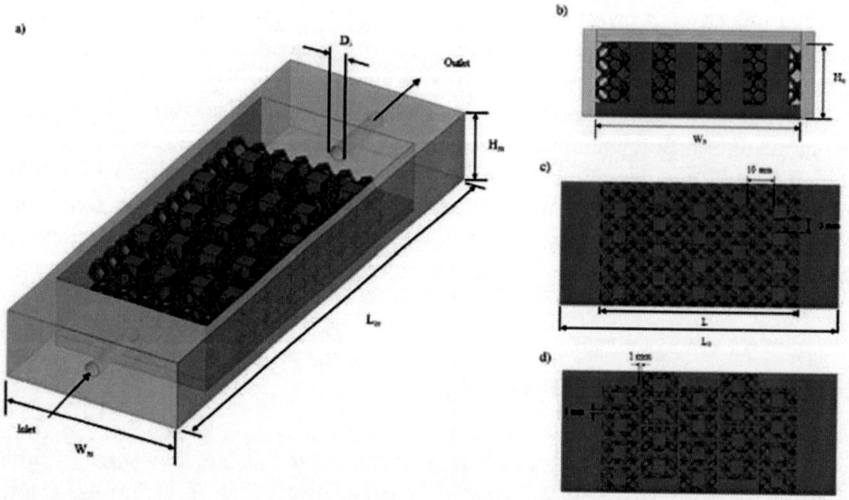

Fig. 1 The developed geometry in **a** isometric, **b** front view, **c** top view of FFMF and **d** top view of PFMF

Table 1 Geometry dimensions and operating parameters of developed model

Parameters	Value	Unit
L_m	145	mm
H_m	19	
W_m	51	
D_i	4	
L_h	105	
H_h	16	
W_h	45	
L	75	
T_{in}	300	K
Porosity, ε	0.90	–
PPI	10	–
Wall heat flux, q	100	kW/m^2
Inlet Reynolds number, Re	100	–
	500	
	1000	
	1500	

$$f = 2\Delta p D_h / \rho_f u^2 L \tag{6}$$

$$Nu = h D_h / k_f \tag{7}$$

$$PEC = (Nu_i / Nu_{base}) / (f_i / f_{base})^{1/3} \tag{8}$$

The boundary conditions at the inlet are constant temperature ($T = T_{in}$) and constant fluid velocity ($\mathbf{u} = \mathbf{u}_{in}$) corresponding to Reynolds number 100, 500, 1000 and 1500. At the outlet, the outlet gauge pressure ($p = p_{out} = 0$) and streamwise gradient of temperature ($n \cdot \nabla T = 0$) are set to zero. As for the wall boundary condition, all of the wall is set to have no-slip condition ($\mathbf{u} = 0$), while the bottom wall is heated with constant heat flux ($q = q_{wall}$).

3 Numerical Implementation

Numerical analysis of the study is conducted using ANSYS FLUENT. This software is specified for computational fluid dynamics analysis and equipped with conserving mass, momentum, and energy equations. Through ANSYS, the models of study meshed and solved with Semi-Implicit-Pressure-Linked equation (SIMPLE) scheme, Least Squared Cell-based method on gradient spatial discretization, and second order for all other spatial discretization. The convergence criteria of the numerical investigation are to have residuals lower than 10^{-6}. Due to the complexity of the models, analysis is carried out using High Performance Computer (HPC). Each run required 20 cores where each core consists of 4 GB RAM. Each run reached convergence at 2300–2500 iterations which acquired up to 6 hours of iterative numerical simulation. Mesh independent study is conducted to omit the influence of mesh number to the result of the simulations. The study is done on a square-pin FFMF heat sink with the mesh number of 8308305, 8440033, 16556119, 33401154, and 173974177. The results are illustrated in Fig. 2.

It can be seen that the results are having minor deviation with only 3.4 and 5.3% for pressure drop and outlet temperature as compared to the highest amount of mesh. Therefore, the impact of the number of mesh to the result is minimal.

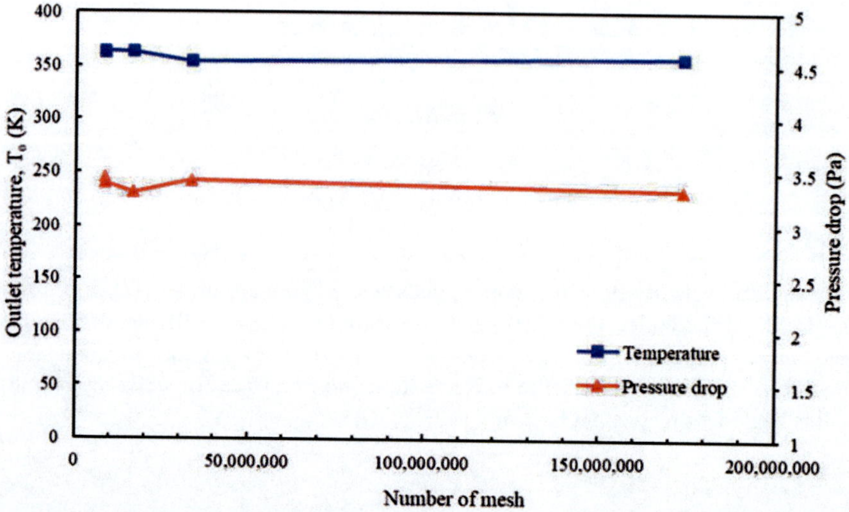

Fig. 2 Mesh independent test showing outlet temperature and pressure drop against number of mesh

4 Result and Discussions

4.1 Validations

To ensure the simulation model prepared can predict the real physical scenario accurately, validation is done before further analysis is conducted. The developed model is compared against experimental data from Li et al. [19]. As mentioned, pinned metal foam heat sink is designed based on the experiment geometry by Li and colleagues. The heat sink is aluminum, working fluid is water and the pin is square. The result is depicted in Fig. 3.

Small margin of error between the experimental model and current study's numerical model can be noticed. The pressure change has an average error of 2.78%, while Nusselt number recorded to have a deviation of 1.25% on average. Hence, the developed computational model is validated to be able to reproduce the experimental works and predict fluid flow and heat transfer phenomena in real life.

4.2 Fluid Flow Characteristic

In general, metal foam will increase the flowing resistance of the fluid due to its structure obstruction. Contours of velocity are illustrated in Fig. 4 to show detailed dielectric coolant flow behavior inside each heat sink model. Figure 5 shows the

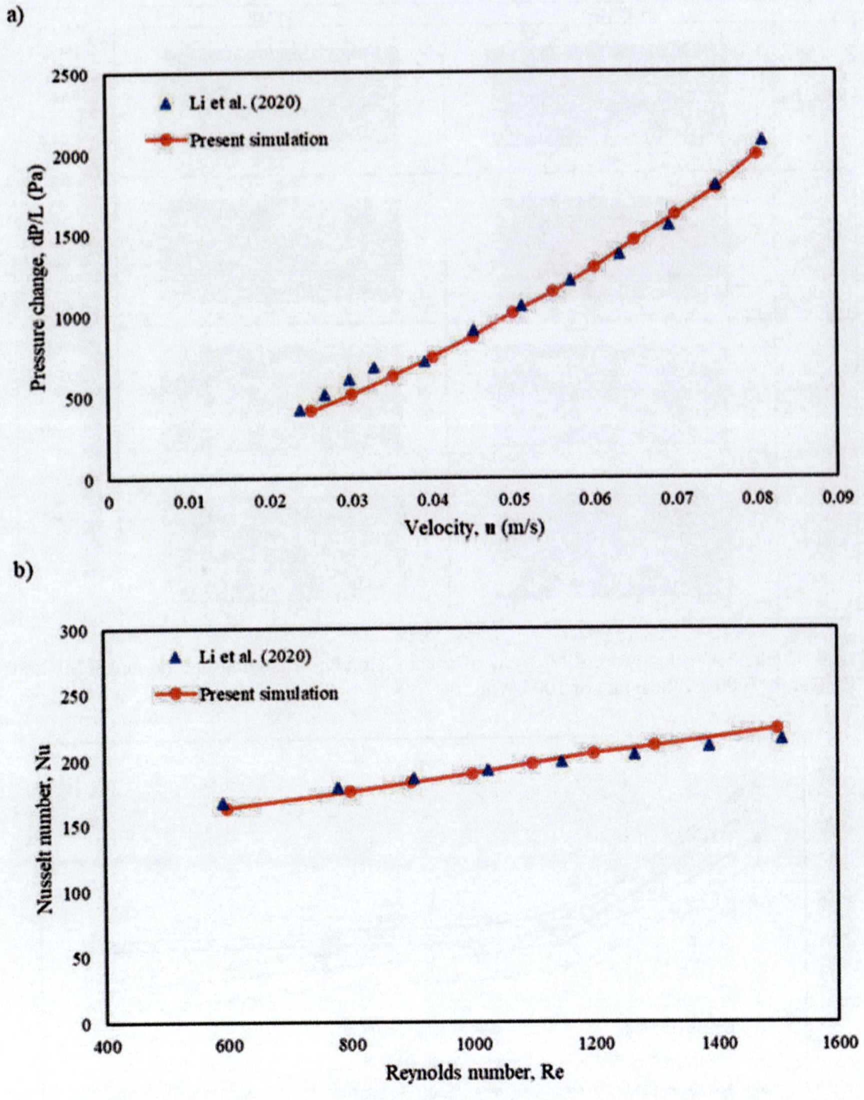

Fig. 3 Validation of present model against Li et al. [19] experimental data of Pressure change and Nusselt number

friction factor, f, of heat sink model under various Reynolds number of FFMF and PFMF with pin design of square, triangular, circular, and no-pin.

The presence of pin also affects the fluid velocity and pressure drop performance of the heat sink models. Pins add obstruction to the coolant flowing through the metal foam. Geometry of the pin affects the flow pattern and velocity as depicted in the velocity contour of Fig. 4. Metal foam without pin resulted in greater fluid flow

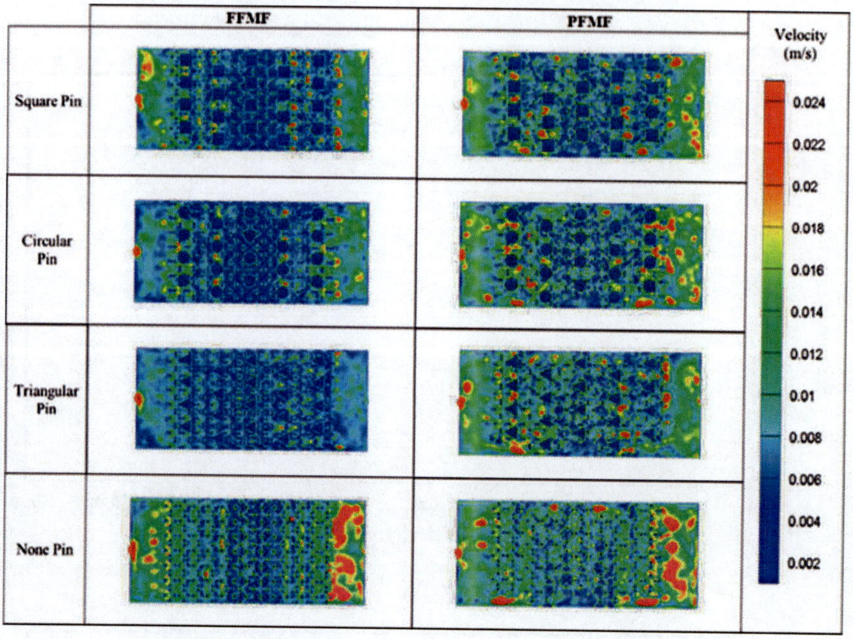

Fig. 4 Contour of velocity from top view of square, circular, triangular and no-pin FFMF and PFMF at *Re* 1000 and heat flux of 100 kW/m^2

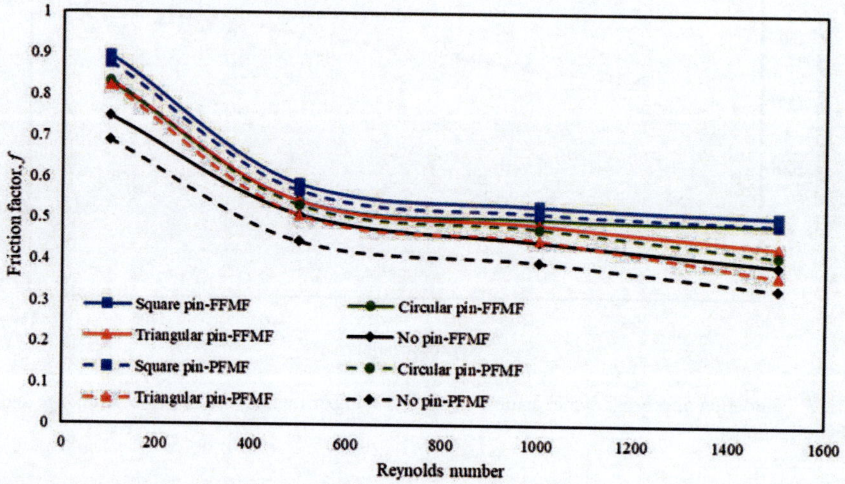

Fig. 5 Graph of friction factor against Reynolds number for square-pin, circular-pin, triangular-pin and no-pin FFMF and PFMF at *Re* 1000 and heat flux of 100 kW/m^2

velocity. Triangular-shaped pin has hydrodynamic advantage to the coolant flows resulting in low pressure drop. Circular-shaped pin also offers similar hydrodynamic streamline, but less significant as compared to triangular. Velocity distribution of square pins model is observed to be similar to circular, but the edgy shape and flat confronting area deteriorates the flow velocity. It is proved in comparison between pin types for FFMF heat sinks in Fig. 5, where the model with square pins yields the highest friction factor. In Contrast, heat sink without pin results in the lowest friction factor. This enforces the earlier statement that pins cause an increase in pressure drop to the flow due to an increase in area of impact which provides more blockage to the flow. When comparing between pin geometries, triangular pins model shows the result of lower friction factor followed by circular pins and square pins models.

On the other hand, friction factor for PFMF is lower as compared to FFMF. This is obvious as reduced metal foam structure reduces the fluid contact surface area with foam wall structure. Figure 4 shows minor reduction in pressure drop in PFMF cases. It can be seen that, smoother dielectric flow is achieved in PFMF due to even gradient contour distribution. Low fluid velocity behavior at the center of FFMF is improved in PFMF. Similar to FFMF, no-pin PFMF heat sink achieves the lowest friction factor. Absence of pin obstruction and less metal foam structure resulting on the great performance in terms of friction factor. In addition, triangular-pin PFMF achieved the lowest friction factor after no-pin metal foams and followed by circular-pin and square-pin finned heat sink models, which the sequence is identical with FFMF. As stated, triangular PFMF excels the other pins with its hydrodynamic structure at the impact side. The difference of pressure drops at *Re* 100 is not significant in pinned model with about 0.38–2% difference, while 7.8% is recorded for no-pin model. The differences in pinned models are more drastic at *Re* 1500 where the pressure drop can reduce up to 16% when the form is partially filled, and the no-pin model also has a reduction of pressure drop of 14.7%.

4.3 Heat Transfer Behavior

Heat transfer performance of heat sink correlates with the total surface area for heat transfer proportionally. Geometry of metal foam offers a high surface area for heat transfer. Figure 6 shows the Nusselt number for FFMF and PFMF under various inlet Reynolds number and pin types.

Based on the result, the increase in Reynolds number improves the heat transfer performance represented by Nusselt number for FFMF heat sink models. Triangular-pin FFMF heat sink allows the working fluid to flow faster, thus resulting in better heat transfer overall compared to other heat sink geometry designs. This can be seen in Fig. 6 where the Nusselt number of triangular pins model is a lot higher than other models. The Nusselt number of square pins model is a little bit higher than circular pin, after the triangular pin. No-pin design is having lowest heat transfer as the absence of pin decrease the heat conduction from the heated surface. Interestingly,

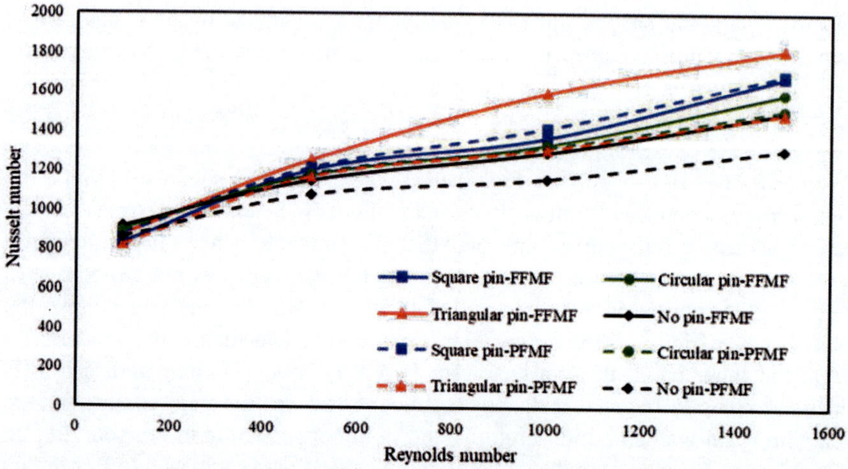

Fig. 6 Graph of Nusselt number against Reynolds number for square-pin, circular-pin, triangular-pin and no-pin FFMF and PFMF at *Re* 1000 and heat flux of 100 kW/m^2

the heat transfer rate of each model in *Re* 100 is totally inverse of the one in higher *Re*.

PFMF heat sink with pins resulted in minor improvement of heat transfer performance when compared to FFMF models as shown in Fig. 6. Square is the highest, followed by circular, triangular and no-pin for PFMF. The Nusselt number of square pins and circular pins models have negligible differences from their counterparts in FFMF. It can be inferred that square-pin and circular-pin FFMF exceed the optimum volume of metal foam where extra metal foam was not assisting to distribute heat transfer effectively at the same time adding pressure drop across the heat sink. No-pin PFMF achieve a lower Nusselt number than no-pin FFMF heat sink. This indicates that pins act as a crucial component in distributing heat energy from the heat sink through conduction when the metal foam is sparse. Absence of pins in this case limits the heat conduction and convection. The Nusselt number of triangular pins model drops drastically in PFMF. This shows that the metal foam in the triangular model is majorly contributed to the heat transfer into the coolant. The Nusselt number in partial foam averagely changes about 2.5%, −1.7%, −14.4% and −9.9% for square, circular, triangular and no-pin respectively.

4.4 Performance Evaluation Criterion (PEC)

Figure 7 shows the PEC of heat the metal foam heat sinks in the same pumping work condition.

It can be seen that reducing the metal foam into partial only benefits the square pins and circular pins models, while declining the heat transfer performance of triangular

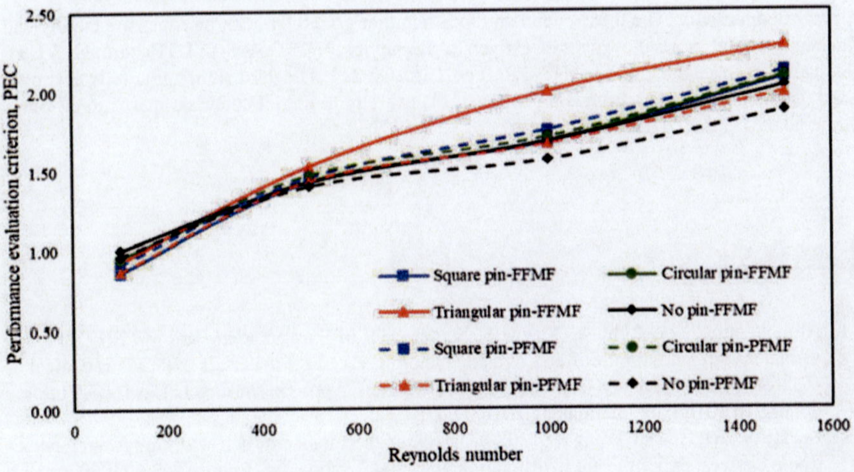

Fig. 7 Performance Evaluation Criterion (PEC) of each model

pins and no-pin models. The highest performance is recorded by triangular-pin FFMF model at *Re* 1500 with 2.28.

5 Conclusion

Throughout this study, heat transfer performance of combined pin fin metal foam heat sinks with dielectric coolant is successfully acquired. The 3D pin fin pore-scale metal foam heat sink model is well validated with existing experimental data. Application of various shaped pin fin (square, triangular and circular) proven to affect the flow characteristics. Triangular pins model is found to have the least friction factor among the pinned heat sink, followed by circular pins and then square pins model. The configuration without pin attained the lowest friction factor overall due to reduction of obstruction in the flow. Based on heat transfer rate in terms of Nusselt number, superior performance was shown by triangular pins heat sink. Triangular pins model acquired the highest Nusselt number followed by square pins, circular pins and no-pin.

Besides, by reducing the metal foam, results show that decreasing of metal foam reduced the pressure drop while able to maintain the heat transfer in square-pinned and circular-pinned metal foam. By truncating the extra metal foam, the flow resistance decreases while having similar heat transfer, hence heat transfer performance is improved. However, triangular pins and no-pin heat sink recorded a decrease in Nusselt number. Overall, triangular pins model has the highest performance in FFMF while square pins model outperformed others in PFMF, as evaluated by PEC.

Acknowledgements The first, second and fourth author gratefully acknowledges the facility and financial support from the Yayasan Universiti Teknologi PETRONAS (YUTP) through YUTP Fundamental Research Grant (YUTP-FRG) no. 015LC0-214. The third author gratefully acknowledge the financial support from Bina Nusantara University through Penelitian International Binus Grant no. 061/VR.RTT/IV/2022.

References

1. Alimohammadi M, Aghli Y, Alavi ES, Sardarabadi M, Passandideh-Fard M (2017) Experimental investigation of the effects of using nano/phase change materials (NPCM) as coolant of electronic chipsets, under free and forced convection. Appl Therm Eng 111:271–279. https://doi.org/10.1016/j.applthermaleng.2016.09.028
2. Liu HR, Li BJ, Hua LJ, Wang RZ (2022) Designing thermoelectric self-cooling system for electronic devices: experimental investigation and model validation. Energy 243:123059. https://doi.org/10.1016/j.energy.2021.123059
3. He Z, Yan Y, Zhang Z (2021) Thermal management and temperature uniformity enhancement of electronic devices by micro heat sinks: a review. Energy 216:119223. https://doi.org/10.1016/j.energy.2020.119223
4. Cengel AY, Ghajar A (2019) Heat and mass transfer: fundamentals and applications, 6th edn. McGraw-Hill Education. [Online]. https://www.mheducation.com/highered/product/heat-mass-transfer-fundamentals-applications-cengel-ghajar/M9780073398198.html. Accessed 30 May 2022
5. Mohammadi M, Taheri A, Passandideh-Fard M, Sardarabadi M (2020) Electronic chipset thermal management using a nanofluid-based mini-channel heat sink: an experimental study. Int Commun Heat Mass Transf 118:104836. https://doi.org/10.1016/j.icheatmasstransfer.2020.104836
6. Feng S, Li F, Zhang F, Lu TJ (2018) Natural convection in metal foam heat sinks with open slots. Exp Therm Fluid Sci 91:354–362. https://doi.org/10.1016/j.expthermflusci.2017.07.010
7. Li H, Hu C, He Y, Tang D, Wang K, Hu X (2021) Visualized-experimental investigation on the energy storage performance of PCM infiltrated in the metal foam with varying pore densities. Energy 237:121540. https://doi.org/10.1016/j.energy.2021.121540
8. Huang S, Lu J, Li Y (2022) Numerical study on the influence of inclination angle on the melting behaviour of metal foam-PCM latent heat storage units. Energy 239:122489. https://doi.org/10.1016/j.energy.2021.122489
9. Moon C, Kim HD, Kim KC (2018) Kelvin-cell-based metal foam heat exchanger with elliptical struts for low energy consumption. Appl Therm Eng 144:540–550. https://doi.org/10.1016/j.applthermaleng.2018.07.110
10. Li Y, Gong L, Xu M, Joshi Y (2021) A review of thermo-hydraulic performance of metal foam and its application as heat sinks for electronics cooling. J Electron Packag 143(3). https://doi.org/10.1115/1.4048861
11. Contento G, Oliviero M, Bianco N, Naso V (2014) The prediction of radiation heat transfer in open cell metal foams by a model based on the Lord Kelvin representation. Int J Heat Mass Transf 76:499–508. https://doi.org/10.1016/j.ijheatmasstransfer.2014.04.058
12. Yang H, Li Y, Yang Y, Chen D, Zhu Y (2020) Effective thermal conductivity of high porosity open-cell metal foams. Int J Heat Mass Transf 147:118974. https://doi.org/10.1016/j.ijheatmasstransfer.2019.118974
13. Morkos B, Dochibhatla SVS, Summers JD (2018) Effects of metal foam porosity, pore size, and ligament geometry on fluid flow. J Therm Sci Eng Appl 10(4):041007. https://doi.org/10.1115/1.4039302

14. Sun M, Zhang L, Hu C, Zhao J, Tang D, Song Y (2022) Forced convective heat transfer in optimized kelvin cells to enhance overall performance. Energy 242:122995. https://doi.org/10.1016/j.energy.2021.122995
15. Zhong Z, Meng L, Li X, Zhang G, Xu Y, Deng J (2020) Enhanced heat transfer performance of optimized micro-channel heat sink via forced convection in cooling metal foam attached on copper plate. J. Energy Storage 30:101501. https://doi.org/10.1016/j.est.2020.101501
16. Almonti D, Baiocco G, Mingione E, Ucciardello N (2020) Evaluation of the effects of the metal foams geometrical features on thermal and fluid-dynamical behavior in forced convection. Int J Adv Manuf Technol 111(3):1157–1172. https://doi.org/10.1007/s00170-020-06092-1
17. Heyhat MM, Valizade M, Abdolahzade Sh, Maerefat M (2020) Thermal efficiency enhancement of direct absorption parabolic trough solar collector (DAPTSC) by using nanofluid and metal foam. Energy 192:116662. https://doi.org/10.1016/j.energy.2019.116662
18. Alfellag MA, Ahmed HE, Jehad MGh, Hameed M (2021) Assessment of heat transfer and pressure drop of metal foam-pin-fin heat sink. Int J Therm Sci 170:107109. https://doi.org/10.1016/j.ijthermalsci.2021.107109
19. Li Y, Gong L, Xu M, Joshi Y (2020) Enhancing the performance of aluminum foam heat sinks through integrated pin fins. Int J Heat Mass Transf 151:119376. https://doi.org/10.1016/j.ijheatmasstransfer.2020.119376
20. Lionello M, Rampazzo M, Beghi A, Varagnolo D, Vesterlund M (2020) Graph-based modelling and simulation of liquid immersion cooling systems. Energy 207:118238. https://doi.org/10.1016/j.energy.2020.118238
21. Qi D, He J, Xu Y, Lin M, Wang Q (2022) Effect of rib diameter on flow boiling heat transfer with staggered rib arrays in a heat sink. Energy 239:122323. https://doi.org/10.1016/j.energy.2021.122323
22. Suresh Patil M, Seo J-H, Lee M-Y (2021) A novel dielectric fluid immersion cooling technology for Li-ion battery thermal management. Energy Convers Manag 229:113715. https://doi.org/10.1016/j.enconman.2020.113715
23. Alhusseny A, Al-Fatlawi A, Al-Aabidy Q, Nasser A, Al-Zurfi N (2021) Dissipating the heat generated in high-performance electronics using graphitic foam heat-sinks cooled with a dielectric liquid. Int Commun Heat Mass Transf 127:105478. https://doi.org/10.1016/j.icheatmasstransfer.2021.105478
24. Chhetri A, Kashyap D, Mali A, Agarwal C, Ponraj C, Gobinath N (2022) Numerical simulation of the single-phase immersion cooling process using a dielectric fluid in a data server. Mater Today Proc 51:1532–1538. https://doi.org/10.1016/j.matpr.2021.10.325
25. (2008) 3M^{TM} Novec^{TM} 7500 engineered fluid: product information. 3M Electronics Markets Materials Division. [Online]. https://multimedia.3m.com/mws/media/65496O/3m-novec-7500-engineered-fluid.pdf

Investigation into Magnetic Drive Sealless Pump Failure at Flare Gas Recovery Unit (FGRU)

M. Farid Yahya, M. Faiz Malek, and Reduan Mat Dan

Abstract Failure investigation was conducted on magnetic drive centrifugal sealless pumps that had only been in operation for six months and were already struggling to rotate. It was discovered that hydrocarbon sludge and other debris had blocked the lubricating channel, which is meant to provide pathways for cooling and lubricating the inner magnetic and containment shell of the pump. Further investigation revealed service water had contaminated the process fluid, precipitating calcium carbonate upon contact with the amine solution and clogging the flushing line. An additional duplex strainer was installed on the repaired pumps, and adequate precautions were taken to prevent water contamination.

Keywords Cleanliness · Internal flushing · Contamination · Clogging · Duplex strainer

1 Introduction

Conventional centrifugal pumps used to deliver fluid in the process refinery plant are normally susceptible to failure of mechanical seal leaks which account for 43% of pump failure modes [1]. This had caused loss of products and created environmental issues due to hydrocarbon spillage. Even under normal operation, the pump

M. Farid Yahya (✉)
Technical Services Department, MRCSB, Persiaran Penapisan, Sg Udang, 76300 Melaka, Malaysia
e-mail: faridyh@petronas.com.my

M. Faiz Malek
Maintenance Department, MRCSB, Persiaran Penapisan, Sg Udang, 76300 Melaka, Malaysia
e-mail: mfaiz.amalek@petronas.com.my

R. M. Dan
Faculty of Mechanical Engineering, Universiti Teknikal Malaysia Melaka, Hang Tuah Jaya, 76100 Durian Tunggal, Melaka, Malaysia
e-mail: reduan.dan@utem.edu.my

F. Ahmad et al. (eds.), *ICREEM 2022*, Lecture Notes in Mechanical Engineering,
https://doi.org/10.1007/978-981-99-5946-4_5

mechanical seal is subjected to minimum permitted leakage rate that is allowed under international standards [2]. In order to completely avoid hydrocarbon leakages that contribute to fugitive emissions into the environment, some refinery plants have switched to using magnetic drive centrifugal pumps or sealless pumps.

In the construction of sealless pump, the process fluid is completely contained within the pump containment shell, preventing product leakage. However, in order to remove the generated heat and provide lubrication, the flushing line makes use of the process fluid as an internal lubricant and cooling medium. Consequently, any restriction to the lubrication flow paths will jeopardize the pump's internal components and result in catastrophic pump failure as a result of lubricant starvation [2]. Additionally, the fluid runs between the inner magnet, the containment shell and through the shaft holes to the rear of the pump shaft where it returns to the pump's suction through the thrust balance hole in the impeller. The pump shaft that is connected to the inner magnet is rotated by the action of the magnetic flux circuit between the inner and outer magnet. The outer magnet is coupled to the driver motor through another shaft to transmit the rotation to the inner magnet [3].

While sealless pump will eliminate leakages, it has very limited solids handling capability. Particles will accumulate in the passages surrounding the magnet and in the close tolerances of bearings such as between the bush and sleeves. This will hinder the pump's performance, resulting in issues and sometimes even failure. In addition, solids will wear out the bearings and other internal pump components [4].

Under the implementation of Plant Change projects, seven sealless pumps (API685) were installed in one of the Malaysia oil refinery plants located in the west coast peninsular. The design of sealless pumps aims to prevent hydrocarbon product leaks into the atmosphere, which otherwise will contribute to fugitive emissions and when the product leakage worsens can escalate to an HSE incident of loss of primary containment. It will also reduce costly periodic pump maintenance due to mechanical seal leaks. However, following the project commissioning, four of the seven installed pumps experienced significant failure. One of the pumps that was installed under Fuel Gas Recovery System was found hard to rotate after seven days of running and the standby pump experienced low discharge pressure and performance after five days of running. The process flow and the location of the pumps are depicted in Fig. 1.

2 Methodology

The magnetic drive pump's internal components were dismantled and examined for the purpose of investigation into the failures. This shall give the clear picture of the extent of damage of the internal component of the pumps. Secondly, the pump suction strainers were examined, and the particles collected inside the pump and at the strainer were sent to an outside lab for further examination. The particles will give clear evidence of the sources of the clogging. Thirdly, X-ray diffraction elemental analysis of the particles collected at the strainer was carried out. Similarly, the particle

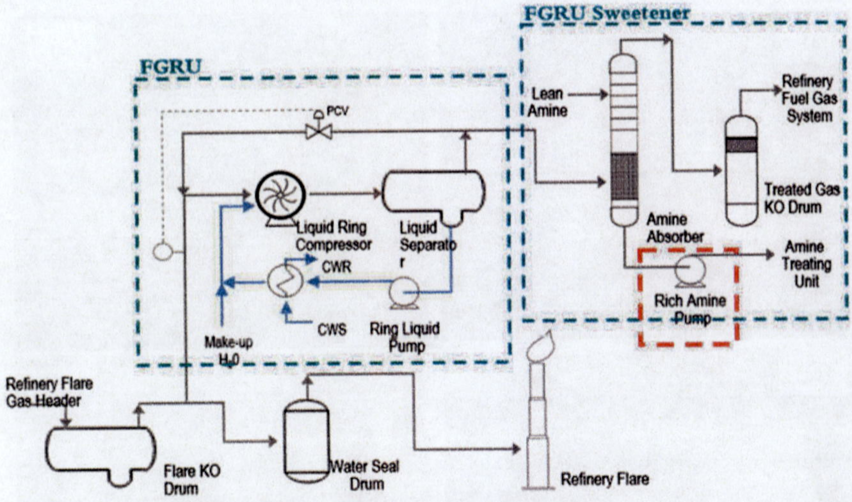

Fig. 1 Process flow diagram of Flare Gas Recovery Unit (FGRU)

sample test is to determine the material of the clogged particle in order to ascertain the origin of the particles. In conclusion, the 5-Why method was used to conduct a Root Cause Analysis (RCA) in order to identify the underlying causes of the failure and offer suggestions for preventing another one of its kind.

3 Result and Discussion

The investigation found that the flow line for internal flushing or internal lubrication was clogged with process solid particles. This had caused lubricant starvation and loss of cooling to the internal parts of the pump. The absence of lubrication had increased friction which caused the mating surfaces to seize. The pump containment shell shows external burnt marks and rubbing and the bushings were found damaged. In fact, it was supported by the temperature records on the containment shell which shows an increase of normal temperature of 60 °C to a high of 103 °C. Figure 2 shows the extent of damage of the internal parts of the pump.

The damage was enhanced with the expanded Outer Magnetic Ring (OMR). Bearing housing could not be removed due to OMR expanded and seized with bearing housing and shell containment. Figure 3 shows the expanded OMR.

In addition, further analysis on the particles filtered, found few types of particles that caused the clogging of the flush line. Figure 4a and b shows images of particles collected at the strainer of the pump. The debris consisted of moist dark and light brownish lumps and had an alkaline PH of 9.5. It contained mainly carbon, oxygen, sulphur and calcium, with other elements such as aluminum, manganese, iron and strontium. The compounds detected contained mainly calcium carbonate (aragonite)

Fig. 2 **a** Clogged flushing line. **b** Process particles. **c** and **d** Burn mark on containment shell. **e and f** bushing damages

with a small amount of calcium carbonate (calcite), iron sulphate hydrate (rozenite) and iron oxide (maghemite).

Based on X-Ray Diffraction analysis as shown in Table 1, the deposited solid particle found at the damaged sealless pump was rich in calcium carbonate. The process theory at upstream FGRU Amine Absorber has demonstrated the governing bicarbonate reaction for tertiary amines (MDEA) on the CO_2 absorption reaction.

Fig. 3 Expanded Outer Magnetic Ring (OMR)

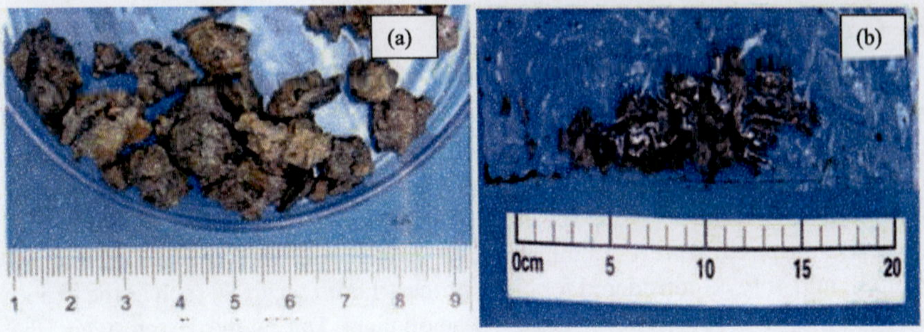

Fig. 4 a and b particles collected at the pump internal and filter

Table 1 X-Ray Diffraction (XRD) analysis result of the particles, approximate element content (wt%)

Element	Wt%
C	32.71
O	35.11
Al	0.1
S	6.04
Ca	23.99
Mn	0.29
Fe	0.36
Sr	1.40

$$CO_2 + H_2O + (HOCHCH_3)_2NCH_3 \leftrightarrow (HOCHCH_3)_2N(CH_3)H^+ + HCO_3^- \quad (1)$$

The solvent lean amine (MDEA) within the operating pH range of 10–11 further dissociates and converts the bicarbonate ion to carbonate ion.

$$HCO_3^- \leftrightarrow H^+ + CO_3^{2-} \quad (2)$$

The origin of calcium ion could only be identified in the service water stream from the upstream process. In this case, the stream acts as a liquid ring at upstream Liquid Ring Compressor and make-up water at 3-Phase Separator Vessel. It was found that the 3-Phase Separator Vessel level had reached a high level (100%) which consequently showed high level build-up (91%) at downstream Amine column. This has indicated a high potential of service water-carryover to the Amine Column. The liquid in the column was then pumped by the two aforementioned FGRU Rich Amine Pumps, with one pump operating continuously and one pump operating on standby. Therefore, by examining these facts, the contamination of the water that resulted in the calcium carbonate rich solid particle generation that clogged the magnetic drive pump's flushing and cooling line may have been the true cause of the failures.

4 Conclusions

Although sealless pumps are the way to go for preventing mechanical seal leaks and lowering emissions of greenhouse gases (GHGs) into the environment, they do have some drawbacks. The following lessons learned should be taken into consideration when selecting a sealless pump for the refinery hydrocarbon application during the project design phase to prevent similar failures in the future. The selection of a sealless pump should only be considered for clean process liquid services, such as the export of the final hydrocarbon product and the base oil plant. During the design stages, the pump datasheet must specify the size and concentration of allowed solid particles in the pumping liquid. To ascertain the solid's size and concentration, a sample of the process liquid must be taken and analyzed at defined intervals. The sealless pump should be designed to accommodate a duplex filter for the internal flush system to further reduce the risk of particle contamination. Additionally, a differential pressure gauge or indicator must be designed and installed across the duplex filter for routine field operator monitoring. The operator will be notified when particles begin to build up inside the duplex strainer as a result of this increase in differential pressure. For high alarm and high-high trip protection, a temperature sensor must be installed in the pump isolation or containment shell. The pump will trip whenever the temperature goes above the normal range, preventing further damage. After the investigation and major pump repair were finished, these suggestions were incorporated and put into action at the Fuel Gas Recovery system. The pumps have been operating normally thus far, with neither a single nor any failures recorded.

Acknowledgements Author acknowledges the contributions from Engineering department at Malaysian Refinery Company Sdn Bhd in the investigation of the failures and contribution from Universiti Teknikal Malaysia Melaka, the member of Center of Advanced Research on Energy (CARE) in preparing the articles.

References

1. Bloch HP, Budris AR (2010) Pump user's handbook: life extension, 3rd edn
2. American Petroleum Institute (API) Standard 682 (2014) Pumps—Shaft sealing systems for centrifugal and rotary pumps, May 2014
3. Al Dossary F, Al Dossary D (2022) Case study: sealless caustic circulation pump failure. Pump & System, April 2022
4. Zimmermann O (2021) Common causes for magnetic drive centrifugal pump failure. March Pumps. https://www.marchpump.com/blog/mag-drive-pump-failure-common-causes/

Investigation on Behavior of New Structures for Airbus A380 Wing Rib Feet Using Fiber Metal Laminates

Mohammed Mahmmud Direa Khairi, Ali Aref Ali Alzanam, Tuan Mahmmad Yuossf, and Ali Ameen Roshan

Abstract Fiber metal laminate (FML) is a lightweight material with outstanding mechanical properties that combines the benefits of metal laminates and fiber reinforced composites. While weight reduction and higher damage tolerance were the primary drivers to develop these novel materials, they have additional benefits that are becoming more relevant to designers, such as cost savings and improved safety. In this work, the design methodology considers the design of the Airbus A380 wing rib, to investigate the mechanical properties of hybrid fiber-metal-laminate (FML) with different orientations composites. The ultimate tensile strengths and strain at failure of the FML composite with varied ply orientation and laminate layup were also investigated and compared to Aluminum 7449 alloy. GLARE was studied as a unique aircraft material. The results reveal high stress (803.52 MPa) and good fatigue life cycles (80* 103). The simulation studies done by ANSYS for the GLARE 4 and the comparison between the existing material of the rib in the Airbus A380, Aluminum 7449, show that through its unique combination of attributes. The study favors GLARE4 as a replacement material. GLARE 4 is a good material for new generation aircraft wing ribs, according to this study.

1 Introduction

Airbus's A320 was the first commercial aircraft to include an all-electric flight control system. Airbus made 626 planes and received 1,503 net orders in 2013 [1]. The wing's aerofoil shape is created by the ribs, which run the length of the wing and act as bridges between the two spars to provide the wing's shape and stiffness in a direction perpendicular to the spars. The wing box's skeleton is made up of the spars and the ribs. When the covers are in place, the wing box construction is completely enclosed. The upper cover is essentially uninterrupted while the lower cover is more

M. M. D. Khairi (✉) · A. A. A. Alzanam · T. M. Yuossf · A. A. Roshan
Mechanical Engineering Department, Universiti Teknologi PETRONAS, 32610 Bandar Seri Iskandar, Perak, Malaysia
e-mail: mohammed_20001015@utp.edu.my

F. Ahmad et al. (eds.), *ICREEM 2022*, Lecture Notes in Mechanical Engineering,
https://doi.org/10.1007/978-981-99-5946-4_6

complex with holes cut in it to provide access ports and connecting points for landing gears, engine pylons and fuel access [2].

There are few breaks in the upper cover, but the lower cover is intricately drilled to permit landing gear pylons, engine pylons, and fuel access [3, 4]. Among them, FML, created by hot pressing and curing a composite of metal plates and fibers due to its high strength-to-weight ratio and interest among scientists [5–9]. Fokker Aircraft's F-27 was the first aircraft to use wing skin made of aramid fiber reinforced aluminum alloy laminates (ARALL) [10] in the 1980s. As a result, Douglas Company in the United States started using the ARALL laminate to make the cargo door of the C-17 big military transport helicopter, which resulted in a 26% reduction in the door's overall weight. Maintaining its anticorrosion performance, impact resistance, and fatigue resistance over its service life, laminate [11, 12]. The aircraft's flat structure or a single curved surface with a minor curvature are the only places where the ARALL can be used because of the material's weak interlaminar shear strength [11]. In the 1990s, glass reinforced aluminum laminates (GLARE) started to be utilized in several huge aircrafts because of their exceptional qualities [13–17]. The A340 passenger aircraft was the first to use GLARE laminates in the construction of the rear bulkhead and three radial crack locking devices [18]. Then, numerous GLARE laminates were also incorporated into the upper fuselage skin, fairing, trailing edge, upper fuselage wall, and upper wall of the A380 aircraft [19, 20]. Furthermore, GLARE laminates were used by American Boeing Aircraft Corporation to construct the cabin floors of the Boeing 757 and Boeing 777 [21]. Although the process of creating and deploying FMLs has come a long way, there are still many issues that need to be addressed. When subjected to shear stress, bending stress, impact stress, or axial stress, FMLs will experience a dramatic decrease in strength and stiffness, leading to interlaminar failure behavior including debonding and delamination. Therefore, analysing the failure mechanism and control methods of FMLs can control the failure of materials to some extent.

The authors conducted a comprehensive study on GLARE with fatigue cracks and through cracks. [15, 22]. Therefore, preliminary research on notches suggested that the quantity of cut fibers significantly affected notch strength by creating holes (for fasteners, e.g.). A through-the-thickness crack or a sharp notch would cause the most harm. This could be the result of fatigue or a foreign item penetrating aluminum. Fatigue would not lead to through-the-thickness cracks in the case of GLARE. It was possible to determine that Glare's residual strength with fatigue cracks was significantly greater than that with cracks. Shear load in an aircraft fuselage is caused by bending and twisting. Therefore, ILSS is also an essential quantity, as the material cannot flex plastically below the limit load [16].

This study has been inspired by previous research [10, 23]. It describes all the procedures to fabricate the GLARE composites and simulate the rib by keeping interfacial bonding strength to investigate the new structure of the rib feet by replacing the Aluminum alloy 7449 which the existing with FML, depending on the extent of damage. Fiber composite patches can be used and there is always the more-costly alternative of part replacement.

2 Methods

During the design process, the Airbus A380 wing rib design is taken into consideration. It was intentional that the tensile elasticity modules of FML composites with different laminate layups should be equivalent to those of FML composites with different ply orientations in order to maintain consistency. In addition to this, the ultimate tensile strengths of the FML composite as well as the strain at failure were investigated and compared to the already established design material, which is aluminum alloy 7449. As a direct consequence of this, it is necessary to make use of design tools (solid work), as well as simulation software (ANSYS). In this study, rock foot made of aluminum alloy was contrasted with FML composites that were reinforced by three layers of natural fiber or glass fibers. In addition to this, the ultimate tensile strengths, and strains at failure of FML composites with variable ply orientation and Aluminum Alloy were investigated.

3 The Wing Rib Feet Simulation Design

The model was built using Solid Work's library and features to draw lines and curves of the geometry, then extruded from 2 to 3D to finish design auditing. Three examples of wing rib foot cracks were studied. According to the Airbus design chose Air ware for the wing ribs, seat rails, and other components of its A350, which will have 16 tonnes of the metal. Constellium's president of aerospace and transportation says Air ware works nicely with existing metal manufacturing, supply chain, and assembly procedures, allowing for high production rates. It's 5% lighter than aluminum and has better corrosion resistance and mechanical qualities. Despite lithium's reactivity, the alloy isn't reactive [24]. The Airbus wing box structure case study showed that the simulated model can accurately anticipate the overall geometrical variations of ribs and skin panels, as well as the positional variations of each rib foot. With the ability to estimate each rib foot's influence and location, skin panel holes can be predrilled. The prediction system will help minimize dimensional variations from previous production processes to assembly and can give a powerful knowledge-based tool for an assembly automation system [25].

4 The Airplane Movement Equations Lift Force

Because a fluid moving past the surface of a body exerts a force on it, a force must be created that is equivalent to or greater than the force that is exerted by gravity in order for there to be a turbulent flow. This flow is in contrast to a laminar flow regime. This force is referred to as lift, and in order to gain a deeper comprehension of how an airfoil generates lift, it is required to utilize significant equations from the field of physical science.

Bernoulli's equation provides the most accurate representation of the pressure fluctuations caused by moving air. Daniel Bernoulli, a Swiss mathematician, was the first person to develop it in order to explain the fluctuation in pressure that is exerted by moving streams of water. The Bernoulli equation is written as:

$$P + {}_{\rho}V^2 = Constant \tag{1}$$

where P is the pressure (force exerted divided by area exerted on), Rho ($\acute{\rho}$) is the density of the fluid being measured, and V is the speed at which the fluid or moving object is traveling.

A wing airfoil does not have a set shape; rather, it is developed based on the purpose of the aircraft that it will be used for. Lift coefficients are a measurement of the amount of lift that can be obtained from a specific airfoil shape. Lift coefficients are used by engineers to assist in the design process. The ratio of the wing's area to its dynamic pressure is directly related to lift. The equation for lift can be written as:

$$\mathrm{Lif_t} = C_L * (1/2\,\rho V^2) * S \tag{2}$$

$$CL = L/(q_{\infty} {}^{*}S) \tag{3}$$

4.1 The Weight of the Rib

When designing an aeroplane, one of the limiting factors that must be considered is the aircraft's weight. A form can be used to determine an individual's weight of Newton's second law:

$$W = mg \tag{4}$$

It can decompose …

$$F = F_{aero} + T + W \tag{5}$$

i.e.

Aero forces Faero = Xaero, Yaero, Zaero

When calculating the angular momentum of an aeroplane, a small fragment of the aircraft with a mass of dm is taken into consideration at r = {x,y,z} wrt the body axes. It is this pressure difference $\Delta P = 34095.7$ Pa, and in fact, F = 56,808.75 N, wingspan 79.75 m, the speed normal rate 300 Km\h, and rang 15200 Km, with that causes the air to flow faster at the plane where the wing of the air plane flying.

Table 1 ANSYS geometry of GLARE4

The properties	Description
Object name	Solid GLARE4
State	Meshed
Stiffness behavior	Flexible
Coordinate system	Default coordinate system
Reference temperature	By environment
Length X	30. mm
Length Y	50. mm
Length Z	100. mm
Volume	20306 mm^3
Nodes	73771
Elements	44722
Physics type	Structural
Analysis type	Static structural
Solver target	Mechanical APDL

5 Results and Discussion

The analysis depicts the material properties at a flying height of 35,000 feet, a temperature of -53° C, and an operating speed of 0.8 Mach. This study focused on the different stress, the equivalent elastic strain, and the total deformation of the wing rib of Airbus A380. The model geometry is tableted in Table 1. The geometry shows the shape, size, relative position of figures, and the properties of space.

6 Normal Stress Characteristic

The difference in stress between two materials with the same dimensions and geometry is caused by the fact that the two objects are made of different materials, which have different elastic modulus of elasticity and so do not deform in the same way. The maximum stress of the GLARE4 composite material is 803.52 MPa based on mild elastic deformation as shown in Fig. 1a. However, after applying the same load to both materials, the maximum stress for aluminum depicted in Fig. 1b reaches 14,196 MPa, which is more than that of the first material due to deformation. Carrillo and Cantwell also conducted quasi-static flexural testing with FMLs that had a variety of fiber orientations. It was discovered that the maximum stresses and strains at failure of the FML are not reliant on the orientation of the fibers, but rather are primarily determined by the volume percentage of the metal layers that are present [26].

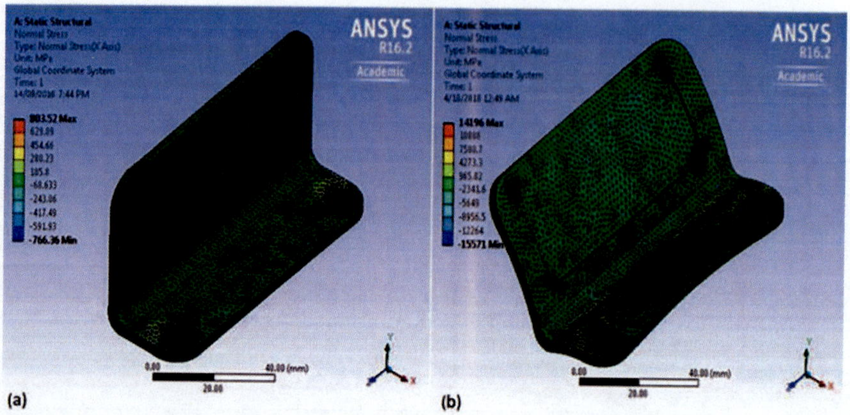

Fig. 1 **a** The stress of the GLARE4 and **b** The stress of the Al 7449 represented in ANSYS

7 Equivalent Elastic Strain Characteristic

In order to account for the strain characteristics of these two material modes, 3D simulations were performed. A finite element study appears to have been performed using GLARE4 and an aluminum alloy. The maximum strain in GLARE4 presented in Fig. 2a is 0.015402, while the maximum strain in Aluminum alloy presented in Fig. 2b is 0.15591 (the larger strain in the latter is obviously related to the plastic behavior of the yield stress), so it is interesting to compare the two materials' strain behavior and determine their equivalent. Additionally, a frictionless condition must be considered to reduce the influence of friction on the deformation behavior of materials.

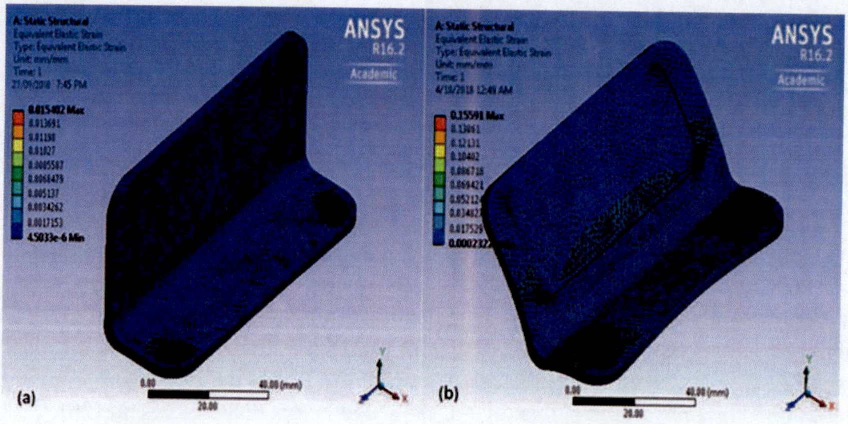

Fig. 2 **a** Equivalent strain of the GLARE4 and **b** Equivalent strain of the Al represented in ANSYS

In order to account for the strain characteristics of these two material modes, 3D simulations were performed. Finite element study appears to have been performed using GLARE4 and an aluminum alloy. The maximum strain in GLARE4 is 0.015402, while the maximum strain in Aluminum alloy is 0.15591 (the larger strain in the latter is obviously related to the plastic behavior of the yield stress), so it is interesting to compare the two materials' strain behavior and determine their equivalent. Additionally, a frictionless condition must be taken into account to reduce the influence of friction on the deformation behavior of materials.

Compared the strain required to perforate GLARE FML, ARAL FML, 2024-T3 aluminum, and a thermoplastic composite (PEEK reinforced with carbon fibers). Comparing areal densities. GLARE FML has a higher perforation strain failure than other materials. High-velocity impact. testing (up to 220 m s-1) to measure the ballistic velocity (impact velocity needed for perforation) of an FML made of layers of glass-fiber-reinforced epoxy composite and aluminum alloy. FMLs made of thin aluminum alloy layers have a moderately higher ballistic limit than a monolithic metal of the same density (i.e. 1.5 and 1.9 mm in thickness) [26].

8 Total Deformation Characteristic

Poisson's ratio between lateral and axial strains causes total deformation in alloy 7449 as shown in Fig. 3a. Strains in three orthogonal directions can assess a material's dilatation in response to multiaxial stress, which is 33.306 mm in GLARE 4 as shown in Fig. 3b. Pressure or hydrostatic stress can modify a material's volume. When the front face of FML panels was hit with a high- velocity GLARE impact, the deformation of the rear (unaffected) aluminum alloy layer was identified using 3D DIC. The impact behavior of FML panels was compared to that of a variety of aluminum alloys in the 3D results. Energy absorption in FML panels is improved by the addition of aluminum alloy and GLAR composite [26].

9 Life Cycle "Fatigue"

Fiber metal laminates' fatigue properties are studied in detail in aviation industry test programmes. The fatigue performance of GLARE4 and Aluminum alloy 7449 is compared while subjected to a simulated rib foot flight in both Figs. 4 and 5. When compared to solid aluminum, glass fibers with high failure stress and strain may withstand more stress before giving out. Consequently, GLARE has a very high fracture toughness. The interfacial debonding between the metal and prepreg layers, the stress redistribution upon crack initiation, and the fracture behavior of the metal and prepreg layers all play a role in determining the fracture toughness of GLARE laminates. Fatigue fractures in fiber metal laminates, as opposed to saw-cut laminates, exhibit greater fracture toughness because the strain length of the fibers

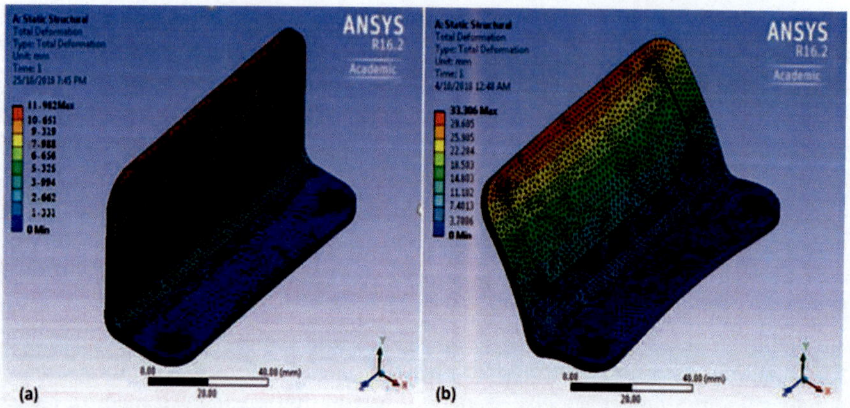

Fig. 3 **a** Total deformation of GLARE4 and **b** Total deformation of Al represented in ANSYS

is effectively increased by the presence of unbroken fibers in the crack's wake and the delamination zone around the break.

A longer fatigue life in the Particulate of GLARE is mostly attributable to the material's greater elastic modulus. In addition to a lower total strain and specifically a lower total plastic strain for a given stress level, a higher elastic modulus also reduces the fatigue process. However, the improvement in tiredness life is not the same for different stress levels. The fatigue life improvement anticipated in the composite due

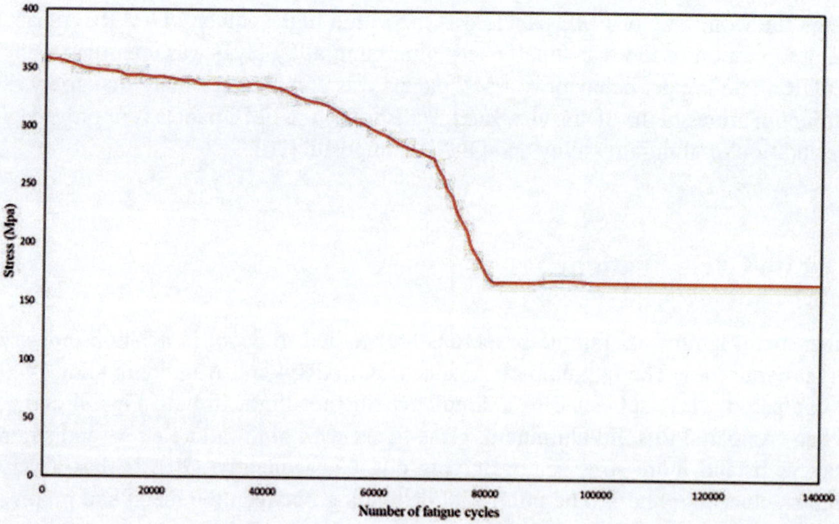

Fig. 4 The fatigue results of GLARE 4

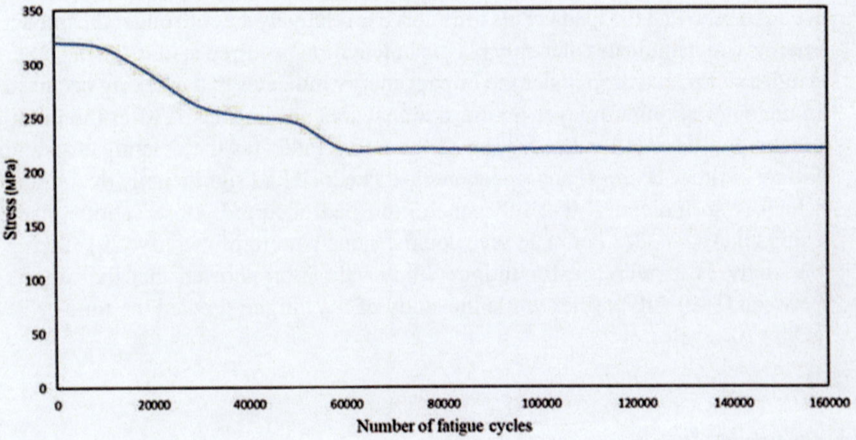

Fig. 5 The fatigue results of Alluminum Alloy 7449

to the reduction in plastic strain was counteracted by the fatigue life reduction caused by the cyclic ductility at high stress levels [27].

10 Validation of the Study

1. Stress levels and stresses associated which is illustrated that the residual stresses in the transverse and shear directions are negligibly small for all off-axis angles. For different angles, therefore, the tensile residual stress GLARE develops in the loading direction to satisfy the overall self-equilibrium of GLARE. Note that for q = 0 and 5° the aluminum–alloy layers yield in compression during unloading of GLARE which shows the maximum stress is 929 MPa. The compressive yielding of the aluminum–alloy layers suppresses further increases in the compressive residual stress of the aluminum–alloy layers and in the tensile residual stress of the GFRP layers for these off-axis angles. [28]. According to previous studies of the FML of GLARE that showed the maximum stress is 929 MPa, and in refer to Table 4.2 which showed the study of the normal stress of GLARE 4 the maximum value the results shows is 803.52 MPa using percentage calculations finds that the different is 12.55% between the two studies.
2. The responses of the results to the local stresses and strains. When looking at the calculated data, we can see that the aluminum alloy breaks at every off-axis angle when the higher stress is originally delivered to the GLARE4 material. This can be seen when the stress is applied to the GLARE4 material. The yielding of the aluminum–alloy layers, on the other hand, occurs exclusively in the off-axis angle range when the loading is increased to a lower stress level [28].

3. It can be seen that the load versus time curve is relatively smooth under the impact energy, which indicates that only plastic indentation occurred at first. On the other hand, a sharp load drop under the impact energy indicates that cracking occurred in the outer aluminum layer on the side that was not impacted. After the initial cracking, a distinct load reduction in the curve indicates the delamination and failure of other layers at an impact energy, The load had significantly decreased, which is an indication that full penetration had occurred. Observations made using GLARE 4–3/2 laminate revealed the same pattern of activity [29]. Taking the study as a reference for fatigue study validation showed that the time is between (8–9) $*10^3$ cycles while the study of the fatigue showed the time cycle is 80 $*10^3$ cycles.

11 Conclusion

ANSYS has been used to analyse the stress constants in FML GLARE4. Composite laminate stress constant change. The elastic characteristics of the Glare 4 composite are superior to those of the aluminum alloy 7449 at any orientation of the fiber composite laminae. G modulus for Glare composite, for any orientation of the fiber composite laminae, is almost higher the modulus of the Aluminum alloy 7449. The study also showed FML's tensile qualities are affected by the materials that make it up. Therefore, the elastic response from the composite laminae and the aluminum strain, as well as the load bearing capacity, are clearly demonstrated good results of the strain behavior of FML. There is a combination of high stiffness and strength from the composite layer and good impact properties from aluminum, resulting in a great performance for aerospace applications. Furthermore, aluminum's toughness and notch sensitivity is linked to its strain plastic region, as for the fiber/matrix interface, so at the same fiber volume fraction the GLARE4 composite tensile strength would be higher than Aluminum 7449 tensile strength.

Fatigue test results of GLARE4 laminates at the maximum stress with a dent inflicted by an impact energy based on a comparison study of the impact energy required to create the first cracking and perforation of different materials on a specified basis, for GLARE 4 material, It appears that the fatigue crack began at the concave dent's edge and made its way slowly outward. However, After the fatigue fracture on GLARE 4 approached the edge of the rib foot, many cracks formed around the dent regions, It led to an exceptionally lengthy fatigue life cycle for the aluminum until failure happened after fatigue cracking had begun near the dent's periphery. There is a discrepancy between the fatigue cycles for crack propagation in GLARE4 laminates and aluminum, with the GLARE4 having longer life cycles.

References

1. Petrescu RV, Aversa R, Akash B, Corchado J, Apicella A, Petrescu FI (2017) Home at airbus. J Aircr Spacecr Technol, 1(2)
2. Marsh G (2010) Airbus A350 XWB update. Reinf Plast 54(6):20–24. https://doi.org/10.1016/S0034-3617(10)70212-5
3. Zheng S, Xu H, Feng J, Zheng Z, Wang Y, Lu L (2011) Lightweight design of automobile drive shaft based on the characteristics of low amplitude load strengthening. Chin J Mech Eng 24(6):1111
4. Bansemir H, Haider O (1998) Fibre composite structures for space applications—recent and future developments. Cryogenics (Guildf) 38(1):51–59
5. Nestler D et al (2017) Continuous film stacking and thermoforming process for hybrid CFRP/aluminum laminates. Procedia CIRP 66:107–112
6. Cantwell WJ (2000) The mechanical properties of fibre-metal laminates based on glass fibre reinforced polypropylene. Compos Sci Technol 60(7):1085–1094
7. Thirukumaran M, Siva I, Jappes JTW, Manikandan V (2018) Forming and drilling of fiber metal laminates–A review. J Reinf Plast Compos 37(14):981–990
8. Dadej K, Bieniaś J, Surowska B (2017) Residual fatigue life of carbon fibre aluminium laminates. Int J Fatigue 100:94–104
9. Sarasini F et al (2019) Effect of temperature and fiber type on impact behavior of thermoplastic fiber metal laminates. Compos Struct 223:110961
10. Pincheira G, Canales C, Medina C, Fernández E, Flores P (2018) Influence of aramid fibers on the mechanical behavior of a hybrid carbon–aramid–reinforced epoxy composite. Proceedings of the Institution of Mechanical Engineers, Part L: Journal of Materials: Design and Applications 232(1):58–66
11. Vermeeren C (2003) An historic overview of the development of fibre metal laminates. Appl Compos Mater, 10(4), pp 189–205
12. Nong T, ElSayed MSA, Biseul C (2019) Experimental optimization and static characterization of modified fiber metal laminates with integrated mechanical interlock bonding system for aerospace applications. Polym Compos 40(4):1510–1525
13. Jia M, Xue P, Yu J (2010) Research on technology and equipment of long fiber reinforced thermoplastics composites. Eng Plast Appl 38(10):83–86
14. Meng Q, Gu Y, Luo L, Wang S, Li M, Zhang Z (2017) Annealing effect on crystalline structure and mechanical properties in long glass fiber reinforced polyamide 66. J Appl Polym Sci, 134(23)
15. Goris S, Osswald TA (2017) Progress on the characterization of the Process-Induced fiber microstructure of long glass Fiber-Reinforced thermoplastics. Plast Eng 73(1):46–47
16. Chai W, Liu X, Shan Y, Wan X, Jiang E (2018) Research on simulation of the bending fatigue test of automotive wheel made of long glass fiber reinforced thermoplastic considering anisotropic property. Adv Eng Softw 116:1–8
17. Airoldi A, Mirani C, Principito L (2020) A bi-phasic modelling approach for interlaminar and intralaminar damage in the matrix of composite laminates. Compos Struct 234:111747
18. Wu G, Yang J-M (2005) The mechanical behavior of GLARE laminates for aircraft structures. Jom 57(1):72–79
19. Cortes P, Cantwell WJ (2004) The tensile and fatigue properties of carbon fiber- reinforced PEEK-titanium fiber-metal laminates. J Reinf Plast Compos 23(15):1615–1623
20. Botelho EC, Silva RA, Pardini LC, Rezende MC (2006) A review on the development and properties of continuous fiber/epoxy/aluminum hybrid composites for aircraft structures. Mater Res 9:247–256
21. Park SY, Choi WJ, Choi HS (2010) A comparative study on the properties of GLARE laminates cured by autoclave and autoclave consolidation followed by oven postcuring. Int J Adv Manuf Technol 49(5):605–613
22. Shayan Asenjan M, Sabet AR, Nekoomanesh M (2019) Long fiber thermoplastic composites under high-velocity impact: Study of fiber length. J Compos Mater. 53(3), pp 353–360

23. Bellini C, di Cocco V, Iacoviello F, Sorrentino L (2019) Performance evaluation of CFRP/Al fibre metal laminates with different structural characteristics. Compos Struct 225:111117
24. Marsh G (2007) Airbus takes on Boeing with reinforced plastic A350 XWB. Reinf Plast 51(11):26–29
25. Saadat M, Sim R, Najafi F (2007) Prediction of geometrical variations in Airbus wingbox assembly. Assem Autom
26. Kaboglu C, Mohagheghian I, Zhou J, Guan Z, Cantwell W (2018) High-velocity impact deformation and perforation of fibre metal laminates. J Mater Sci 53(6):4209–4228. https://doi.org/10.1007/s10853-017-1871-2
27. Arivukkarasan S, Dhanalakshmi V, Aruna M (2013) Performance study on fatigue behaviour in aluminium alloy and alumina performance study on fatigue behaviour in aluminium alloy and alumina silicate particulate composites. February 2014, https://doi.org/10.6180/jase.2013.16.2.03
28. Kawai M, Hachinohe A (2002) Two-stress level fatigue of unidirectional fiber—metal hybrid composite : GLARE 2. 24, pp 567–580
29. Wu G, Thomas ÆJYÆH (2007) The impact properties and damage tolerance and of bi-directionally reinforced fiber metal laminates, pp 948–957. https://doi.org/10.1007/s10853-006-0014-y

Thermal Analysis of Helical Pin Fins at Different Pitch Steps Through Numerical Technique

Syed Waqar Ahmed, Adeel Tariq, Khurram Altaf, Sadaqat Ali, Ghulam Hussain, and Masri B. Baharom

Abstract Many modern-day applications rely on heat sinks for heat dissipation. Pin fins are one of the heatsinks that are used for cooling and the convection coefficient of a pin fin depends on many parameters that need optimization. The pin fin in its plain form isn't efficient enough and there are ways to not only enhance its heat transfer rate but also reduce its weight simultaneously. This study aims to geometrically modify a plain pin fin to enhance the effective area for convective heat transfer. This study contains a comprehensive comparison between simple pin fin heatsinks and their corresponding heatsinks with helical-shaped profiles on the outer surface. A helical pattern was made using different pitches 2, 4, 6, and 8 mm and they were tested at air speeds of 2, 4, 6 and 8 m/s. Computational fluid dynamic analysis was performed to measure the performance of heatsinks. The tetrahedral mesh was used for simulation, and it was concluded that the helical fin having a 2 mm pitch performed the best and therefore had the highest convection coefficient of 202.69 W/m^{2}K at 8 m/s airspeed.

S. W. Ahmed (✉) · A. Tariq · K. Altaf · M. B. Baharom
Department of Mechanical Engineering, Universiti Teknologi PETRONAS, Seri Iskandar, Malaysia
e-mail: syed_19000908@utp.edu.my

A. Tariq
e-mail: adeel_22007073@utp.edu.my

K. Altaf
e-mail: khurram.altaf@utp.edu.my

M. B. Baharom
e-mail: masrib@utp.edu.my

S. Ali
School of Mechanical and Manufacturing Engineering, National University of Science and Technology, Islamabad, Pakistan
e-mail: sadaqat.ali@smme.nust.edu.pk

G. Hussain
Department of Mechanical Engineering, University of Bahrain, Zallaq, Bahrain
e-mail: ghussain@uob.edu.bh

F. Ahmad et al. (eds.), *ICREEM 2022*, Lecture Notes in Mechanical Engineering,
https://doi.org/10.1007/978-981-99-5946-4_7

Keywords Computational fluid dynamics · Heatsinks · Pin fins · Convection · Heat transfer

1 Introduction

According to the caloric theory presented by Lavoisier, heat is a fluid that contains no mass and is termed as caloric [1]. Moreover, Joule indicated through experiments that heat can travel from one state to another when the temperature gradient is provided [2]. For some applications, high temperatures are required however for most engineering systems proper heat dissipation has been a pain point. Therefore, optimal heat dissipation is required for the proper functioning of heat-emitting engineering systems. For example, internal combustion engines convert only 25% of their produced energy to useful work whereas about 30% of the generated heat is dissipated into the environment through engine walls [3, 4]. Therefore, to maintain the heat generation systems at their optimal performance, heat sinks are applied onto the heat-emitting surfaces to effectively help in heat dissipation into the environment. For this purpose, researchers are investigating different geometries to enhance the heat dissipation capabilities of the systems.

Double circular pin fins were used by Feng et al. to enhance thermal performance. The study concluded that pin fin height and adjacent spacing are important factors to improve heat transfer rate. Simulations were carried out and it was observed that a pin fin diameter of 12 mm, a height of 0.08 mm, and a pin fin spacing of 0.5 mm enhanced the heat transfer rate by 42% [5]. Adeel et. al increased the surface area of slotted perforated plate fins and found that perforations and slotting of plate fins increase the convective heat coefficient of the fins significantly [6–8]. Trapezoidal-shaped plate fins were introduced by Ozdilli and Sevik with incorporated rectangular and corner-rectangular channels. When compared to standard plate fins, their modified plate fins presented much improved convective heat transfer [9]. A numerical approach was used by Al-Sallami that compared different geometries containing notches, slots, and various holes in the plate-fin heat sink. It was observed that fins with slots performed best among the tested configurations [10]. As it is established that heat fins improve the overall thermal performance of the system, however, they also add material weight to the system. Therefore, there must be a trade-off between the thermal efficiency of the heat fins and the amount of weight they add to the system. It is also critical to mention that for systems that are small in size, such as, electronic systems, the weight of the fins must be minimal, and the design must be compact. For this purpose, interruptions and perforations are introduced in the fin design. The objective of enhancing convective heat transfer can be achieved through perforating the fins which would result in a weight reduction of fins with the increased surface area as well [11, 12]. Heat transfer characteristics and flow movement observations were carried out by Narato et al. for circular pin fins at various inclination angles from 30° to 150°. It was observed that pin fin arrays that were inclined at 120° and 135° significantly enhanced the heat transfer when compared to the vertical pin fin [13].

According to a study conducted by Jassem, it was observed that an increase in the diameter of perforations also increases the heat transfer rate. Different geometries of perforations were applied, and it was also concluded that the heat performance of the perforated fin is also dependent on the perforation geometry [14]. Therefore, elliptical-shaped perforations were done on the rectangular plate fins by Gurav et al. The effect was studied through numerical simulations concluding that elliptical perforations on rectangular fins enhance the overall heat transfer rate [13]. Therefore, in the current study, different slots are being introduced on an array of pin fins to analyze their thermal performance.

2 Numerical Analysis

A. *Problem Description*

Conventional pin fins carry a lot of weight with less thermal performance. To curtail this problem, helical slots of 1mm^2 area are formed on the simple pin fins. The helical slots are also varied at different pitches to study their effect. This would shred off the material from the pin fins and in turn, also enhance the convective heat coefficient due to an increase in the surface area. Each heat sink is comprised of an array of 4 pin fins with different helical slot configurations on each array as shown in Fig. 1. The material of heat sink for the current numerical simulations is Aluminum with thermal conductivity of 239 W/m °K.

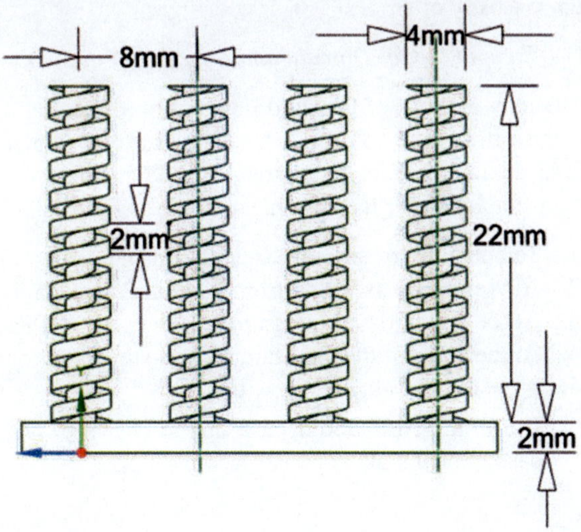

Fig. 1 Schematic diagram of Helical Fin with 2 mm pitch

B. *Heat Transfer Model*

When a heat sink is applied to the heat source, it extracts the heat from the source through conduction. Whereas, when air is passed through the heat fin, the convection phenomenon occurs. This phenomenon forms a conjugate heat transfer model. For the current study, air enters the domain at 26 °C. While air velocity is varied from 2 to 8 m/s with an interval of 2 m/s air velocity.

Fourier's conduction law, given in Eq. (1), was applied to solve the heat conduction model for all the considered simple pin fins and helical pin fins.

$$\nabla \cdot (k_s \nabla T_s) = 0 \tag{1}$$

where,

k_s = conduction coefficient for aluminum, 239 W/mK

T_s = Fin temperature.

For turbulent flow in the heat sink model, Reynold's averaged Navier Strokes (RANS), Eqs. (2) and (3), were applied. The time-averaging of the momentum, energy, continuity, and decay variables from mean and variable components were utilized to attain these equations.

$$\frac{\partial \rho}{\partial t} + \nabla.\rho \underline{U} = 0 \tag{2}$$

$$\frac{\partial \left(\rho \underline{U}\right)}{\partial t} + \nabla.\left(\rho \underline{U}\right)\underline{U} = \nabla.\left(\underline{\underline{\sigma}} - \rho \overline{\underline{U}'\underline{U}'}\right) \tag{3}$$

where $\underline{\underline{\sigma}} = -\mathrm{p}\underline{\underline{\mathrm{I}}} + \mu\left(\nabla\underline{\mathrm{U}} + |\nabla\underline{\mathrm{U}}|^{\mathrm{T}}\right)$ and $-\rho\overline{\underline{U}'\underline{U}'} = \mu_{\mathrm{t}}\left(\nabla\underline{\mathrm{U}} + |\nabla\underline{\mathrm{U}}|^{\mathrm{T}}\right) - \frac{2}{3}\left(\rho k \underline{\underline{\mathrm{I}}}\right)$ are the Newtonian and Reynolds stress tensors. p is the pressure density and ρ, μ is the viscosity of air.

C. *Computational Domain and Boundary Equations*

The inlet and out of the fluid domains were kept at a fairly away distance so that reverse flow may not occur. Heat sink array was placed perfectly in center of the fluid domain. The computational fluid dynamics (CFD) boundary conditions that were employed for the current study are as follows:

1. Temperature of air at the inlet is considered at room temperature of 26 °C
2. Air velocity was varied from 2 to 8 m/s with an increment of 2 m/s air velocity
3. At the outlet, the air pressure is $\mathrm{P} = \mathrm{P}_{\mathrm{gauge}} = 0$ Pa
4. At the walls of the fluid domain, no-slip conditions are applied
5. $\frac{\mathrm{dT}}{\mathrm{dx}} = 0$, $\frac{\mathrm{dT}}{\mathrm{dy}} = 0$, and $\frac{\mathrm{dT}}{\mathrm{dz}} = 0$
6. Uniform distribution of heat flux at $\mathrm{Q} = 39{,}777.25\ \mathrm{W/m^2}$ was provided at the base of heat sink

7. By considering the no-slip condition at the fluid domain i.e. air and heat sink interface, the heat flux is conserved.

D. *Solution Methods and Convergance Criteria*

ANSYS fluent was used for energy and momentum equations. Utilizing the second order up-winding solution, the solution was considered converged after the sum of residuals was <10^{-6}. However, for solving continuity and momentum equations, residuals sum of <10^{-4} was considered.

E. Grid Independence Test

The geometry of the heat sink is discretized into small elements up to a value that mesh independence is achieved. The mesh geometry of the independent elements of the mesh is tetrahedral in nature as shown in Fig. 2. A mesh independence test was conducted, which concluded that the number of elements chosen for the study is appropriate and any further increase in the elements does not alter the results significantly as shown in Table 1.

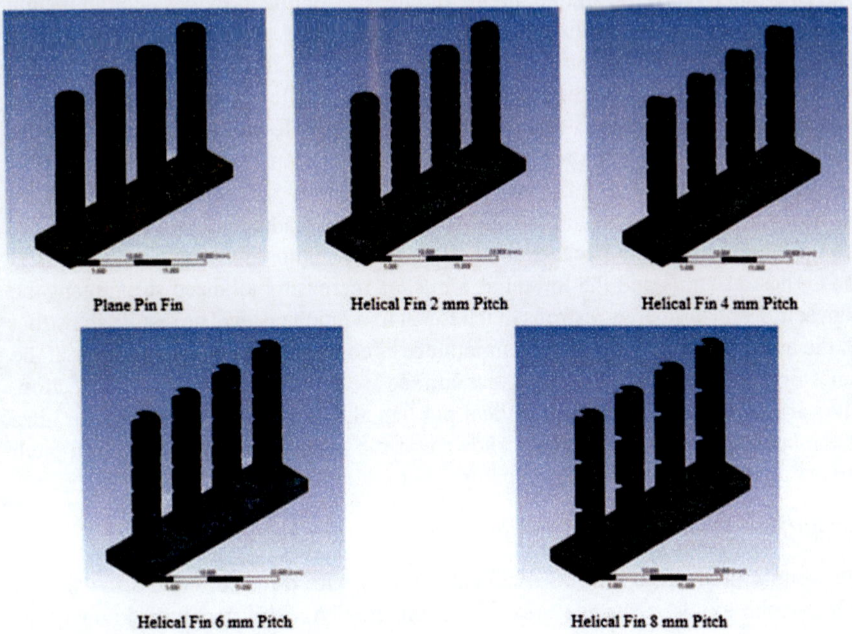

Fig. 2 Tetrahedral meshing grids

Table 1 Mesh independence test

Heatsink	Number of elements	Base Temperature (°C)
Plane fin	209,598	82.8
	215,087	83.1
	226,349	83.3
Fin with 2 mm pitch	84,801	72.7
	85,164	72.9
	87,383	73.2
Fin with 8 mm pitch	71,427	77.4
	73,564	77.6
	75,965	77.7

3 Results and Discussion

The CFD analysis of the simple pin fin and helical pin fins provides enough insight on the thermal aspects so that we can draw conclusions about the performance of the simple pin fin and helical fins. Moreover, a comparison is also presented in this study. It was observed that the modifications in the simple pin fin are significant and the introduction of the helix-shaped cutout in the simple pin fin has yielded better thermal performance. The profiles of the steady-state temperature of all the fins at 2, 4, 6, and 8 m/s air velocities are given in Figs. 3, 4, 5, 6 and 7. The base of all the heatsinks is the hottest and the temperature drops gradually as we move towards the tip due to conjugated heat transfer. The overall temperature of the heatsinks is the highest at 2 m/s and the lowest at 8 m/s, as increasing airspeed strengthens the convection and temperature drops much lower and much faster. The results show that all the helical pin fins have lower temperature as compared to the plain pin fins at the same airspeed, because of having larger surface areas subjected to forced convection. The variation of the pitch of the helical pin fins also affects the overall temperature of the heatsink. The CFD analysis shows that the helical pin fin with a 2 mm pitch drops the temperature the most as compared to all other pin fins.

A. *Surface Area Versus Fin Pitch Size*

The primary reason to introduce helical grooves with different pitch sizes was to increase the exposed surface area of the pin fins. As it is generally known that greater the surface area of the heat sink, greater will be the heat dissipation of the fin. Therefore, Fig. 8, presents the graph between the exposed surface area to the environment corresponding to the different pin fins at different pitch sizes. This will ultimately help us understand why any pin fin is performing better than the other ones.

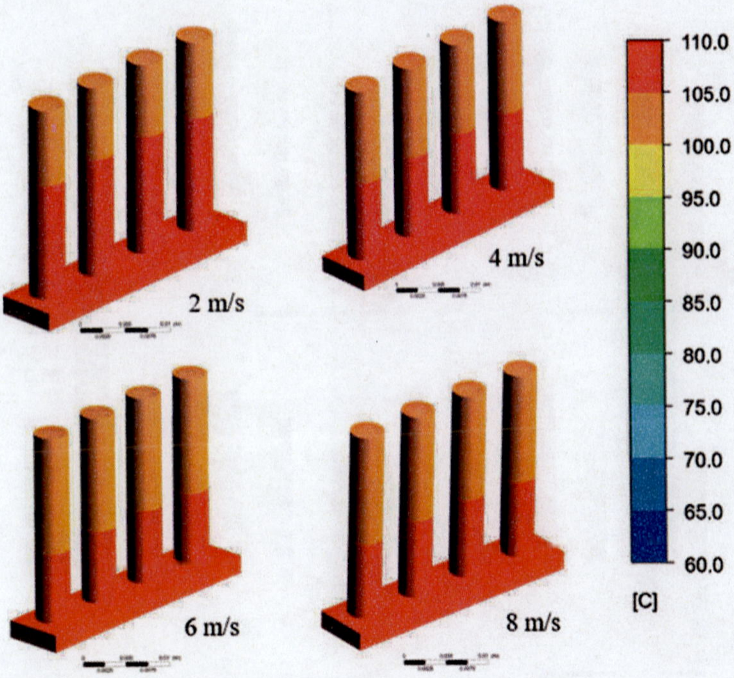

Fig. 3 Plane pin fins temperature contours

B. *Fin Base Temperature Versus Air Velocity*

When considering temperature at the base of the heat sink as a parameter for comparison among all the proposed designs, it was noted that the helical fin having a 6 mm pitch has the best performance. Whereas the simple pin fin without any modifications was ranked as the worst performing fin. The helical fin with 6 mm pitch reached a maximum temperature of 92.4 °C while, the simple pin fin reached a maximum temperature of 102 °C at 2 m/s air velocity, respectively. Likewise, for helical fin with 6 mm pitch and simple pin fin at the air velocity of 8 m/s, the temperature reached 65.4 and 79.2 °C, respectively. This clearly indicates the improved performance of helical pin fins as compared to the simple pin fin. Figure 9 below shows the comparison of all the fins at different wind speeds.

C. Fin Tip Temperature Versus Air Velocity

When the tip temperature of the fins is considered as a parameter for comparison, the helical fin with 2 mm pitch has the best performance. At a speed of 2 m/s, the temperature was 63.6 °C whereas simple pin fin had a temperature of 92.8 °C. For temperatures at 8 m/s air velocity, the helical fin with 2 mm pitch and simple pin fin had 42.5 and 70.85 °C, respectively. Figure 10 demonstrates a comprehensive idea of the performance of all the fins. The reason that fin with 2 mm pitch performed best is that it has the highest surface area of 1453 mm^2 that is responsible for convection.

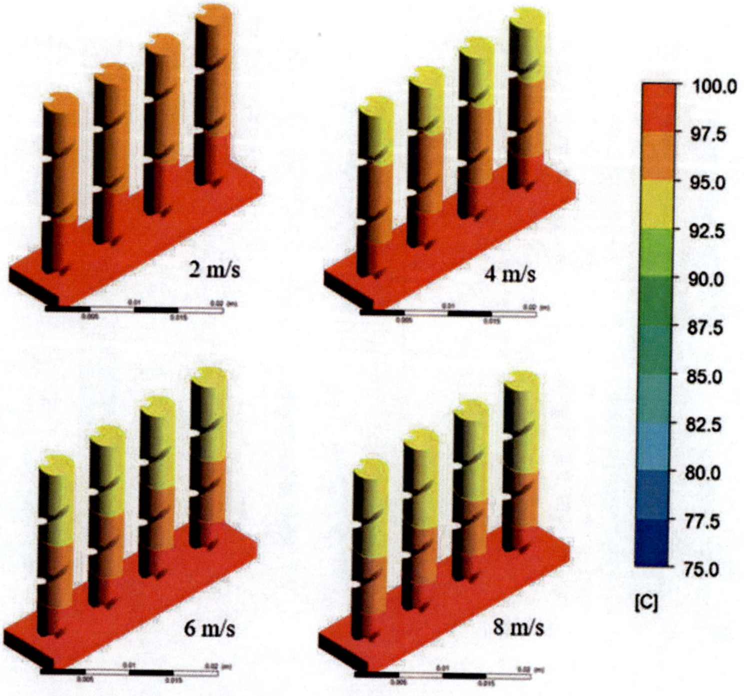

Fig. 4 Temperature contours of Helical Pin with 8 mm pitch

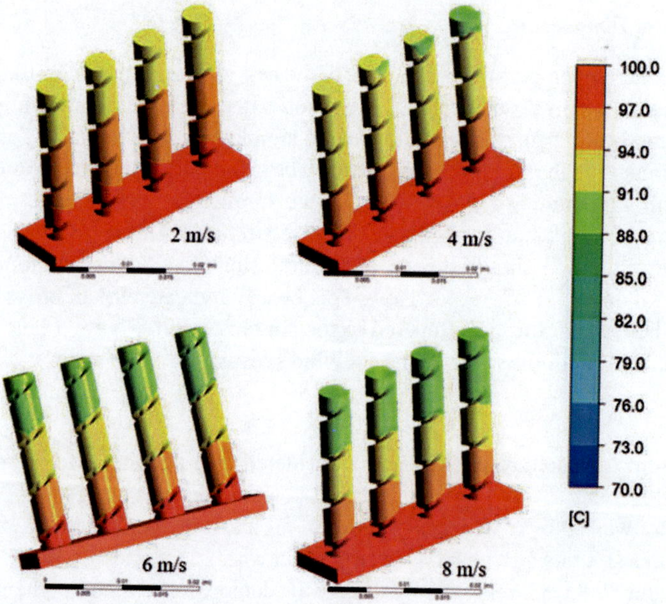

Fig. 5 Temperature contours of Helical Pin with 4 mm pitch

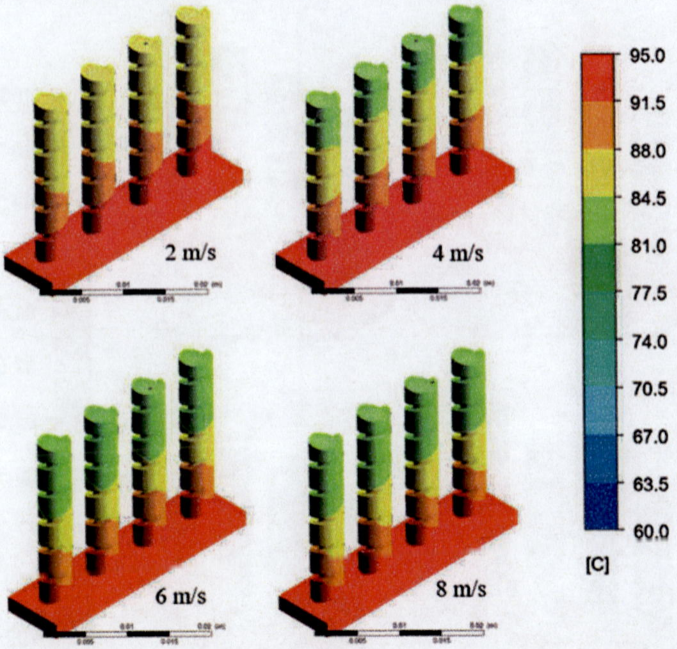

Fig. 6 Temperature contours of Helical Pin with 6 mm pitch

D. *Convection Coefficient Versus Air Velocity*

The convection coefficient of all the fins was calculated using Eq. (4). Helical fin with 2 mm pitch has the highest convection coefficient corresponding to all the air velocities as evident from Fig. 11. Whereas, helical fin with 8 mm pitch has the worst convection coefficient of 67.18 W/m^2K at 2 m/s air velocity. The reason for worst performance of 8 mm pitch among rest of the helical fins is that it has the lowest surface area of 1725 mm^2.

$$h = \frac{Q}{A_s(T_s - T_\infty)} \tag{4}$$

where,

T_s = Average surface temperature of heat sink
T_∞ = Average temperature at the inlet and outlet of test section
A_s = Surface area of heat sink responsible for convection.

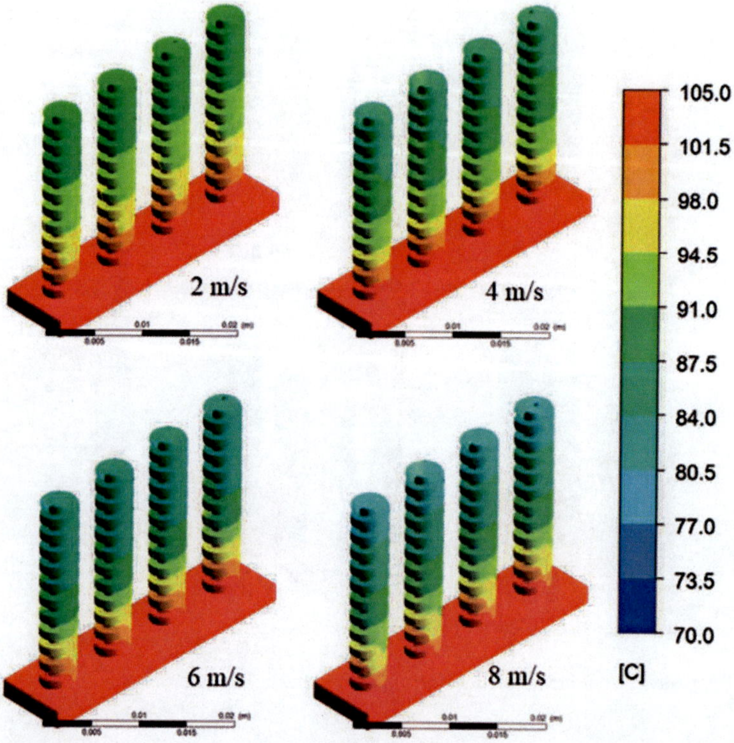

Fig. 7 Temperature contours of Helical Pin with 2 mm pitch

Fig. 8 Surface area versus pitch size

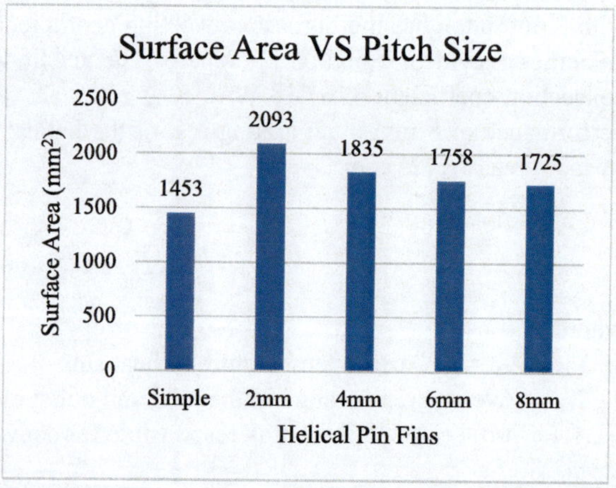

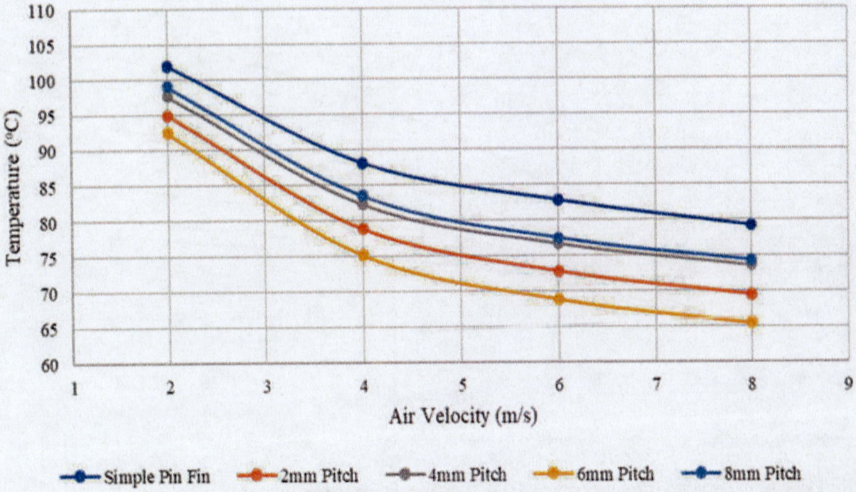

Fig. 9 Base temperature comparison at different air velocities

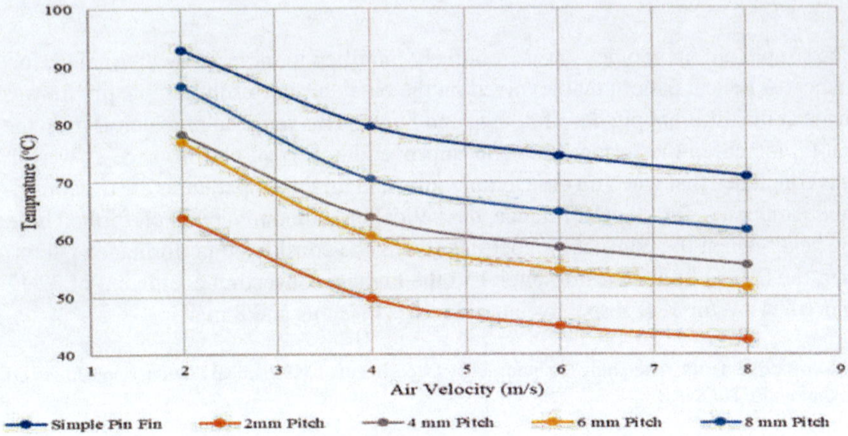

Fig. 10 Tip Temperature comparison at different air velocities

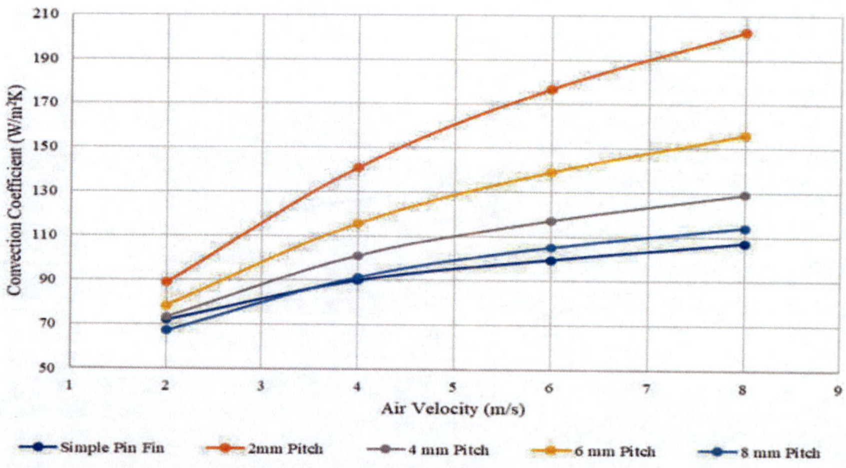

Fig. 11 Convection coefficient comparison at different air velocities

4 Conclusion

The simple pin fin geometry was iteratively modified to increase its thermal performance. A helical pattern that revolved on the outer surface of the simple pin fin was made at the different pitches of 2, 4, 6, and 8 mm. This removal of material from the pin fin increased the surface area and improved its thermal performance as well. It was concluded that when fin base temperature and fin tip temperatures are the considered parameters for fin performance, the helical fin with 6 mm pitch performed best. However, when the convection coefficient was the considered performance parameter, the helical fin with 2 mm pitch had the highest convection coefficient of 72.12 and 107.41 W/m^2K at respective air velocities of 2 m/s and 8 m/s.

Acknowledgements This study is supported by Department of Mechanical Engineering, Universiti Teknologi PETRONAS.

References

1. Mareš JJ, Hubík P, Šesták J, Špička V, Krištofik J, Stávek J (2008) Phenomenological approach to the caloric theory of heat. Thermochim Acta 474(1–2):16–24. https://doi.org/10.1016/j.tca.2008.05.001
2. Bejan A, Kraus AD (2003) Heat transfer handbook, 1st edn. Wiley
3. Pulkrabek WW (2013) Engineering fundamentals of the internal combustion engine, 2nd edn.
4. Tariq A, Altaf K, Rani AMA, Baharom M, Aziz ARA (2019) Study of heat transfer attributes of custom fins for Crank-Rocker engine block using ANSYS. J Adv Res. Fluid Mech Therm Sci 62(2):235–243

5. Li HY, Chen KY (2007) Thermal performance of plate-fin heat sinks under confined impinging jet conditions. Int J Heat Mass Transf 50(9–10):1963–1970. https://doi.org/10.1016/j.ijheatmasstransfer.2006.09.024
6. Tariq A, Altaf K, Ahmad SW, Hussain G, Ratlamwala TAH (2021) Comparative numerical and experimental analysis of thermal and hydraulic performance of improved plate fin heat sinks. Appl Therm Eng 182(August 2020):115949. https://doi.org/10.1016/j.applthermaleng.2020.115949
7. Altaf K, Tariq A, Ahmad SW, Hussain G, Ratlamwala TAH, Ali HM (2022) Thermal and hydraulic analysis of slotted plate fins heat sinks using numerical and experimental techniques. Case Stud Therm Eng 35(May):102109. https://doi.org/10.1016/j.csite.2022.102109
8. Tariq A, Altaf K, Hussain G, Reezwan M (2021) No Title. In: Clean energy opportunites in tropical countries. Springer, Singapore, pp 115–131
9. Wei XQ, Zhang YH, Hu WL, Wang LB (2016) The fluid flow and heat transfer characteristics in the channel formed by flat tube and dimpled fin. Int J Therm Sci 104:86–100. https://doi.org/10.1016/j.ijthermalsci.2015.12.018
10. Chang SW, Wu HW, Guo DY, Jie Shi J, Chen TH (2017) Heat transfer enhancement of vertical dimpled fin array in natural convection. Int J Heat Mass Transf 106:781–792. https://doi.org/10.1016/j.ijheatmasstransfer.2016.09.094
11. Shaeri MR, Yaghoubi M, Jafarpur K (2009) Heat transfer analysis of lateral perforated fin heat sinks. Appl Energy 86(10):2019–2029. https://doi.org/10.1016/j.apenergy.2008.12.029
12. Bar-Cohen A, Iyengar M, Kraus AD (2003) Design of optimum plate-fin natural convective heat sinks. J Electron Packag Trans ASME 125(2), SPEC:208–216. https://doi.org/10.1115/1.1568361
13. Gaurav RB, Patil JD (2013) Finite volume analysis of convective heat transfer augmentation from horizontal rectangular fin by elliptical perforations. Int J Glob Technol Inst 2:40–44
14. Shaeri MR, Yaghoubi M (2009) Numerical analysis of turbulent convection heat transfer from an array of perforated fins. Int J Heat Fluid Flow 30(2):218–228. https://doi.org/10.1016/j.ijheatfluidflow.2008.12.011

A Conceptual Design of Sustainable Solar Photovoltaic (PV) Powered Corridor Lighting System with IoT Application

John Yuan En Tin, Woan Wen Tan, Asfarina binti Abu Bakar, Mohd Syukur bin Mahali, Florence Francis-Lothai, Nurul Farahana Mohammad, Siti Syafinah Ahmad Hassan, and Kui Fern Chin

Abstract Electricity consumption keeps rising every year as the world's population and economic expansion increase. Conventional energy sources such as petroleum, coal, and natural gas continue to be the most widely used energy sources on the planet. They are used in heating, power generation, and as motor fuel. However, fossil fuels are not a suitable source of energy in the long run since they are depleting rapidly. The conventional energy system is incapable of managing and monitoring energy usage. Most of the research that has been done in the past has been on energy monitoring systems, especially on simulation models where the results obtained do not always match the real-time measurements. Hence, the initial investment in creating solar energy with an IoT system is crucial in order to overcome this issue. This paper

---

J. Y. E. Tin (✉) · W. W. Tan · A. A. Bakar · M. S. Mahali · F. Francis-Lothai · N. F. Mohammad · S. S. A. Hassan · K. F. Chin
Faculty of Science and Technology, Department of Electrical Engineering (Power), CATS University College, Kuching, Malaysia
e-mail: john@icats.edu.my

W. W. Tan
e-mail: tanww@icats.edu.my

A. A. Bakar
e-mail: asfarina@icats.edu.my

M. S. Mahali
e-mail: m.syukur@icats.edu.my

F. Francis-Lothai
e-mail: florence@icats.edu.my

N. F. Mohammad
e-mail: farahana@icats.edu.my

S. S. A. Hassan
e-mail: syafinah@icats.edu.my

K. F. Chin
e-mail: ckfern@icats.edu.my

F. Ahmad et al. (eds.), *ICREEM 2022*, Lecture Notes in Mechanical Engineering,
https://doi.org/10.1007/978-981-99-5946-4_8

is going to present the conceptual design of a sustainable solar photovoltaic (PV) powered corridor lighting system with IoT application. The overall system consists of six major components, which include a solar PV array, an inverter with maximum power point tracking (MPPT), battery storage, load (corridor lighting system), utility supply, and monitoring system. The inverter with MPPT is used to select the types of input for the load, either in PV mode, battery mode or utility mode. There are three switching methods to turn on the load in this design: either via IoT mode, bypass mode, or photocell mode. IoT mode requires internet technology to operate, while bypass mode is performed manually by the users according to their needs. On the other hand, photocell mode will operate based on the ambient light intensity. The proposed design is expected to have positive impacts on the environment, especially on energy usage and sustainability issues.

Keywords Energy · Solar · IoT · Lighting · Sustainability

1 Introduction

The sustainable power source application in Malaysia is still considered new and in a progressing state. The current power production and demand trends indicate that Malaysia has a reserve margin that will only last for a few years [1]. Therefore, it is vital to conduct extensive research work, particularly on sustainable resources such as renewable energy. This kind of energy has the potential to help reduce pollution and support strategies that can help with sustainable development [2].

The cost of energy consumption is facing upward pressure with the intensification of the world energy crisis and the advent of the era of the fossil fuel energy crisis. Countries across the world are looking for solutions to solve the energy crisis, such as exploring the use of new energy and renewable energy or finding new energy-saving technologies in order to reduce energy consumption and improve energy efficiency concurrently.

University buildings generally demand a lot of energy to power the lights, air conditioners, and even heavy equipment and machines. This contributes to an increase in the institution's overall operational costs. Alternative energy resources can be used to achieve the institution's long-term efficiency and sustainability goals while minimizing energy waste. Some other practices that can be implemented include installing energy-efficient lighting systems [3] and intelligent systems for energy management in buildings [4, 5].

Currently, all lighting in i-CATS University College buildings is controlled manually using switches. There are days when the lights are not required, and there are evenings when they must be switched on. It is a challenge to ensure that these lights are switched on in the presence of users during dark weather conditions while avoiding energy wastage when such lights are not required. A lighting control system that uses an IoT application and allows the user to control the switches with their smartphone is useful [6–8]. Photocell sensors can also be installed to facilitate the effective use of light in the darkness.

This paper presents a conceptual design for a sustainable solar photovoltaic (PV) powered corridor lighting system [9] for two blocks of buildings on the i-CATS University College campus. The solar PV system is expected to power at least 16 units of T8 20W LED tube lights with an average daily usage of 12 h. This system can also be monitored with an IoT application. It is user-friendly and capable of resolving issues with evening classes while also reducing the university's overall operational costs. The proposed system can also be used as a backup power supply for the buildings in case the main power supply fails.

2 Literature Review

The application of solar cells is growing rapidly as a source of sustainable energy. The system design can be diversified into standalone, grid-tied, and hybrid systems. Adaption with the Internet of Things (IoT) helps consumers monitor and analyse energy consumption automatically with electronic devices. Manual screening of energy consumption is time-consuming and needs to be examined by a trained electrician. Hence, there is a great interest in creating systems that can automatically monitor and analyse energy.

Since this project is engaged with the photovoltaic component of solar energy, a broad range of research has been conducted to study not only the integration with IoT but also other technologies such as passive infrared sensors (PIR), methods of identifying anomalies and inefficiencies for energy cost savings, and issues and challenges of relevant applications and controllers.

As mentioned by Moura [4], transforming buildings into nearly zero-energy buildings (nZEB) enables a high reduction in net electricity. They implemented smart energy services on a campus with the deployment of an IoT platform. Their developed system reduces grid feed-in, an energy storage system with lithium-ion batteries with a total capacity of 480 kWh with a minimum state of charge of 20% and a round trip efficiency of 85%. They measured the available area and orientation on the campus using Google Earth and Sunny Design to install and design the PV system. This proposed method uses smart CEI Moncloa and is based on a wireless sensor network to monitor and control energy. The system used the energy record of the building in the year 2019 to evaluate self-consumption of the generated energy and managed to achieve self-consumption of 86.4%.

Thomas [10] proposed a method to evaluate the performance of laboratory built-in perovskite solar versus commercial state-of-the-art technologies. They have developed a simple device powered by perovskite solar cells consisting of sensors and communication beacons. The system was applied to three locations with an IoT device and measured the light intensity logged during a number of days using a pyranometer. The proposed method is able to prove that perovskite solar cells are a potential power source for self-powered devices where their work has shown that perovskite solar cells can suffer from low shunt resistance at low intensity and carbon perovskite solar cells, in particular, suffer from a slow response to changing light intensity.

Araujo [11] used the Stochastic Petri Nets (SPN) availability model for the smart photovoltaic management system. Their method evaluated implementation costs between grid tie and hybrid solar by calculating the estimations of other measures besides steady-state availability, such as reliability. Their system was run on a building with 15 floors containing 4 apartments per floor, where each apartment has 3 smart plug installations (sensor plug). The system used sensors in the building to capture information and to monitor the amount of energy with adaption to cloud-computing infrastructure and one switch power. The proposed system was able to achieve 99.4494% and 99.4496% availability analysis of the autonomous energy management system. Meanwhile, the system was also able to analyse reduction cost or financial return between grid-tied and hybrid systems, where the grid-tied showed energy produced in a year was 4,621.78 kWh with a cost of $462.17 as compared to the hybrid system which was 4,562.91 kWh with a cost of $1,277.61.

A system designed by Abuzairi [12] managed to utilize the solar module more efficiently with a combination of solar charge controller (SCC) and Maximum Power Point Tracking (MPPT). The proposed system uses an Arduino microcontroller (Arduino Nano) to control the circuit and battery and to produce PWM signals to regulate the DC-DC converter. It was conducted with several tests to validate the efficiency of the MPPT algorithm and managed to achieve the highest efficiency of 78% with MPPT and 26% without MPPT. Meanwhile, at low irradiance levels, their proposed system increased SCC efficiency by up to 52%.

According to Somefun [13], they have developed an electrical system that automatically controls the working operation of electrical appliances within a house to maintain comfort, minimize energy consumption levels, and save money over a period of time. Their proposed system uses a 32-bit ESP8266 system on chip (SoC) wireless fidelity (Wi-Fi) module which is linked to some sensors through a microcontroller for data communication. They have designed a user-friendly remote access which monitors the operation and management of the system with the Cayenne IoT platform and an ESP-01 Wi-Fi module. Their system showed a 17% savings in energy consumption, which in turn reduces the cost of energy usage for the same comfort level.

Meanwhile, Khaleel [14] proposed a low-cost data acquisition system for monitoring electrical and environmental parameters in a solar station. To measure and monitor the solar panel system, they used a Liquid Crystal Display LCD 4 20 connected to an I2C interface module, a serial monitor connected via a USB cable to the Adriano board, and a Wi-Fi module (ESP3266) on a smartphone or PC. The system results showed the characteristics of the solar cell completely identical to the practical results and data collected by several sensors, yielding errors of less than 1%.

Politi [15] has designed a simple and reliable method that provides a model-based evaluation of the harvestable energy from any real indoor light environment. The prototype system was a micro source of a classic wireless e-ink Wi-Fi connected device, with an average power consumption of around 10 mW. The developed system started with applying the Shockley-Queisser (SQ) model and comparing it to some of the usual real-life PV converters. A commercial flexible thin-film Gallium-Arsenide (GaAs) PV solar cell from Alta Devices Inc. Kayes [16] under several controlled single sources of light was calculated to present an energy harvesting system. The proposed system showed a mean absolute percentage error (MAPE) between the model curve and the experimental power curve of each of the 21 days was lower than 6%.

Sales [17] proposed a monitoring and tracking PV system for photovoltaic water pumping systems with a Li-Ion battery. Their system used a 2.4 kWp photovoltaic field, a 3.6 kVA hybrid inverter, a 3.3 kWh/3 kW lithium-ion battery, a 2.2 kVA variable speed driver, and a 1.5 kW submersible pump, which were connected with the Modbus-RTU protocol over an RS485 bus and open software for monitoring the system. The proposed system was compared between PVWPS + LIB (Lithium-ion batteries) and direct photovoltaic water pumping systems (DPVWPS), which shows that the system demonstrates perfect tracking of the MPP for all levels of irradiance, without the irradiance threshold common in DPVWPS schemes.

Yang [18] has evaluated and designed a smart campus service discovery algorithm and a perceptual data fusion method, as well as examined the challenges associated with integrating various forms of service data into smart campuses. However, Jing [19] demonstrates how resource waste in higher education institutions can be eliminated by performing intelligent control for energy savings on a campus using multidimensional situational information fusion approaches, while the incorporation of modern technologies such as on-cloud and big data architecture into a traditional university campus is proposed by Villegas-Ch [20]. In order to promote sustainable energy consumption, Hannan [21] provided an in-depth examination of various types of Internet-of-energy-based building energy management systems, such as energy routers, storage systems, and materials, along with renewable technologies. Mylonas [22] demonstrated how IoT platforms can be used in institutes to encourage energy-saving behaviours supported by data collected by such platforms, as well as possible avenues for identifying anomalies and inefficiencies based on preliminary findings from a study of school-building data.

Kekre and Gawre [23] developed a low-cost IoT application based on an integrated solar PV monitoring system that uses a GPRS module and a microcontroller to send measured data that can then be retrieved from anywhere on the planet, assisting us with maintenance and fault detection, as well as keeping a record of all data at predetermined intervals. Nevertheless, Tejwani [24] developed a novel technique for remote monitoring of solar photovoltaic systems in rural applications using GSM voice channels, so as to overcome the limitations of other strategies in the form of an analogue signal between transmitter and receiver, while Lopez-Vargas [25] proposed a real-time monitoring system for solar home systems based on Arduino microcontroller with a 3G connection is discussed, with an emphasis on the technical challenges faced and solutions applied, as well as the platform's functionality, services, and potential. Badave [26] proposed a wireless remote monitoring system for solar power plant with a validation concept that uses eight solar panels to track string voltage, string current, temperature, and humidity via a CC3200 microcontroller with an ARM Cortex-M4 architecture.

Sahraei [27] proposed a low-cost and gradually deployable occupancy detection system based on battery-powered wireless sensor nodes. Throughout their investigation, they used PIR to determine occupancy. They showed that by putting this information about occupancy into a building simulation model, they could reduce the amount of energy used for heating, ventilation, and air conditioning (HVAC) by 10% to 15%.

Lakra [28] developed an automated method for controlling electric equipment based on the number of people in the lecture room. In their research work, they created an automated energy-saving solution for university lecture halls that uses PIR to control electrical gadgets automatically. After deploying the technology, they were able to manage the university's electricity waste.

Table 1 presents a comparison of previous research and studies for the development of solar panels with the adaption of IoT applications.

3 Methodology

The conceptual design of this project is comprised of six main components: solar PV array, inverter with Maximum Power Point Tracking (MPPT), battery storage, load, utility supply, and monitoring system. The correlation between these components is illustrated in Fig. 1.

The solar PV array, battery storage, and utility supply serve as the input power supply for the load via inverter with MPPT. If the inverter is in PV mode, then the solar array will become the main power supply for the load. Concurrently, the battery will be charged by the solar array as well. If the solar PV array is not available, the inverter will be switched to battery mode, and the load will operate using the power stored in the battery storage. Utility mode will be turned on only if both the solar PV and battery are unavailable. The utility supply will then serve as the main power supply for the load and charge the battery simultaneously. A monitoring system is connected to the inverter to monitor the conditions and readings of the solar PV array, battery, utility supply, and load.

The overall flow chart of the system is divided into Part I and Part II as shown in Fig. 2 and Fig. 3. Part I shows the three types of modes in the system. The system will operate either in PV mode, battery mode, or utility mode.

Three conditions have to be fulfilled for the inverter in PV mode to operate. First, the solar PV array must have sufficient energy gain. Secondly, the amount of current (ampere) supplied has to be lower than the fuse rated value in order to fulfill the safety purpose. Third, the PV isolator has to be switched on. If these three conditions are fulfilled, then the inverter will be in standby mode for the load in part II. The battery will also be charged by the solar PV array when the PV mode is activated.

If either one of the aforementioned three conditions is not achieved, then the inverter will be switched to battery mode. In this mode, the battery will serve as

Table 1 Development of solar panels with adaption of iot application

Name	Year	Method	Major findings
Somefun et al. [13]	2022	The proposed method uses the 32-bit ESP8266 system on chip (SoC) wireless fidelity (Wi-Fi) module, which is linked to some sensors through a microcontroller for data communication	The main contribution of this paper is a new approach that was deployed to achieve a demand-side energy management scheme which does not compromise the comfort of the users
Khaleel et al. [14]	2022	To measure a solar panel system, the proposed method employs a Liquid Crystal Display LCD 4 20 connected to an I2C interface module, a serial monitor connected via a USB cable to the Arduino UNO board, and a Wi-Fi module (ESP3266) on a smartphone or PC	The main contribution of this paper is to develop a low-cost data acquisition system for monitoring electrical and environmental parameters in a solar station
Thomas et al. [10]	2022	This proposed method uses perovskite solar cells, consisting of sensors and communication beacons, to measure the energy yield for a given PV technology in various situations	The main contribution of this paper is to evaluate the performance of laboratory-built perovskite solar cells versus that of commercial state-of-the-art technologies
Moura et al. [4]	2021	This proposed method uses smart CEI Moncloa based on a wireless sensor network to monitor and control energy	The main contributions of this paper are an integrative perspective that includes the energy option to achieve a sustainable energy campus and the deployment of an IoT platform that ensures the necessary conditions to implement smart energy services
Araujo et al. [11]	2021	The proposed method uses cloud computing and the Internet of Things (IoT) to perform the building's energy management	The main contribution of this paper is designing a dependability model to evaluate an IoT system
Politi et al. [15]	2021	The proposed method uses the micro source of a classic wireless e-ink wifi connected device with an average power consumption of 10 mW to harvest energy and the Shockley-Queisser (SQ) model to compare with some usual real-life PV converters	The main contribution of the paper is to develop a simple and reliable method that provides a model-based evaluation of the harvestable energy from any real indoor light environment

(continued)

Table 1 (continued)

Name	Year	Method	Major findings
Sales et al. [17]	2020	The proposed method uses a 2.4 kWp photovoltaic field, a 3.6 kVA hybrid inverter, a 3.3 kWh/3 kW lithium-ion battery, a 2.2 kVA variable speed driver, and a 1.5 kW submersible pump, which are connected with Modbus-RTU protocol over an RS485 bus and open software for monitoring system status	The main contribution of this paper is to develop a small-scale prototype of an autonomous solar water pumping system intended for the constant supply of water without the use of fossil fuels
Abuzairi et al. [12]	2019	The Arduino Nano microcontroller is used to control the circuit and battery to make PWM signals that control the DC-DC converter	The main contribution of this paper is that the designed SCC with MPPT technology was able to manage maximum power tracking and control battery charging
Lopez-Vargas et al. [25]	2019	The proposed method uses Arduino microcontroller with a 3G connection for solar home systems	The main contribution of this paper is that the suggested technique showed the system's resilience and the reliability
Badave et al. [26]	2018	The proposed method uses a CC3200 microcontroller with an ARM Cortex-M4 architecture for solar power systems that monitor respective parameters	The main contribution of this paper is that it improves overall system performance while taking up less space and makes it easier to check system metrics every time interval
Sahraei et al. [27]	2018	This proposed method uses a low-cost and progressively deployable occupancy detection system based on wireless sensor nodes that are operated by batteries	The main contribution of this paper is that they could reduce the amount of energy needed for HVAC by 10% to 15%
Kekre and Gawre [23]	2017	This proposed method uses an IoT application that employs GPRS module and a microcontroller	The main contribution of this paper is to help in maintenance, defect detection, and data recording
Tejwani et al. [24]	2014	The proposed method uses GSM voice channels for data transfer for remote monitoring of solar PV systems in rural settings	The main contribution of this paper is the capacity to overcome the restrictions of existing strategies in the form of an analogue signal between transmitter and receiver

Fig. 1 Simplified block diagram of the overall system

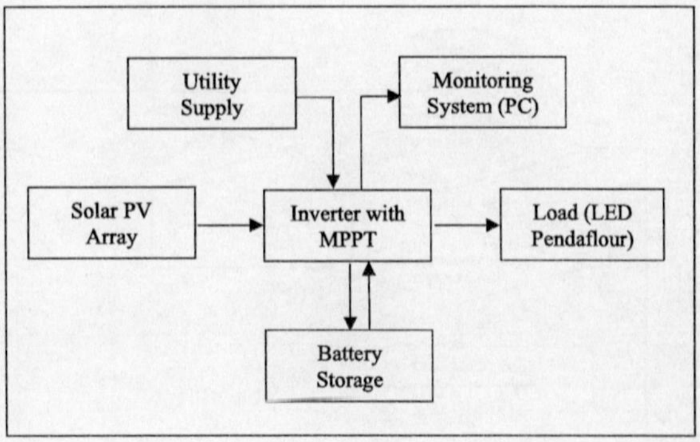

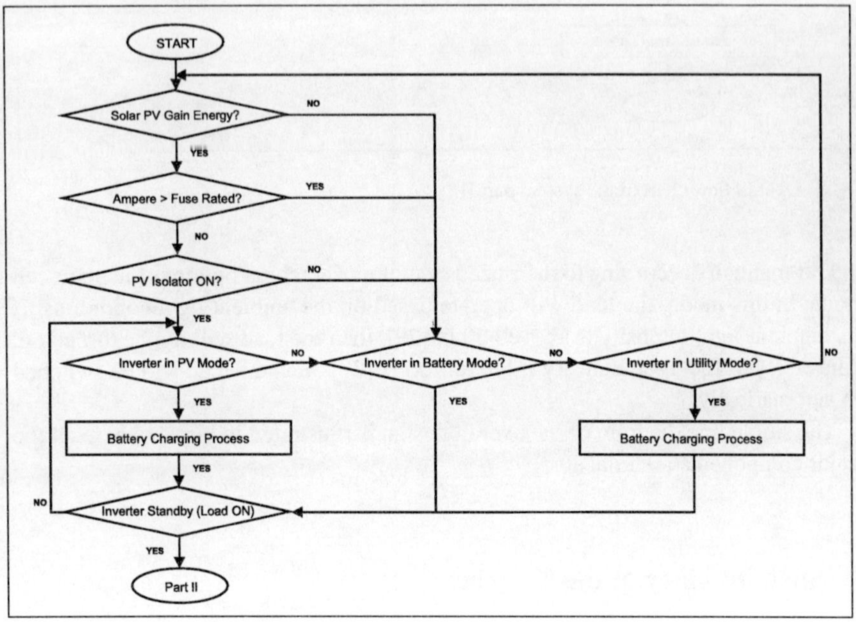

Fig. 2 Overall flow chart of the system part I

the main power supply for the load. If the PV mode is not available and the battery storage is not sufficient enough to power the load, then the inverter will be switched to utility mode. In this mode, the load draws power directly from the utility source, and the battery is charged at the same time.

Part II of the flow chart shows the three switching options available for the load. The load can be turned on by the selector switch either through IoT mode, bypass mode, or photocell mode. If the switch for IoT mode is selected, then the users can turn on and off the load remotely as long as they are connected to the internet via IoT platform. If the switch is in bypass mode, then the users can turn the load on

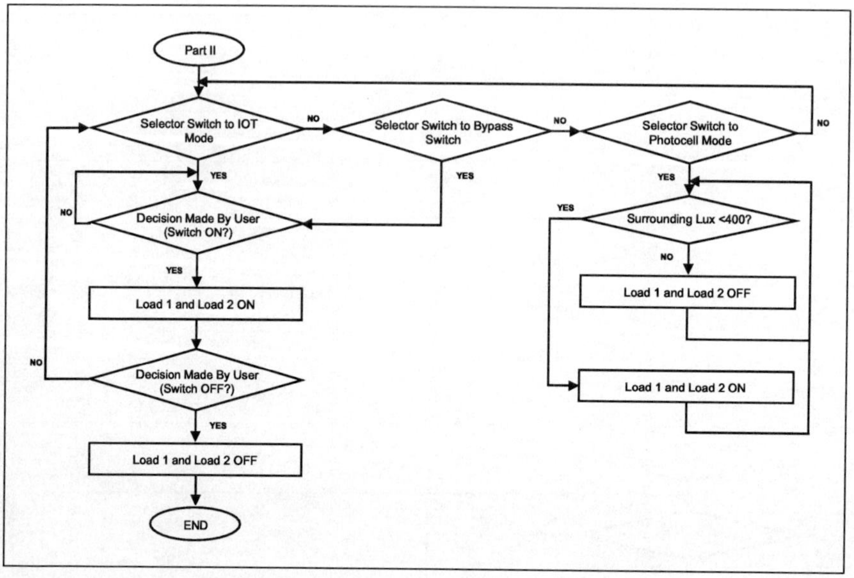

Fig. 3 Overall flow chart of the system part II

and off manually according to their needs. Another switching option is the photocell mode. In this mode, the load will operate based on the ambient light conditions. If the ambient light intensity is above 400 lx [29], then the load will not be turned on. Conversely, if the light intensity is below 400 lx [29], then the load will be switched on automatically.

The single line diagram of the overall design is illustrated in Fig. 4. where all the major components are indicated.

4 Sustainability of the Project

This project is going to be a long-term, progressive-based project to capitalize on its expected positive outcomes. In order to preserve the sustainability aspect of the project, constant maintenance and calibration will be carried out to ensure that all the components of the system are in optimum condition. If the project produces the expected results, the project's scope can be expanded further to diversify the applications of the designed system, such as by incorporating a basic ventilation system or even an air-conditioning system into the overall system design. The sustainability of the project can also be enhanced by the implementation of different IoT attributes into the system. To improve the project's feasibility, additional features such as smart remote control on solar PV array, monitoring of dust accumulation in PV system,

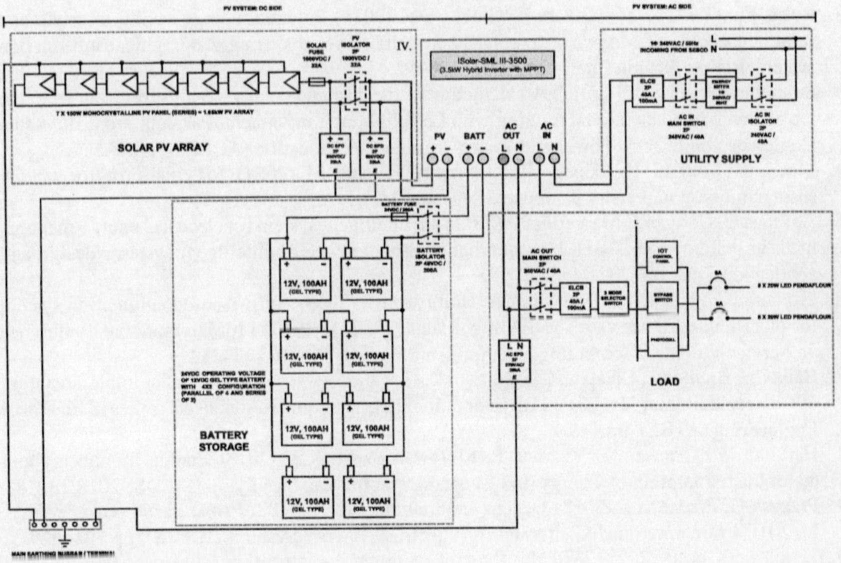

Fig. 4 Single line diagram of the overall design

monitoring of PV power, array inspection, tracking system, fault and control, and sunlight intensity [30] can be added to the system.

5 Conclusions

In conclusion, a conceptual design of a sustainable solar photovoltaic (PV) powered corridor lighting system with IoT application is proposed and discussed. The system consists of three operation modes: PV mode, battery mode, and utility mode. Only one of these modes will operate each time according to the settings of the system. The load (lighting system) can be turned on and off via three switching methods: either through IoT technology, bypass switch, or photocell mode. The proposed project is expected to have positive impacts on the environment, especially in the sustainability aspect as well as energy usage.

Acknowledgements Special thanks to i-CATS UC for providing the funding for this paper.

References

1. Basri NA, Ramli AT, Aliyu AS (2015) Malaysia energy strategy towards sustainability: a panoramic overview of the benefits and challenges. Renew Sustain Energy Rev 42:1094–1105

2. Rosas-Flores JA, Zenón-Olvera E, Gálvez DM (2019) Potential energy saving in urban and rural households of Mexico with solar photovoltaic systems using geographical information system. Renew Sustain Energy Rev 116:109412
3. Siddiqui MF, Soleja TJ (2017) A calculation of the cost and energy savings resulting from the replacement of conventional ligthing with LED lighting for Karachi, Pakistan. In: 2017 Asian Conference on Energy, Power and Transportation Electrification (ACEPT). pp 1–3
4. Moura P, Moreno JI, López GL, Alvarez-Campana M (2021) IoT platform for energy sustainability in university campuses. Sensors 21(2):1–22
5. Anastasi G, Corucci F, Marcelloni F (2011) An intelligent system for electrical energy management in buildings. In: 2011 11th international conference on intelligent systems design and applications. pp 702–707
6. Motlagh NH, Khajavi SH, Jaribion A, Holmstrom J (2019) An IoT-based automation system for older homes: a use case for lighting system. In: 2018 IEEE 11th international conference on Service-Oriented Computing and Applications (SOCA). pp 247–252
7. Haba CG, Breniuc L, Gălăţanu CD, Petrişor D, Herţanu R (2019) IoT system for implementation of connected lighting. In: 2019 11th international symposium on Advanced Topics in Electrical Engineering (ATEE). pp 1–5
8. Hartman WT, Hansen A, Vasquez E, El-Tawab S, Altaii K (2018) Energy monitoring and control using Internet of Things (IoT) system. Syst Inf Eng Des Symp (SIEDS) 2018:13–18
9. Prakash G, Pradeepa S (2018) Design and modeling of MPPT for solar based power source. In: 2018 4th International Conference on Electrical Energy Systems (ICEES), pp 624–629
10. Thomas SK et al (2022) Will the Internet of things be perovskite powered? Energy yield measurement and real world performance of perovskite solar cells in ambient light conditions. IoT 3:109–121
11. Araujo E, Pereira P, Maciel P (2021) Dependability impact in the smart solar power systems: an analysis of smart buildings. Energies 14:1–24
12. Abuzairi T, Ramadhan WWA, Devara K (2019) Solar charge controller with maximum power point tracking for low-power solar applications. Int J Photoenergy, Hindawi 1–11
13. Somefun TE, Abdulkareem A, Awosope CO, Akanji O (2022) Smart home comfort and energy conservation using internet of things. TELKOMNIKA Telecommun Comput Electron Control 20(2):357–365
14. Khaleel KK, Atyia TH, Al-Naib AMI (2022) Design and development of real-time data acquisition of photovoltaic panel parameters via IoT. NTU J Renew Energy 3(1)
15. Politi B et al (2021) Practical PV energy harvesting under real indoor lighting conditions. Sol Energy 224:3–9
16. Kayes BM et al (2011) 27.6% conversion efficiency, a new record for single-junction solar cells under 1 sun illumination. In: 2011 37th IEEE photovoltaic specialists conference. pp 000004–000008
17. Sales FJG et al (2020) PV monitoring system for a water pumping scheme with a lithium-ion battery using free open-source software and IoT technologies. Sustainability 12(24):10651
18. Yang AM, Li SS, Ren CH, Liu HX, Liu L (2018) Situational awareness system in the smart campus. IEEE Access 6:63976–63986
19. Jing X, Li S, Cheng J, Guo J (2020) Multidimensional situational information fusion method for energy saving on campus. J Intell Fuzzy Syst 38:4793–4807
20. Villegas-Ch W, Palacios-Pacheco X, Lujan-Mora S (2019) Application of a smart city model to a traditional university campus with a big data architecture: a sustainable smart campus. Sustainability 11(2857):1–28
21. Hannan MA et al (2018) A review of internet of energy based building energy management systems: issues and recommendations. IEEE Access 6:38997–39014
22. Mylonas G, Amaxilatis D, Chatzigiannakis I, Anagnostopoulos A, Paganelli F (2018) Enabling sustainability and energy awareness in schools based on IoT and real-world data. IEEE Pervasive Comput 17:53–63
23. Kekre A, Gawre SK (2017) Solar photovoltaic remote monitoring system using IOT. In: International conference on recent innovations in signal processing and embedded systems (RISE). pp 619–623

24. Tejwani R, Kumar G, Solanki C (2014) Remote monitoring for solar photovoltaic systems in rural application using GSM voice channel. Energy Procedia 57:1526–1535
25. Lopez-Vargas A, Fuentes M, Vivar M (2019) IoT application for real-time monitoring of solar home systems based on Arduino with 3G connectivity. IEEE Sens J 19(2):679–691
26. Badave PM et al (2018) Health monitoring system of solar photovoltaic panel: an internet of things application. Adv Smart Grid Renew Energy 347–355
27. Sahraei N, Looney EE, Watson SM, Peters IM, Buonassisi T (2018) Adaptive power consumption improves the reliability of solar-powered devices for Internet of things. Appl Energy 224:322–329
28. Lakra M, Kiran KV, Chinara S (2016) Design of smart and intelligent power saving system for Indian universities. In: Proceedings of first international conference on information and communication technology for intelligent systems, vol 1. pp 245–253
29. Alshikh Z, Abufarha S, Zannerni R, Mushtaha E, Altan H (2018) Analysis of lighting performance in residential unit converted into nursery. In: Zemch 2018 international conference. pp 1–15
30. Priharti W, Rosmawati AFK, Wibawa IPD (2019) IoT based photovoltaic monitoring system application. J Phys Conf Ser 1367(1)

Studies of $BaTiO_3$/PVDF-Based Nanocomposites as Nanogenerator Application

Ahmad Firdaus Che Omar, Tunku Ishak Tunku Kudin, Ainnur Izzati Kamisan, Ainnur Sherene Kamisan, Mohamad Fariz Mohamad Taib, Oskar Hasdinor Hassan, and Ahmad Sukri Ahmad

Abstract The performance of energy harvesting nanogenerator based on $BaTiO_3$-PVDF has been investigated. The as-prepared $BaTiO_3$-PVDF (BaPV50%), showed a maximum voltage peak to peak of 22 V and a short current of 0.164 nA at an applied force of 15 N, BaPV50% displayed an output energy power of 2.2194 nW and power density of 0.7065 nW/cm^2. All fabricated energy harvester devices have excellent stability even after a 500 pressing releasing cycle where the output was recorded without any external poling process; indicating that BaPV50%-NG is a promising flexible energy harvester.

Keywords Piezoelectric · Polymer · Reduced graphene oxide · Nanogenerator

---

A. F. C. Omar (✉) · A. S. Kamisan · M. F. M. Taib
Faculty of Applied Sciences, Universiti Teknologi Mara, Shah Alam, Malaysia
e-mail: afcheomar10@gmail.com

A. S. Kamisan
e-mail: ainnur@uitm.edu.my

M. F. M. Taib
e-mail: mfariz@uitm.edu.my

T. I. T. Kudin
Faculty of Defense Science and Technology, Universiti Pertahanan Nasional Malaysia, Sungai Besi, Malaysia

A. I. Kamisan · O. H. Hassan
Ionic Materials and Devices (iMADE) Research Laboratory, Institute of Science, Universiti Teknologi Mara, Shah Alam, Malaysia
e-mail: oskar@uitm.edu.my

A. S. Ahmad
PETRONAS Group Research & Technology, PETRONAS Research Sdn.Bhd. (PRSB), Bangi, Malaysia
e-mail: ahmadsukri.ahmad@petronas.com

F. Ahmad et al. (eds.), *ICREEM 2022*, Lecture Notes in Mechanical Engineering,
https://doi.org/10.1007/978-981-99-5946-4_9

1 Introduction

Currently, the emerging development and application of AIoT (Artificial Intelligence of Things) have increased the demand for energy generation. Non-renewable and renewable energy sources are commonly used in energy generation that supports human social and economic development, welfare, and health. Subsequently, most of these non-renewable sources have a drawback due to energy crises as well as environmental pollution. Renewable energy sources such as the sun, ocean, and wind, on the other hand, are examples of green technology, however, the main disadvantages of these technologies are dependent on environmental conditions. Immensely, energy harvesting using a piezoelectric nanogenerator (PENG) has attracted the attention of researchers as one alternative way to harvest energy [1]. PENG has a feasible approach for capturing energy from the environment or kinetic/mechanical energy and converting it into electrical energy [2]. The piezoelectric effect or piezoelectricity is a phenomenon in which an electric potential is formed across a piezoelectric material when external pressure is applied to it, and this potential can generate electricity [3, 4]. To date, several PENG made of different organic piezoelectric materials such as Zinc Oxide (ZnO), lead zirconate titanate (PZT), barium titanate ($BaTiO_3$), polyvinylidene fluoride (PVDF), aluminum nitride (AlN) and lithium niobate ($LiNbO_3$) commonly has been used in energy harvesting system [5–10].

Polymer piezoelectric PVDF has been widely used in flexible PENG applications due to its excellent mechanical flexibility, low cost, chemically stable, and ease of fabrication. However, PVDF has a low dielectric constant (k), and low piezoelectric charge constant (d_{33}) and needs an external poling process to increase efficient conversion [11, 12]. Therefore, to increase piezoelectricity performance in PENG application, the employment of ceramic-based material would increase the K and d_{33} value of the composite. $BaTiO_3$ is a lead-free material which exhibits a high piezoelectric coefficient value range of 350–460 pC/N [13]. Despite having a high piezoelectric coefficient, $BaTiO_3$ is highly fragile to stress and strain effects due to its ceramic nature [14]. Thus, $BaTiO_3$ material in energy harvesting applications would be limited. Integrating $BaTiO_3$ ceramic with PVDF polymer would resolve the mechanical flexibility, further improving the energy output performance of the PENG application [15].

The initial part of this investigation is to determine the optimal PVDF content in $BaTiO_3$-PVDF composites that yield high energy production. Five different samples with PVDF content of 10, 20, 30, 40, and 50 wt% in the $BaTiO_3$-PVDF nanocomposite are prepared and characterized. All PENG nanocomposites were then fabricated by spin-coating onto indium tin oxide/polyethylene terephthalate (ITO/PET) substrate and depositing a gold (Au) electrode on top of the thin film. The energy output of PENG measurement was assessed using the open-circuit voltage, short-circuit current of the fabricated PENG nanocomposite.

2 Experimental Section

2.1 Preparation of $BaTiO_3$-PVDF Nanocomposite

$BaTiO_3$ was synthesized using the sol–gel technique and calcined at 1100 °C [16] and commercial medium molecular weight (543 K) Polyvinylidene fluoride was obtained from Sigma Aldrich for nanocomposite preparation. Firstly, 1 g of $BaTiO_3$ nano-powder with a high crystalline structure was dispersed in 5 ml of N–N Dimethyl-formamide (DMF) for 2 h at 30 °C. Then 0.5 g PVDF was dispersed in 10 ml (DMF) for 4 h at 40 °C. $BaTiO_3$ solution then was mixed with PVDF solution and sonicated using ultrasonic for 1 h at room temperature to obtain a homogenous nanocomposite dispersion. A series of $BaTiO_3$ with different weight percentages from 10 to 50 wt% of PVDF was prepared.

2.2 Fabrication of $BaTiO_3$/PVDF PENG

To fabricate the $BaTiO_3$/PVDF nanocomposite PENG, the ITO plastic substrate was first peeled and verified the conductive part using a multimeter. Then the conductive part was cleaned with methanol to remove any residue. All five samples were spin-coated on the ITO substrate followed by heat treatment at 80 °C for 5 min. Then gold (Au) was coated on top of the $BaTiO_3$/PVDF nanocomposite using a sputter coater for 30 s. The coated area was sealed with Kapton tape to encapsulate the active material. This encapsulation protects active material from damage or peels from the ITO substrate. The fabricated nanocomposite was labeled as BaPV10%, BaPV20%, BaPV30%, BaPV40% and BaPV50% for nanocomposite with PVDF concentrations 10, 20, 30, 40, and 50 wt% respectively. The surface of the active layer of the PENG nanocomposite is 2 × 2 cm (length × width).

2.3 Characterization

Phase transition analysis was carried out using an X-ray diffractometer (XRD; XpertPro, Malvern Panalytical, UK) with a Cu Kα ($\lambda = 1.542$ Å) radiation source (45 kV, 40 mA). The sample was scanned from 0° to 90° with normal rate scanning. The surface morphologies of the composite thin film were examined via field emission scanning electron microscopy (FESEM; JSM-7600F Joel, Japan). Electrical characterization of open-circuit voltage and short-circuit current of nanogenerator was conducted using an oscilloscope (InfiniVision, DSOX3014T, Keysight, Malaysia) and source meter (2400, Keithley, Singapore) connected to a PC by utilizing the Keysight BenchValue software for voltage data storage and Kickstart2 software for current data collection.

3 Result and Discussion

(1) *Crystal Structure and Morphology Study*

XRD analysis of $BaTiO_3$, PVDF, and $BaTiO_3$/PVDF nanocomposite is carried out to determine the detailed crystallographic structure of the materials. Figure 1 shows the XRD pattern of the $BaTiO_3$ indicating the tetragonal structure of the $BaTiO_3$ phase having peak diffraction at (002), (102), (112), and (202). Meanwhile, the XRD pattern of the pristine PVDF is presented in Fig. 2. It can be observed that there were three prominent peaks at $2\theta = 18.4°$, 19.8°and 26.4° present confirming that the PVDF used in this work was α-phase [17]. There are two broad peaks observed in the XRD located approximately between 17.9° and 26° while the second peak appeared at a 2θ range of 30°–45°. $BaTiO_3$/PVDF nanocomposites were successfully obtained from the ultrasonic homogenizer method. Figure 3 shows the XRD pattern observed at 10°–80° range showing the BaPV nanocomposite with different weight percentages of PVDF. The BaPV nanocomposite XRD pattern shows an identical peak with $BaTiO_3$ (BTO) value at the plane (100), (110), (111), (200), (211), (220), and (300) respectively, this implies that the BaPV nanocomposite still preserves the BTO phase characteristic even though PVDF polymer was added in this composition.

As the PVDF polymer composition in the nanocomposites increases, BTO tetragonal characteristics in BaPV composite show significantly decreased intensity and the tetragonal secondary peak pattern did not appear at the plane (001), (101), (002), (102) and (202) as the only primary peak of BTO appears. This suggested the sample in the BaPV nanocomposite phase changed from tetragonal into cubic phase. Meanwhile, by increasing the amount of PVDF, diffraction peaks for BaPV XRD patterns matched well with JCPDS No. 05-06262728 supporting the existence and uniform dispersion of BAPV nanoparticles. From the observation above, the peak intensity for BaPV nanocomposite decreases dramatically at 50 wt% of PVDF due to the

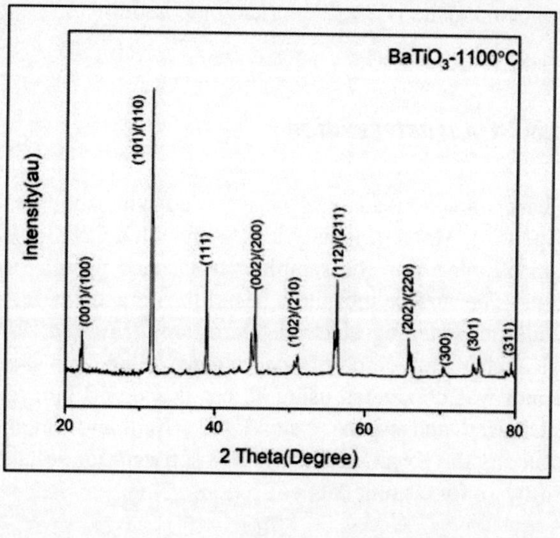

Fig. 1 XRD pattern of $BaTiO_3$ tetragonal nanopowder

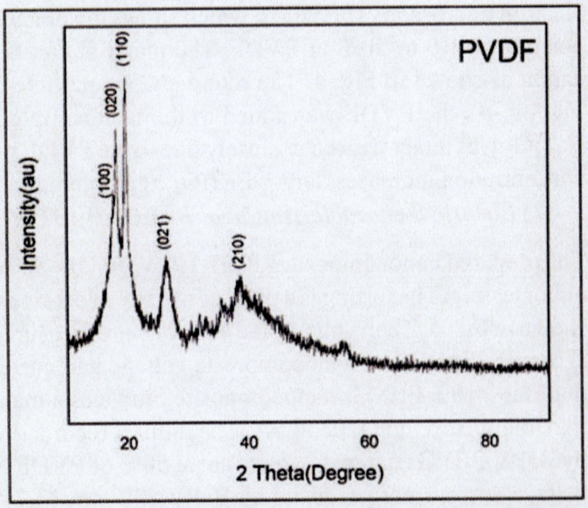

Fig. 2 XRD pattern of PVDF powder

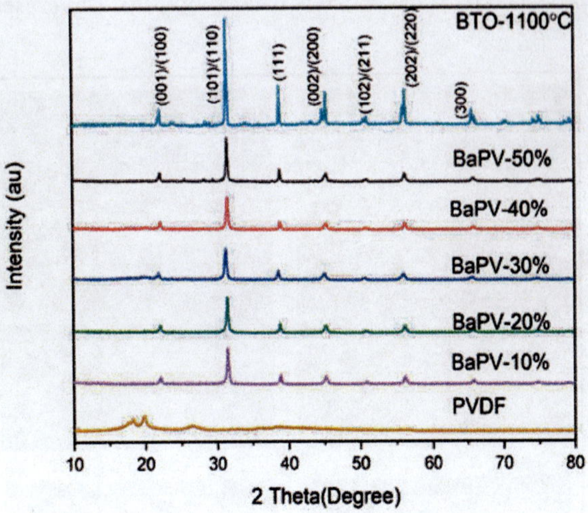

Fig. 3 XRD pattern BaPV nanocomposite with different weight percentages of PVDF

increasing amount of the amorphous PVDF in the nanocomposite. However, a change in phase for the PVDF is more obvious, as shown in Fig. 2. The pure PVDF shows the strongest peak plane at (100), (120), (110), (021), and (210) but the diffraction peaks of PVDF in the BaPV nanocomposites are negligible due to high crystallinity and strong diffraction peaks of $BaTiO_3$ in the nanocomposites. This is because the $BaTiO_3$ nanoparticles dispersed in the PVDF might have an important effect on the crystallization of the PVDF polymer [18].

The surface morphologies of PVDF and $BaTiO_3$/PVDF nanocomposite thin film with a different weight percentage of PVDF are analyzed by using field scanning electron microscopy (FESEM), which shows the images of $BaTiO_3$/PVDF nanocomposite from 10 to 50% of PVDF. The material can be seen under 80,000 magnification as shown in Fig. 4. The round-shaped particle of pure PVDF can be seen in Fig. 4a. As the PVDF was added to the nanocomposite, from the image reported, $BaTiO_3$ was intact with each closely due to the PVDF matrix. Thus we can see PVDF concentration increases, larger $BaTiO_3$ agglomeration can be seen in Fig. 4.

(2) *Electrical characterization of $BaTiO_3$/PVDF PENG*

The prepared nanocomposites $BaTiO_3$/PVDF (10–50%) were subjected to nanogenerator testing. The testing is a process of 15 N of external load applied to the PENGs as shown in Fig. 5. Then voltage and current were measured during the process. Figure 6 shows BaPV PENGs nanocomposite voltage and current output. It can be observed that BaPV10% PENGs nanocomposite produces a maximum voltage of 4.9 V and a maximum current at 0.0344 nA. The highest recorded voltage and current produced by BaPV PENGs nanocomposite are at 50% of PVDF concentration. The maximum voltage and current produced by BaPV-50% are 13.7 V and 0.162 nA. This shows an increment of 36% of voltage output and an increment of 21% of current output produced by BaPV PENGs nanocomposite. The presence of the BTO ceramic filler

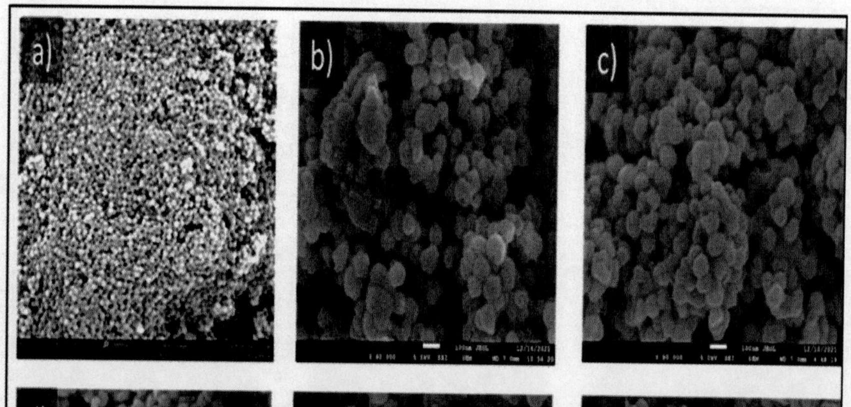

Fig. 4 Surface morphologies of **a** PVDF powder **b** BaPV10% nanocomposite **c** BaPV20% nanocomposite **d** BaPV30% nanocomposite **e** BaPV40% nanocomposite **f** BaPV50% nanocompsite

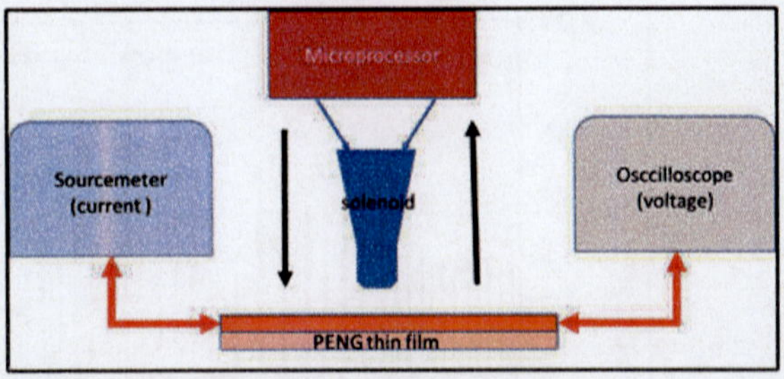

Fig. 5 Schematic of electrical diagram measurement system

that occupies more than 50% of PVDF polymer matrix provides a higher piezoelectric effect, enhances electrical displacement, and high breakdown strength which is in agreement with the literature [19, 20].

While the linear trends were observed for maximum voltage-current as the PVDF in BaPV nanocomposite increases. As mentioned above, the role of PVDF polymer improve conversion efficiency in PENG nanocomposite and increase coupling coefficient for higher energy output [19].

Thus, voltage-current increases as the concentration of polymer increases in PENG nanocomposite. Similar behavior was also reported by Nunes et al. 2015 where the concentration of PVDF polymer increases the coefficient efficiency of dielectric response on PENG nanocomposite [21].

Table 1 shows the output voltage at different PVDF concentrations and power of BaPV PENG nanocomposite respectively. The power density was calculated by measuring the output power of PENG across the unit area of the active layer of PENG. From this table, the highest power density produced is at BaPV50% at 0.7065 nW/cm^2.

The addition of PVDF polymer into $BaTiO_3$ nanoparticles not only increases elasticity but also enhances the energy output of BaPV nanocomposite due to PVDF piezoelectric characteristics that improve conversion efficiency in nanocomposite and increase coupling coefficient for higher energy output [22]. The prepared BaPV nanocomposite shows a higher voltage value, but the current produced by it is relatively smaller.

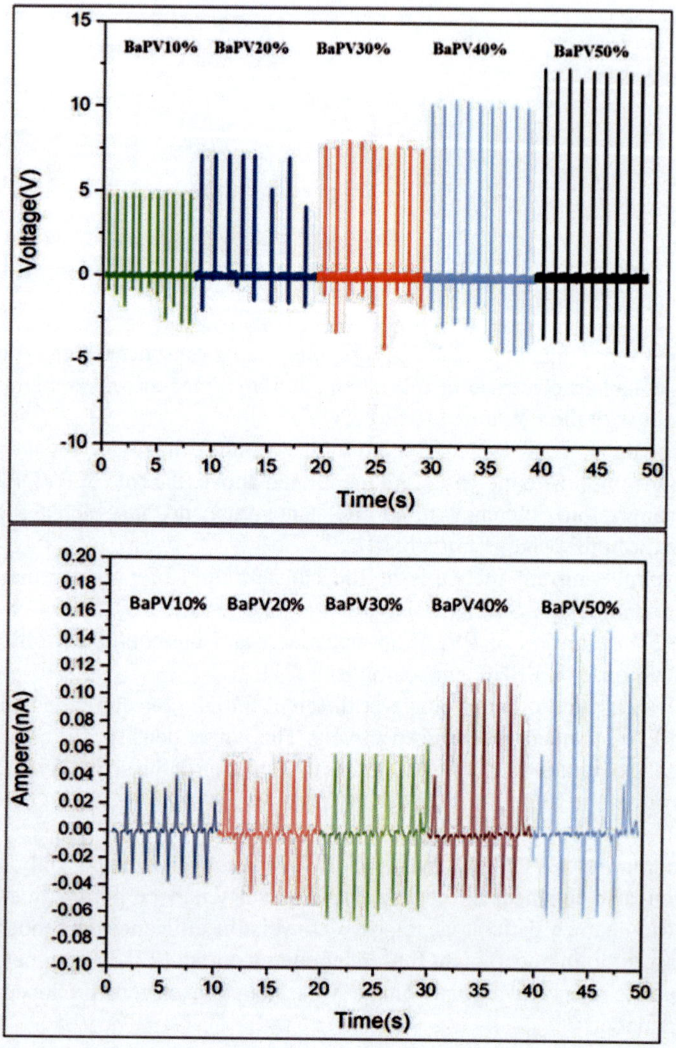

Fig. 6 BaPV nanocomposite energy output of voltage value over time (above) and current over time (below)

Table 1 Energy output of nanocomposite

PENG		Energy output			
	Vp-p (V)	Max (V)	Max (nA)	Power (nW)	Power density (nW/cm^2)
BaPV10%	6.0	4.9	0.0344	0.1686	0.054
BaPV20%	7.4	6.0	0.0379	0.2273	0.072
BaPV30%	8.0	7.3	0.0484	0.3533	0.1124
BaPV40%	10.0	7.8	0.0906	0.7063	0.2248
BaPV50%	22.0	13.7	0.164	2.2194	0.7065

4 Conclusion

BaPV nanocomposites with 50 wt% (BaPV50%) produces maximum energy output at 13.7 V with 0.162 nA. Furthermore, the device exhibited the highest power at 0.7065 nW/cm^{-2}. This shows that $BaTiO_3$/PVDF nanocomposite is a good candidate for flexible energy harvester application.

Acknowledgements Authors would like to thank Ministry of Science, Technology and Innovation Malaysia (MOSTI) for the ICF grant awarded.

References

1. Covaci C, Gontean A (2020) Piezoelectric energy harvesting solutions: a review. Sensors (Switzerland) 20(12):1–37. MDPI AG. https://doi.org/10.3390/s20123512
2. Khaligh A, Zeng P, Zheng C (2010) Kinetic energy harvesting using piezoelectric and electromagnetic technologies state of the art. IEEE Trans Industr Electron 57(3):850–860. https://doi.org/10.1109/TIE.2009.2024652
3. Nguyen V, Kelly S, Yang R (2017) Piezoelectric peptide-based nanogenerator enhanced by single-electrode triboelectric nanogenerator. APL Mater 5:74108. https://doi.org/10.1063/1.4983701
4. Ghosh SK, Sinha TK, Mahanty B, Mandal D (2015) Self-poled efficient flexible 'ferroelectretic' nanogenerator: a new class of piezoelectric energy harvester. Energ Technol 3(12):1190–1197. https://doi.org/10.1002/ente.201500167
5. Sukumaran S, Chatbouri S, Rouxel D, Tisserand E, Thiebaud F, ben Zineb T (2021) Recent advances in flexible PVDF based piezoelectric polymer devices for energy harvesting applications. J Intell Mater Syst Struct 32(7):746–780. https://doi.org/10.1177/1045389X20966058
6. Kamaruzaman D, Ahmad N, Rosly MA, Mamat MH (2021) Piezoelectric energy harvesting based on ZnO: a review. AIP Conf Proc 2332. https://doi.org/10.1063/5.0044476
7. Ambrosio R, Jimenez A, Mireles J, Moreno M, Monfil K, Heredia H (2011) Study of piezoelectric energy harvesting system based on PZT. Integr Ferroelectr 126(1):77–86. https://doi.org/10.1080/10584587.2011.574989
8. Clementi G, Margueron S, Suarez MA, Baron T, Dulmet B, Bartasyte A (2019) Piezoelectric and pyroelectric energy harvesting from lithium niobate films. J Phys: Conf Ser 1407(1). https://doi.org/10.1088/1742-6596/1407/1/012039

9. Jeong CK et al (2013) Virus-directed design of a flexible $BaTiO_3$ nanogenerator. ACS Nano 7(12):11016–11025. https://doi.org/10.1021/nn404659d
10. Progress in the applications of smart piezoelectric materials for medical devices enhanced reader
11. Baur C, Apo DJ, Maurya D, Priya S, Voit W. Downloaded via 103 [Online]. https://pubs.acs.org/sharingguidelines
12. Cauda V, Stassi S, Bejtka K, Canavese G (2013) Nanoconfinement: an effective way to enhance PVDF piezoelectric properties. ACS Appl Mater Interfaces 5(13):6430–6437. https://doi.org/10.1021/am4016878
13. Simon-Seveyrat L, Hajjaji A, Emziane Y, Guiffard B, Guyomar D (2007) Re-investigation of synthesis of $BaTiO_3$ by conventional solid-state reaction and oxalate coprecipitation route for piezoelectric applications. Ceram Int 33(1):35–40. https://doi.org/10.1016/j.ceramint.2005.07.019
14. Nunes-Pereira J et al (2015) Energy harvesting performance of $BaTiO_3$/poly(vinylidene fluoride-trifluoroethylene) spin coated nanocomposites. Compos B Eng 72:130–136. https://doi.org/10.1016/j.compositesb.2014.12.001
15. Fu J, Hou Y, Zheng M, Wei Q, Zhu M, Yan H (2015) Improving dielectric properties of PVDF composites by employing surface modified strong polarized $BaTiO_3$ particles derived by molten salt method. ACS Appl Mater Interfaces 7(44):24480–24491. https://doi.org/10.1021/acsami.5b05344
16. Omar AFC, Hatta FF, Kudin TIT, Mohamed MA, Hassan OH (2019) Calcination effect on structural transformation of barium titanite ferroelectric ceramic by Sol-Gel method. Int J Eng Adv Technol 9(1):5893–5896. https://doi.org/10.35940/ijeat.A3023.109119
17. Newman BA, Yoon CH, Pae KD, Scheinbeim JI (1979) Piezoelectric activity and field-induced crystal structure transitions in poled poly(vinylidene fluoride) films. J Appl Phys 50(10):6095–6100. https://doi.org/10.1063/1.325778
18. Dang ZM, Shen Y, Nan CW (2003) Dielectric behavior of threephase percolative Ni-$BaTiO_3$/polyvinylidene fluoride composites. Appl Phys Lett 81(25):4814–4816. https://doi.org/10.1063/1.1529085
19. Song Y, Shen Y, Liu H, Lin Y, Li M, Nan CW (2012) Enhanced dielectric and ferroelectric properties induced by dopamine-modified $BaTiO_3$ nanofibers in flexible poly(vinylidene fluoridetrifluoroethylene) nanocomposites. J Mater Chem 22(16):8063–8068. https://doi.org/10.1039/c2jm30297g
20. Goyal RK, Katkade SS, Mule DM (2013) Dielectric, mechanical and thermal properties of polymer/$BaTiO_3$ composites for embedded capacitor. Compos Part B Eng 44(1):128–132. https://doi.org/10.1016/j.compositesb.2012.06.019
21. Nunes-pereira J, Sencadas V, Correia V, Cardoso VF, Han W, Rocha JG (2014) Energy harvesting performance of $BaTiO_3$/poly(vinylidene fluoride–trifluoroethylene) spin coated nanocomposites. Compos: Part B 1–7
22. Ruan L, Yao X, Chang Y, Zhou L, Qin G, Zhang X (2018) Properties and applications of the β phase poly(vinylidene fluoride). Polymers (Basel) 10(3):1–27. https://doi.org/10.3390/polym10030228

Transient Modeling of Two-Phase Flows Using OLGA

Seshu Kumar Vandrangi, Tamiru Alemu Lemma, and Syed Muhammad Mujtaba

Abstract The leakages in pipelines have been raising serious environmental concerns; various systems have been developed to detect leakages in a pipeline. Hence, there is a need for early leakage detection, as well as a leakage prevention system. A detailed analysis of two-phase flows using transient modeling prior to leak detection using transient modeling is much required. This paper focuses on validating the two-phase flows using transient modelling with the help of OLGA, a multiphase software. The software was able to validate both steady-state pressure drop calculations and transient simulations in two-phase flows using a black oil model. A pipeline with 18,000 ft length with varying various other parameters such as inlet flow rates, diameters, and Gas Liquid Ratio (GLR) are studied and validated. The validation was based on the study which was analyzed for the purpose of operating and planning of new oil field in Saudi Arabia. The current paper was able to validate the study both in steady-state and transient flow simulations.

Keywords Two-phase flows · OLGA · Equation of state · Pressure · Mass flow rate · GLR · Water cut

1 Introduction

One of the biggest obstacles the sector is experiencing right now is extracting hydrocarbons at a reasonable cost. Regarding this, transient multiphase simulation has been given a lot of focus to understand the flow behavior in long and deep flow line risers of various topologies. There is a lot of uncertainty about the application of the huge diameter to the situation described above. The available experimental and field data only cover much smaller dimensions, therefore this is the case [1].

Major mishaps in multiphase flow systems usually result from improper operation or incorrect calculations during design. The incapacity to assess real-world flow

S. K. Vandrangi (✉) · T. A. Lemma · S. M. Mujtaba
Department of Mechanical Engineering, Universiti Teknologi PETRONAS, Tronoh, Malaysia
e-mail: seshu1353@gmail.com

F. Ahmad et al. (eds.), *ICREEM 2022*, Lecture Notes in Mechanical Engineering,
https://doi.org/10.1007/978-981-99-5946-4_10

patterns is the root cause of all of these acts. Therefore, the determination of flow resistance, the determination of flow stability, including the engineering application of heat transfer/pressure drop theory rely heavily on the identification and theoretical prediction of flow.

Every year, millions of dollars worth of oil and gas are carried through land and sea, and even a fractional loss can have a significant economic impact. Even the most advanced leak detection systems are very inexpensive compared to the total cost of building and installing pipelines, and operators are wise to prioritize efficiency and effectiveness above minimization of expenses. The most significant problem is leakage, which is an unfortunate fact of life in the pipeline industry today.

For efficient management and strategic planning of a large Saudi oilfield, accurate data on well performance is essential. Only by building massive facilities for routinely testing every output in the field can complete design data be gathered [2]. The three possible setups of the testing pipeline were carefully evaluated. Dimensions such as oil production rate, gas/liquid ratio, water cut, and pipeline length were considered. Approximately 10 miles (16.1 km) in length were examined [2].

Brill et al. [2] in their work have used various combinations of mass flow rate changes which were simulated using the Taitel, Lee and Dukler model [3]. And also few runs using the Scoggins model [4] gave similar results. Further, calculations were performed with the Taitel, Lee and Dukler model [3].

There were three potential setups for the testing pipeline, and we've labeled them A, B, and C. For each possibility, three distinct types of two-phase flow computations were carried out. The lengths, diameters, flow rates, gas-liquid ratios, and water cuts for steady-state pressure losses were the initial variables to be determined. In the second set of computations, we simulated the transient behavior of the flow as the inlet flow rate changed. The last kind of computation included forecasting the length and other properties of the liquid slugs that are expected to generate. To model PVT behavior, we employed a black oil model [2].

This current research validates the work done by Brill et al. [2], with varying parameters such as mass flow rate, diameter, GLR, water cut and pipeline lengths. The paper validates the steady-state pressure loss calculations and transient simulation flow behavior using OLGA software for two-phase flows. Till date, there is no proper literature available for two-phase flow transient modeling, hence this work of validation helps in doing further in the field of leak detection and two-phase flows.

2 Literature Review

Important to the oil and gas sector is the study of multiphase pipe flow, or the simultaneous passage of two or more phases in a single pipe. Two- and three-phase flows are seen in many technical applications, not just the petroleum industry. This includes chemical facilities, nuclear power plants, and drainage systems [5].

2.1 Flow Regimes

Several different types of two-phase flows exist. Stratified smooth flow, stratified wavy flow, slug flow, bubble flow, and annular flow are the most common types of oil and gas flows [6]. (1) stratified flow, wherein the liquid is continuously flowing at the pipe's base and gas is continuously flowing above it (occasionally stratified-wavy flow is distinguished from stratified smooth flow); (2) slug flow, a type of stratified flow with slugs of very turbulent liquid (plug flow is a type of slug flow that happens at lower velocities); (3) a gas stream with entrained liquid droplets, an adherent liquid film, and an annular flow; and (4) bubble flow refers to a continuous flow of liquid that contains gas bubbles. The overall characteristics of a two-phase flow are established by the flow regime in which the flow is operating. Pipe diameter and the transport qualities of gas and liquid determine the regimes. Pipelines used to transport oil can only support two types of flows: stratified and slug.

2.2 Flow Assurance

Flow assurance is the prevention of production delays caused by problems anywhere in the production chain between the reservoir and the control facility. This could include (1) alteration of the phase structure (hydrates and paraffin), (2) precipitation (scale and asphaltenes), (3) reaction (corrosion), (4) solids (production fines, hydrate crystals, and wax particles) [7]. Moreover, there are flow assurance difficulties that are a direct consequence of multiphase flow. These would include (1) terrain or riser slugging, (2) ramp-up slugs, and (3) pigging/sphering slugs. Therefore, multiphase flow is the backdrop against which all flow assurance challenges arise. Before evaluating any other flow assurance issue, a multiphase model must be built in most cases. Figure 1 depicts the flow assurance framework.

2.3 Governing Equations for Transient Flow

Continuity Equation (Conservation of Mass): Through the pipeline, mass conservation is stated by the continuity equation [8, 9],

$$\frac{\partial(\alpha_k \rho_k)}{\partial t} + \frac{\partial}{\partial x}(\alpha_k \rho_k v_k) = \Gamma_{ki} + \Gamma_{kw} \tag{1}$$

Momentum Equation (Newton's Second Law of Motion): Newton's second rule of motion, expressed as the momentum equation, describes the instantaneous force balance on the fluid within a cross section of the pipeline [8, 9].

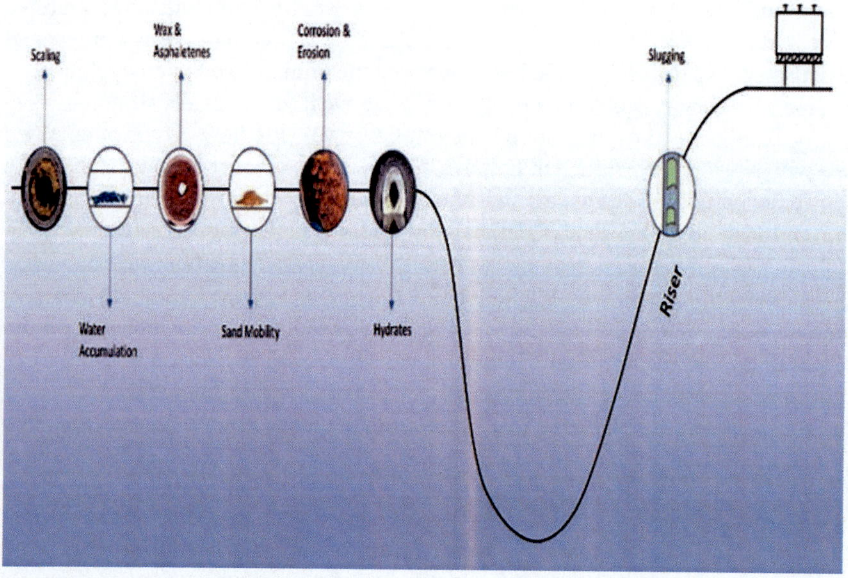

Fig. 1 Flow assurance in a pipeline

$$\frac{\partial(\alpha_k\rho_k v_k)}{\partial t}+\frac{\partial\left(\alpha_k\rho_k v_k^2\right)}{\partial x}=-\alpha_k\frac{\partial p_k}{\partial x}dx+(S_{ki}+S_{kw})+v(\Gamma_{ki}+\Gamma_{kw})-\alpha_k\rho_k g \tag{2}$$

Energy Equation (Conservation of Energy): In fluid elements, energy is conserved as shown by the energy equation [8, 9].

$$\frac{\partial}{\partial t}(\alpha_k E_k)=-\frac{\partial}{\partial x}[\alpha_k v_k(E_k+p_k)]+q_{ki}+q_{kw}+w_{ki}+w_{kw}+\Gamma_{ki}h_{ki}+\Gamma_{kw}h_{kw} \tag{3}$$

2.4 Equation of State (EOS)

An equation of state is used to explain the most important property of a fluid. The relationship between density, pressure, and temperature is given by an equation of state and can be written in a broad sense as $\rho = f(p, T)$. The equation of state is often extremely intricate and dependent on the dynamical changes in fluid characteristics. Cubic equations are the most well-known type of EOS; they are based on Van der

Waal's (vdW) EOS, the first model that can explain properties in both liquid and vapr phases [10]. With regards to parameter estimation and extending Peng-Robinson EOS to mixes, there have been over 220 variations and many investigations. Due to their simplicity in mathematical formulation and the display of reasonable correlations and forecasts of thermodynamic parameters, cubic EOS have gained widespread use in the oil and gas sector [11]. Equation (4) provides the most generic form of cubic equations of state for the aforementioned equations of state, and Table 1 provides the corresponding constants and terms.

$$\mathrm{P} = \frac{RT}{V-b} - \frac{a\alpha(T)}{(V-bm_1)(V-bm_2)} \tag{4}$$

3 Fluid Database

The fluid's composition and, by extension, its properties, might be difficult to determine, especially in the beginning stages of a flow assurance project. Various petroleum fluids, from dry gas to heavy oil, are roughly categorized in Table 2 [8, 9].

The behavior of different fluids is illustrated in Fig. 2. The C-marks denote the crucial spots. Fluids can exist in the diagram as pure gas, pure liquid, or a mixture of the two. It also demonstrates that black oil is more likely to take the form of a liquid, while gas is more likely to take the form of a gas, as may be expected. Hence a black oil model was used to simulate all PVT behavior and it is also used in Brill's [2] work. However, the detailed composition was not given in the work of Brill, hence Black oil model available in OLGA was used.

4 OLGA Modeling Methodology

IFE and SINTEF, Norway, collaborated to create the OLGA (OiL and GAs) code. The simulator is used for a wide variety of design and operational issues, including startup and shutdown transients, terrain slugging, variable production rates, pigging, gaslift, and many more. In addition to a steady-state point model (OLGAS) and a full transient computational code, this program can be used to simulate a one-dimensional, extended two-fluid system. Although OLGA's popularity has grown steadily since its 1983 introduction to the market, there are few freely available sources that analyze the drug's efficacy.

Among the many methods of flow modeling, the use of conservation equations such as those for mass, momentum, and energy, as well as equations of state (EOS) for the fluids, is widespread. To model the two-phase flows, the current study employed OLGA, a multiphase simulator developed by the Schlumberger group. The default black oil model available in OLGA was used.

Table 1 Constants and α terms in various EOS [12]

EOS	a	b	m_1	m_2	α
Van der Waals (vdW)	$\frac{27}{64}\frac{(RT_c)^2}{P_c}$	$\frac{1}{8}\frac{RT_c}{P_c}$	0	0	1
Redlich-Kwong (RK)	$0.42748\frac{R^2T_c^{2.5}}{P_c}$	$0.08664\frac{RT_c}{P_c}$	−1	0	$\frac{1}{\sqrt{T}}$
Soave–Redlich–Kwong (SRK)	$0.42748\frac{(RT_c)^2}{P_c}$	$0.08664\frac{RT_c}{P_c}$	−1	0	$[1+(0.48508+1.55171\omega-0.15613\omega^2)(1-\sqrt{T_r})]^2$
Peng–Robinson	$0.45724\frac{(RT_c)^2}{P_c}$	$0.07780\frac{RT_c}{P_c}$	$-1+\sqrt{2}$	$-1-\sqrt{2}$	$[1+(0.37464+1.54226\omega-0.26992\omega^2)(1-\sqrt{T_r})]^2$

Table 2 Types of naturally occurring petroleum fluids

Fluid	Gas-oil ratio	Occurrence in reservoir
Dry gas	∞	Gas
Wet gas	>100	Gas + tiny amount of liquid
Condensate	3–100	Gas + some liquid
Volatile oil	1.5–3	Liquid + considerable gas
Black oil	0.1–1.5	Liquid + some gas
Heavy oil	~0	Almost no gas

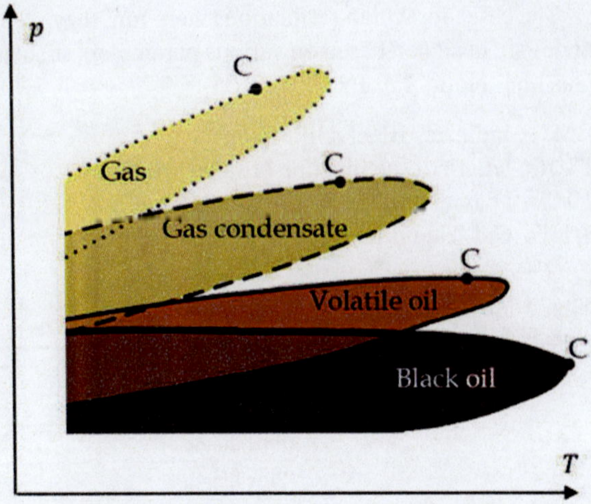

Fig. 2 Typical two-phase flow envelopes for different fluid types

4.1 Pipe Modeling

The valve in the middle of the pipe was permanently open, with a mass rate-driven inlet node at one end and a pressure-driven output node at the other. The pipe was then segmented into two sections, since the no of sections was not specified in Brill's [2] work. Nominal pipeline diameters of 8 and 10 inches were used for all calculations. The sample model of the pipe is shown in Fig. 3.

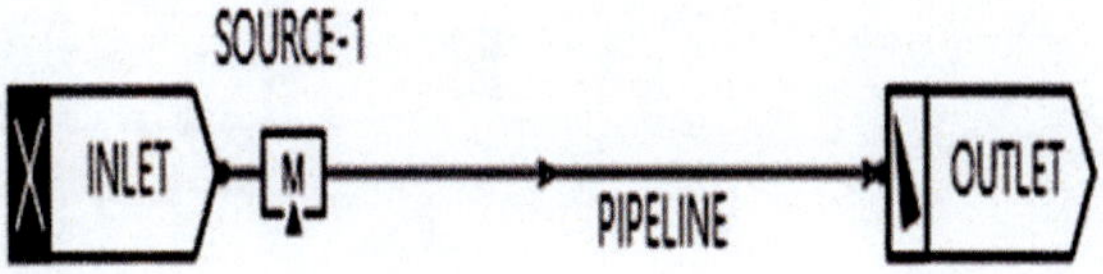

Fig. 3 The simulated OLGA model—5 km pipeline

4.2 Boundary Conditions

The boundary conditions used for running simulations for steady-state and transient flow conditions for two-phase pipelines are tabulated in Table 3. Though there are three configurations discussed in Brill's [2] work, only a straight pipeline is used in showing the transient simulation results. Hence, a straight pipeline was considered in this research as well for transient simulation.

However, for steady-state results, they used one of the three configurations from Brill's [2] work as shown in Fig. 4. The pipeline length was divided in equal sections and simulations were performed extracting the data from every section.

The cases for which simulations were run, they were named with codes. The case codes are designed based on various parameters such as length of pipeline, diameter, Gas liquid ratio. For Ex: BM18105

BM Indicates Black Oil Model
18/10 length of the pipeline (18,000, 10,000 ft)
10/08 Diameter of the pipeline (10, 8 inch)
5/1 Gas Liquid Ratio (500, 100 GLR)

Table 3 Boundary conditions

Component	Details
Fluid	Black oil
Pipe material	Steel (assumed)
Pipe diameter	8, 10 inches
Pipe length	18,000 ft
Outlet pressure	65 psig
Inlet temperature	65.6 °C
Outlet temperature	65.6 °C
H ambient	1.5–2.0 BTU/h t^2
Flow rate	2000–8000 STBL/D
Gas liquid ration	100, 363, 500

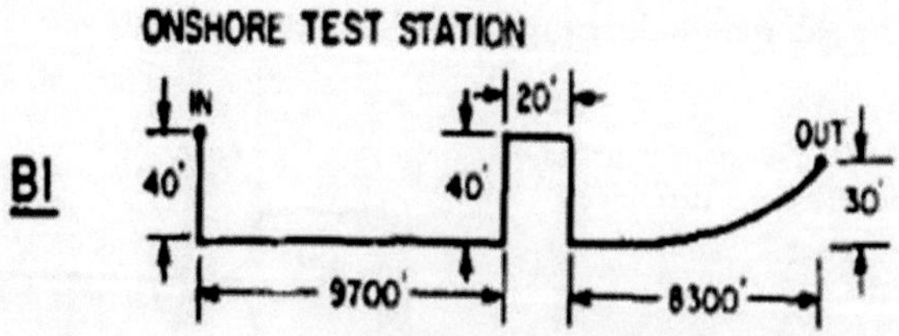

Fig. 4 Pipeline geometry used for steady-state simulations

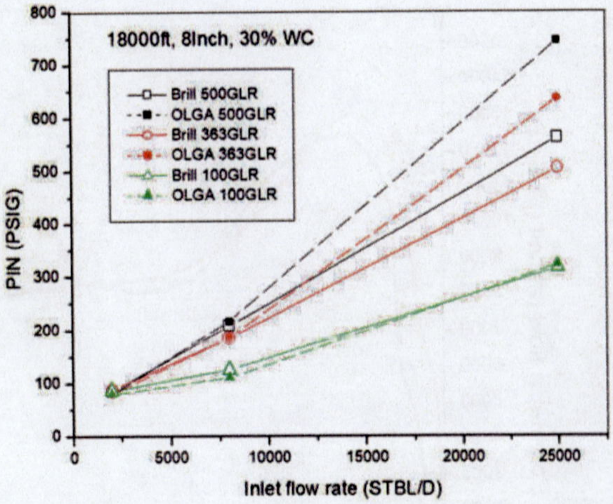

Fig. 5 PIN (pressure at inlet) versus inlet flow rate for steady-state simulations

5 Pipe Modeling—OLGA Results

5.1 *Steady-State—Pressure Loss Calculations*

All pressure loss calculations used a pipeline pressure at the separator of 65 psig. The length of the pipeline is 18,000 ft with a diameter of 8 inches. A water-cut of 30% was used in the simulations. The PIN (Pressure at inlet) was plotted against Inlet flow rate as shown in Fig. 5.

For inlet flow rates of 2000, 8000 STBL/D, the deviation of PIN with Brill's value is less than 8% for 500, 363 GLR. However, for 25,000 STBL/D, the deviation is more than 25%. Whereas for 100 GLR, for all values of flow rates, the deviation is less than 12%. Based on these observations, further simulations can be performed for flow rates below 8000 STBL/D.

5.2 *Transient Simulations*

The time required to reestablish steady-state flow after a disturbance in inlet boundary conditions can be calculated with transient two-phase flow models. When inlet flow rates are changed in a pipeline, the time required for outlet flow rates and pressures to stabilize varies with diameter, flow rates, and pipeline length. Stabilized flow must occur before accurate well tests can be conducted.

Various combinations of flow rate changes were simulated including both increasing and decreasing flow rates. The effects of pipeline length, diameter, elapsed time to change flow rate, and gas-liquid ratio were analyzed for both increasing and decreasing flow rate changes.

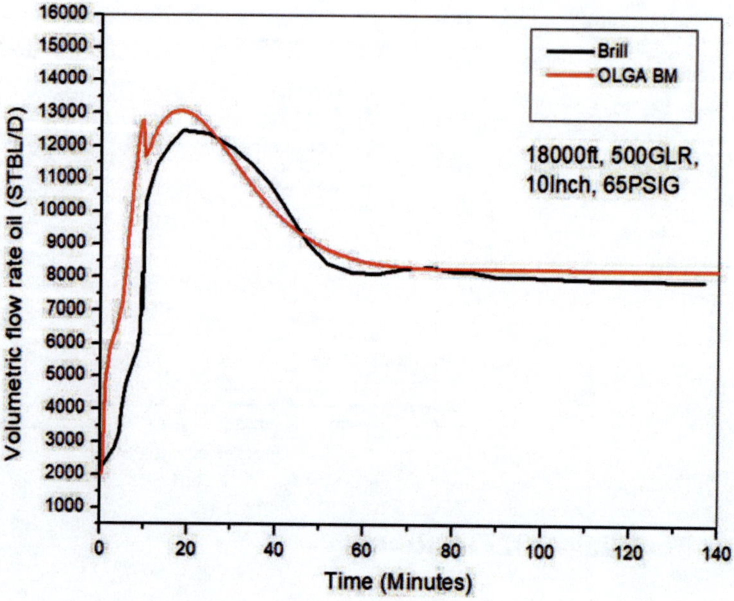

Fig. 6 BM18105 outlet flow rate versus time

The outlet volumetric flow has been plotted against time for increasing flow rate 2000 to 8000 STBL/D at the inlet. The pipeline has a length of 18,000 ft for 10 inch diameter and 500 GLR has been maintained. The flow rate stabilizes around 8000 STBL/D after 40–50 min of the start. The OLGA values are almost in-line with Brill's values as shown in Fig. 6.

For similar increasing flow rates of 2000–8000 STBL/D at the inlet, the outlet flow rate has been plotted against time for a 10,000 ft length pipeline. The stabilization time is much less as shown in Fig. 7 and can be attributed length of the pipeline. The values obtained from OLGA are in good acceptance with Brill's values.

For similar increasing flow rates of 2000–8000 STBL/D at the inlet, the outlet flow rate has been plotted against time for 18,000 ft length pipeline with 8 inch diameter. The stabilization time occurs after 50 min of the start as shown in Fig. 8. The values obtained from OLGA are in good acceptance with the Brill's values after the stabilization.

On the other hand, the Pressure at inlet values is plotted against time for a constant pipeline length of 18,000 ft with varying diameter of 8, 10 inch. The results are quite predictable with lesser the diameter higher the pressure as shown in Fig. 9. The inlet flow rates were increased 2000–8000 STBL/D and the effect of diameter has been observed. Also, the values obtained from OLGA is in good match with Brill's values.

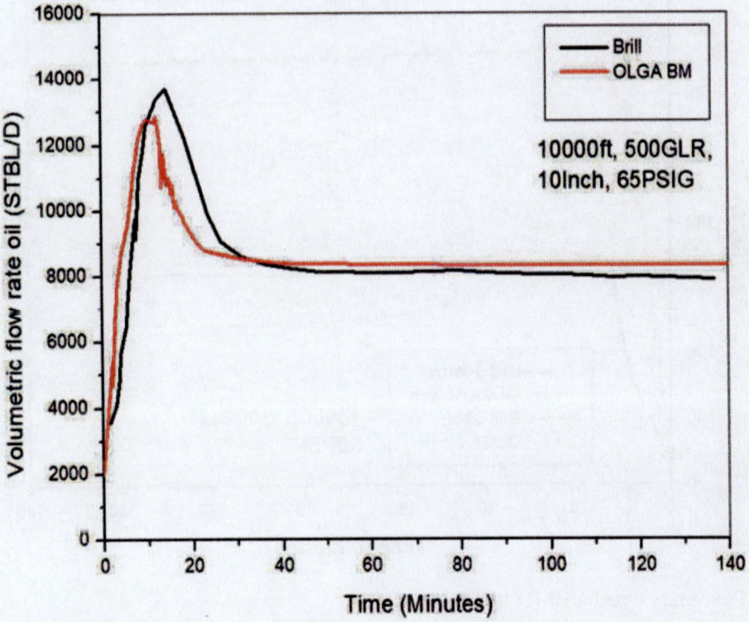

Fig. 7 BM10105 outlet flow rate versus time

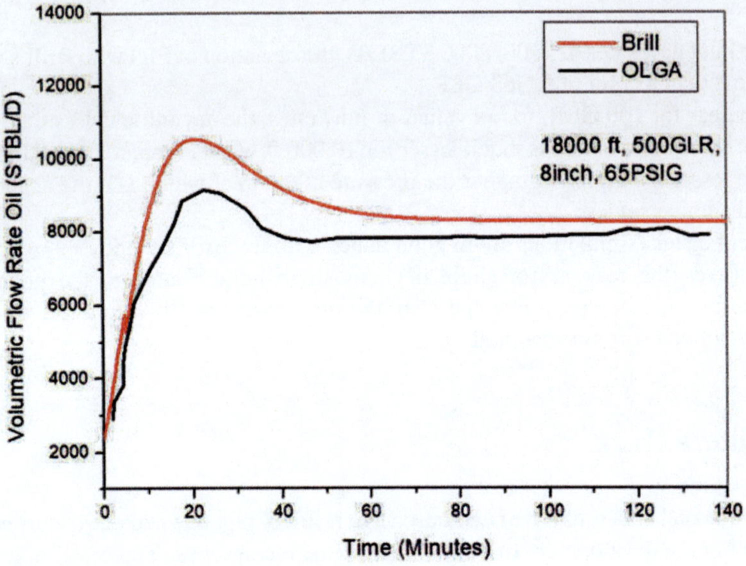

Fig. 8 BM18085 outlet flow rate versus time

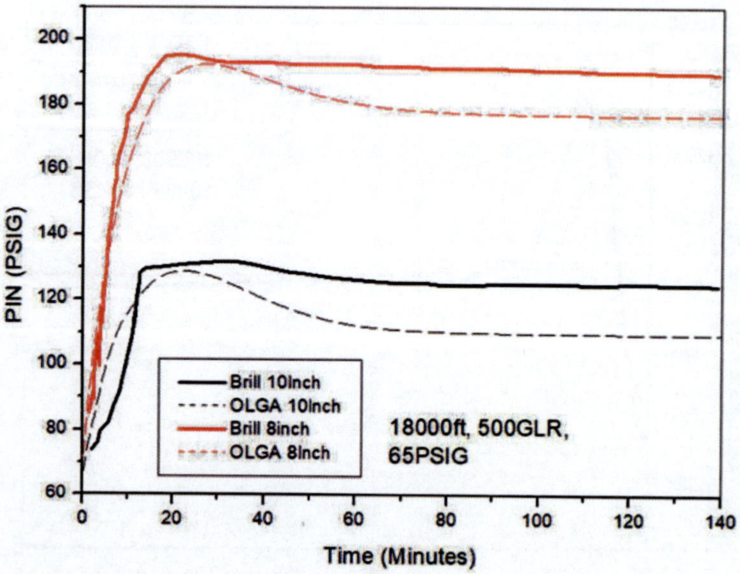

Fig. 9 PIN versus time for BM18105 and BM18085

6 Conclusions

- For inlet flow rates of 2000, 8000 STBL/D, the deviation of PIN with Brill's value is less than 8% for 500, 363 GLR
- Whereas for 100 GLR, for all values of flow rates, the deviation is less than 12%
- The stabilization time is much lesser for 10,000 ft when compared to 18,000 ft
- The lesser the diameter higher the pressure at the inlet which is quite according to the universal law
- The transient simulations are in good match with the Brill's work
- However, the reasons for slight deviations can be attributed to the unknown detailed fluid compositions and also, the no. of sections to which the pipeline was divided was not specified

7 Future Work

Environmental concerns have been raised as a result of pipeline leakage occurrences, prompting the development of various leak detection systems. Therefore, a system for early leak detection and leak prevention is required. These simulations can be used as no leak conditions and leak conditions can be modified and used in leak detection techniques.

Acknowledgements The authors acknowledge the financial help from the Universiti Teknologi PETRONAS (UTP) under YUTP-FRG, Cost Center 015LC0-251 and are thankful for the opportunity.

References

1. Ali SF, Yeung H (2010) Experimental investigation and numerical simulation of two-phase flow in a large-diameter horizontal flow line vertical riser. Pet Sci Technol 28(11):1079–1095
2. Brill J, Sifferman T, Schmidt Z, Sharma Y, Dutta-Roy K (1983) Evaluation of two-phase flow test pipeline configurations for an offshore oil field. In: Middle east oil technical conference and exhibition. OnePetro, pp 607–610
3. Taitel Y, Lee N, Dukler A (1978) Transient gas-liquid flow in horizontal pipes: modeling the flow pattern transitions. AIChE J 24(5):920–934
4. Scoggins Jr MW (1977) A numerical simulation model for transient two-phase flow in a pipeline. The University of Tulsa
5. Danielson TJ (2012) Transient multiphase flow: past, present, and future with flow assurance perspective. Energy Fuels 26(7):4137–4144
6. Liu Y, Yang W, Wang J (2008) Experimental study for the stratified to slug flow regime transition mechanism of gas-oil two-phase flow in horizontal pipe. Front Energy Power Eng Chin 2(2):152–157
7. Bai Y, Zhang T, Li Y, Ruan W, Zhu K (2017) A new leak detection method for subsea pipelines. Ships Offshore Struct 12(sup1):S144–S152
8. Bratland O (2010) Pipe flow 2: multi-phase flow assurance. Ove Bratland Flow Assurance Consulting, Chonburi, Thailand
9. Bratland O (2009) Pipe flow 1: single-phase flow assurance, vol 2, pp 21–92
10. Van der Waals JD (1873) Over de Continuiteit van den Gas-en Vloeistoftoestand, vol 1. Sijthoff
11. Papadopoulos A, Tsivintzelis I, Seferlis P, Linke P (2018) Computer-aided molecular design: fundamentals, methods, and applications. In: Reedijk J (eds) J Elsevier Ref Modul Chem Mol Sci Chem Eng
12. Patricio RA, Sondermann CN, Figueiredo A, Viggiano R, Bodstein GC, Rachid FBdF, Baptista RM (2018) Numerical simulation of pig motion in two-phase flow pipelines in the presence of a leak. In: International pipeline conference. American Society of Mechanical Engineers

CFD-DEM Validation of Sand Retention Testing

Aimi Zahraa Zainal, Javed Akbar Khan, and Mohd Azuwan Maoinser

Abstract The production of sand in unconsolidated reservoirs has always been a problem to tackle where it affects oil production when plugged. With that, sand retention testing has been conducted to select the sand screen based on its performance and permeability. Numerous experimental studies have outlined ways to select the optimal screen size. However, not many numerical studies have been conducted to simulate the experiment as it comes to the limitations of computational power. In this study, it was shown that it is possible to simulate the experimental conditions into the simulation platform and obtain results that would suffice in selecting the screen size with other factors to consider. The results obtained from the experiment were used to compare with the results obtained from the simulation. The obtained numerical simulation results were not accurate as there were several factors to consider, namely the duration of the simulation difference, a comparative study between the particle size distribution of the simulated particles and the actual ceramic beads, and to conduct of an independent mesh study.

Keywords CFD-DEM · Particle–fluid flow · Plugging · Jamming · Sand retention · Slurry test

A. Z. Zainal (✉) · J. A. Khan · M. A. Maoinser
Institute of Hydrocarbon Recovery, Universiti Teknologi PETRONAS, Bandar Seri Iskandar, Malaysia
e-mail: aimi.zainal@utp.edu.my

J. A. Khan
e-mail: javedkhan_niazi@yahoo.com

M. A. Maoinser
e-mail: mazuwan.maoinser@utp.edu.my

J. A. Khan
Department of Mechanical and Electrical Engineering, China University of Petroleum, Qingdao 266000, China

F. Ahmad et al. (eds.), *ICREEM 2022*, Lecture Notes in Mechanical Engineering,
https://doi.org/10.1007/978-981-99-5946-4_11

1 Introduction

The selection of the optimal screen is essential to ensure that the opening size is not too small, which could lead to production restriction, or that the screen is too large that there is too much sand being produced with the consideration of the type of screen and pore structure for the application of the specified reservoir and wellbore [1]. Literature has highlighted the necessity to develop measurement techniques and analysis for screen testing to have the predictive properties for the expected screen performance and to have more available data capacity to improve future predictions [2]. So that it would be simplified to determine the critical properties to measure the screen performance testing [2]. Generally, the purpose of the sand retention test would be to measure the mass of sand produced as a function of the time of mass of sand injected, the pressure developed across the screen and sand pack, and the particle size distribution of the produced solids [3]. There are several empirical criteria used for screen performance [4]. But still, there exists an issue with the efficiency of the screen due to plugging. The plugging typically occurs at the annulus between the screen and the wellbore wall if this space is filled or partly filled by something with considerably less permeability than the reservoir sand [5]. On the other hand, plugging can be described as the movement of particles that migrate to the pores and, therefore, accumulate (or jam) when the particles migrate to the narrow channels [6]. Figure 1 shows the schematic of the particles plugging in the shale pores where the schematic describes the flow decreases after the plugging occurs as the channel decreases in size.

The unconsolidated formation has higher elasticity as compared to consolidated formation [7, 8]. And sand production is more in the unconsolidated formation which caused plugging. The phenomenon of plugging and jamming have been studied widely across many applications but there have been limited studies related to the sand bridging across the mechanical sand screen, especially on numerical simulation, as many experimental studies have been made over the years. shows the different jamming states from dilute, dense, and jammed (Fig. 2).

Various CFD studies have been conducted to transport sand particles in the fluid and to investigate permeability reduction in the reservoir [10–13]. There have been

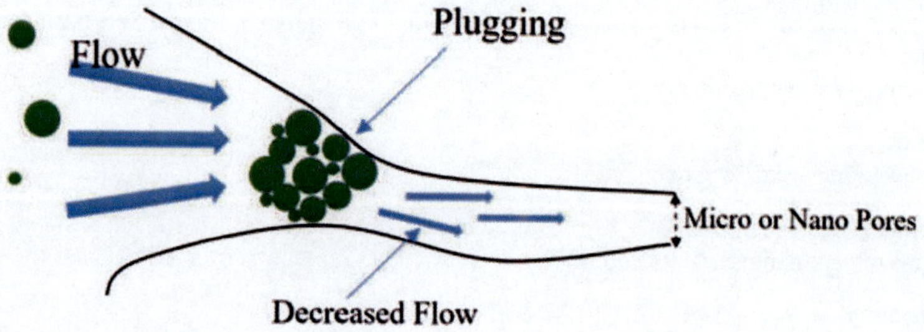

Fig. 1 Schematic of particles plugging [6]

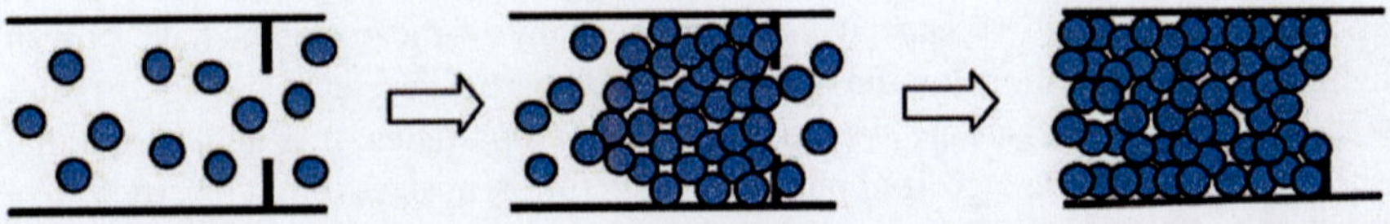

Fig. 2 Different states of particles flow; from dilute flow (left most), to dense flow (middle), to jammed state (right most) [9]

limited studies on the numerical simulations that involved the coupling of Computational Fluid Dynamics (CFD) and Discrete Element Modelling (DEM) to study the interaction between the particles and the fluid that the particle is suspended in. Based on the understanding of the movement and collision of the particles suspended in the fluid, this phenomenon becomes essential to enable the sand retention to occur at the sand screen to the point where the sand particles do not plug the opening entirely and are able to allow the fluid to still flow in the channel. Figure 3 shows the comparison of a previous study that shows the particle bridging and compares it to the experimental side of the particle bridging as part of their simulation validation. A study shows both experimental and numerical investigations for the sand retention [4]. These numerical simulations were focused on the effect of liquid velocity, effect of slot size, and the effect of particle concentration on the amount of sand produced. However, the particle size has been kept constant.

The study of jamming or plugging had many considerations on which parameters were deemed precise to mimic the real-life conditions. A study found that the screen selection is dependent on the particle size distribution of the formation grains and also considers other reservoir and production parameters [3]. They had considered the grain size distribution and uniformity, screen type, strength, and adequate sizing as part of the factors to consider in designing the screen, as well as screen permeability

Fig. 3 Comparison of the experimental particle bridging and the simulated particle bridging [1]

(as an indicator of inflow capacity). However, many of those studies have proven that there have been limits on the mono-dispersed particle systems and do not cater to the poly-dispersed particle systems, such as sand particles. It is understood that sand particles can be categorized physically into different sizes through particle size distribution (PSD). A CFD study of particles mentioned in a fluid has been performed using different particle sizes, it shows a small change in particle diameter and has significant change in the settlement or suspension of the particle [14]. The results obtained from the PSD show the different particle sizes that can be found throughout a sample. This would follow a lognormal distribution of a poly-dispersed particle system [15]. These simulation results show that the probability of jamming depends on the diameter of the larger particles in the distribution, especially the characteristic particle diameter, d_{90} [15].

2 Methodology

A.A.A. Development of Geometry

The geometry had been created to mimic the visual cell of the sand retention testing rig, specifically at the cell where the slurry will be injected via Autodesk Fusion 360. Based on Fig. 4a, the height of the visual cell is 92 mm with a diameter of 33 mm. The thickness of the sand screen is 3 mm. The inlet, visual cell, sand screen, and outlet are labelled in Fig. 4. The sand retention testing rig, SRT-350 is shown in Fig. 4b along with a diagram of the visual cell as used in the SRT system. Figure 4c shows the wire-wrapped screen used as a reference to replicate the 3D modelling of the sand screens. The CFD-DEM domain is constrained to the visual cell along with the inlet at the top, followed by the sand screen and the outlet section at the bottom.

B.B.B. CFD-DEM Simulation

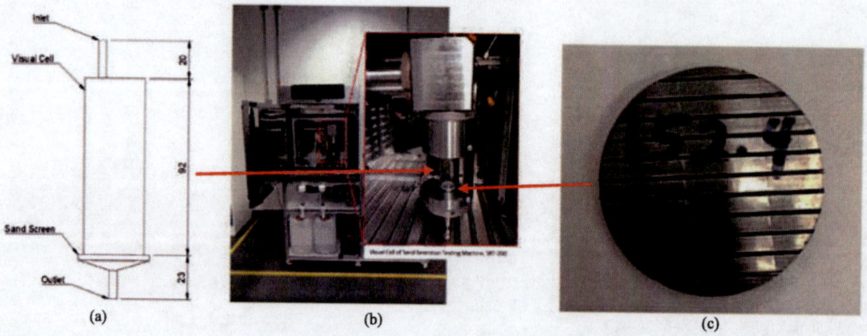

Fig. 4 Sand retention test setup **a** dimension of SRT cell **b** Sand Retention Testing Machine, SRT-350, and the Visual Cell of SRT-350, **(b)** the wire-wrapped screen used in the sand retention testing, 150 μm

CFD will be used to study the fluid flow, whereas DEM will be for the particle motion. The coupling of the two software will allow the effects of particle motion affecting the fluid in the CFD domain and vice versa for the DEM software to consider the fluid effect on the particles during the simulation. Figure 5 shows the methodology flow in conducting the CFD-DEM simulation to evaluate the particle flow through the sand screen of different opening sizes.

There were several considerations on the CFD-DEM simulation input as well as the boundary conditions. The simulation had been done in Altair AcuSolve and Altair

Fig. 5 Methodology of the CFD-DEM process

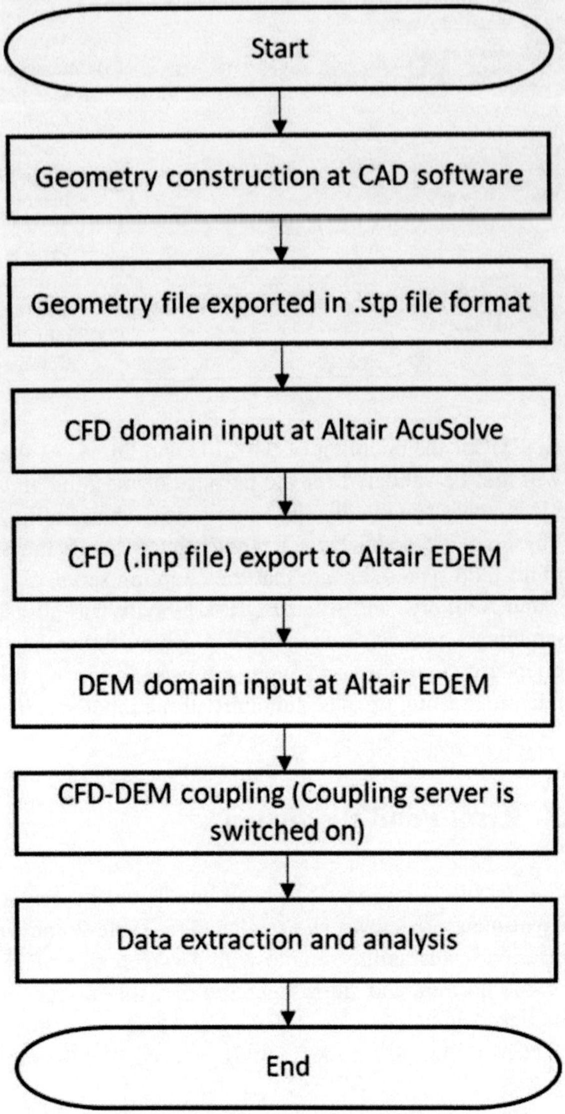

Table 1 Summary of input for CFD and DEM domains

CFD domain (Fluid Flow)	DEM domain (Particle Motion)
(1) Turbulence model: Spalart–Allmaras (2) Inlet: 0.0067 kg/s (Profiled Inlet, Mass Flow Rate) (3) Carrier Fluid Volume Fraction: 0.5 (4) Outlet: Pressure Profile (5) Walls: No slip (v = 0 m/s) (6) Time Step Size: 0.1 s (7) Final Time: 7 s (8) Multifluid Type: EDEM Bidirectional Material (Water) (9) Meshing size: • Volume Mesh: 0.005	(1) Particle Diameter: 150 μm (Spherical-shaped) (2) Size Distribution: Lognormal (3) Bulk Material Properties: • Poisson's Ratio: 0.25 • Solids Density: 2500 kg/m^3 • Shear Modulus: 1 × 10^8 Pa (4) Particle Properties: • Mass: 4.41786 × 10^{-9} kg • Volume: 1.767158 × 10^{-12} m^3 (5) Interactions: • Coefficient of Restitution: 0.5 • Coefficient of Static Friction: 0.5 • Coefficient of Rolling Friction: 0.1 (6) Particle Factory Generator: • Factory Type: Dynamic • Total Mass: 0.04 kg • Particle Velocity: 3 m/s (negative Z-direction) • Generation Rate: 8.3 × 10^{-5} kg/s (7) Simulator Grid: • Cell Size: 1 mm

EDEM for the coupling of the CFD and DEM. At the CFD domain, the geometry will first be validated for the purpose of meshing and to ensure that the generated 3D model is suitable for CFD simulation. This is followed by the specification of the Physics of the model. Here, it is essential to specify the Bidirectional EDEM Coupling multi-fluid type to ensure that the coupling server can be established through both Altair AcuSolve and Altair EDEM. Gravity was not taken into consideration as the jamming was considered to be fluid-driven, rather than gravity-driven jamming. The boundary conditions and input are used following the conditions set on the sand retention testing rig. The summary of the input is shown in Table 1.

3 Results and Discussion

Constant flow rate test had been conducted experimentally. The results from the experiment are shown and tabulated in Table 2 and Fig. 6d. The experiment used ceramic beads as the sample with a screen size of 150 microns, a fluid injection of 400 ml/min, and slurry injection of 5 ml/min. From the experiment, the sample produced and retained were 9.87 g and 5.75 g, respectively. The sample produced is defined as the mass of the sample produced after the sand screen, whereas the sample

Table 2 Experimental results

Sample	Sample produced (g)	Sample retained (g)
Ceramic beads	9.87	5.75

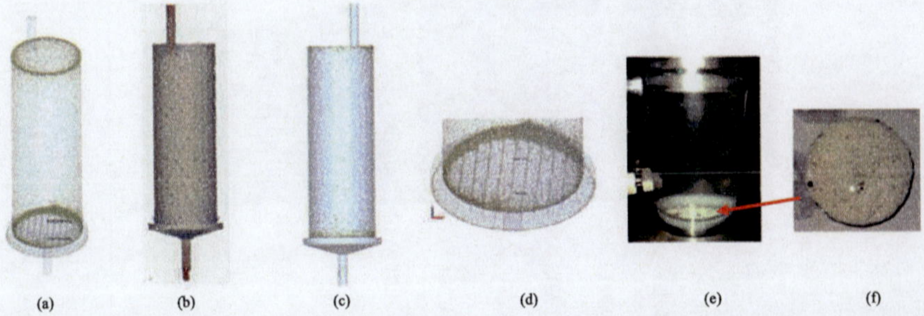

Fig. 6 Simulation and experimental results of ceramic retention on the screen **a** Total mass sensors added before and after the screen to record the sand produced and retained, respectively. **b** The flow of the fluid throughout the cell **c** 3D modelling of the particle simulation within the visual cell **d** Simulation of ceramic accumulated at the screen. **e** Ceramic accumulation during the experiment **f** Thick ceramic layer retained on the sand screen

retained was justified as the mass of the sample on the screen. Figure 6(c) shows the motion of the particles in the DEM simulation.

The results of the amount of sand produced (Table 2) will be used to compare with the results obtained numerically. Figure 6 shows the mass of the sample produced and the mass of the sample retained on the screen.

Table 3 compares the experimental and numerical simulations. It was shown that the ceramic beads produced and retained during the experiment were 9.87 g and 5.75 g, respectively.

There were some factors that had been taken into consideration for the likeness of the simulation to the experimental setup that could contribute to the difference in results. One of the factors include the duration of the sand retention testing. The duration of the experiment lasted between 15 – 25 min. The duration of the simulation took about 7 s. With a longer duration of the simulation, a higher percentage of sample retained and produced could be achieved to match the experimental setup. However, with the simulation duration of 7 s, the flow of the fluid was shown in Fig. 6c from the inlet to the outlet with the fluid passing through the screen. Figure 6d shows the accumulation of the sand on top of the sand screen which is also seen in the experiment, as shown in Fig. 6e, f with the fluid carrying the ceramic beads flowing down onto the screen to be filtered. This emphasizes that it is possible to

Table 3 Comparison of experimental and simulation results

Sample	Ceramic Beads	Simulation
Sample produced (g)	9.87	0.00000056
Sample retained (g)	5.75	0.228469

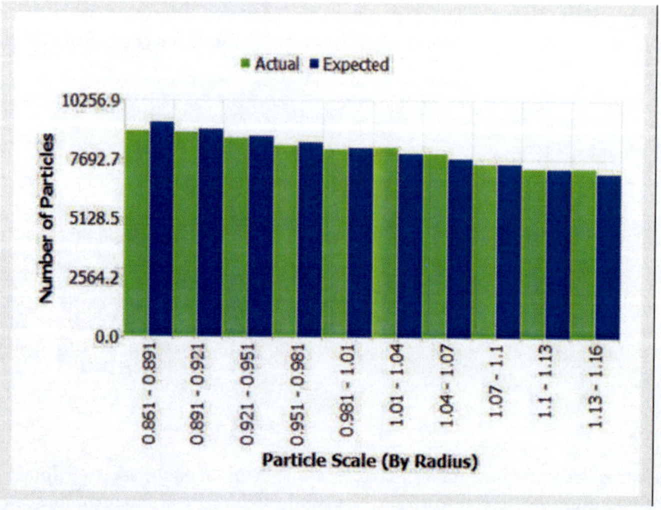

Fig. 7 Particle size distribution

simulate the movement of the particles via CFD-DEM. The current settings for the particle properties are set according to the default settings (namely on the particle density, Poisson's ratio and shear modulus). The particle size distribution could also contribute to the difference in sample production and retention. Figure 7 shows the particle size distribution that shows the number of particles produced by the particle scale. This is based on the particle size input of 150 μm that is lognormally distributed. A comparison of the particle size distribution of the simulated particles with the actual ceramic beads could ensure the properties of the particles to be properly defined.

To also ensure the simulation is able to run for longer periods of time (to match the experiment duration), a lower meshing number could be implemented to do so. With this, a mesh independent study could be implemented to ensure that the physics of the model can be validated to follow the actual experimental setup.

4 Conclusion

The production of sand in unconsolidated reservoirs has always been a problem to tackle on where it affects the production of oil when plugged. With that, sand retention testing has been conducted to select the sand screen based on the performance and permeability. Numerous experimental studies have outlined ways on how to select the optimal screen size. However, not many numerical studies have been conducted to simulate the experiment as it comes to the limitations of computational power. In this study, it was shown that it is possible to simulate the experimental conditions into the simulation platform and obtain results that would suffice in selecting the screen size with other factors to take into consideration. The results obtained from the experiment were used to compare with the results obtained from the simulation. The obtained numerical simulation results were not accurate as there were several factors to consider, namely the duration of the simulation difference of 7 s during the simulation and the range of 15 – 25 min using the experimental setup. It is believed that the accuracy may be achieved with a longer simulation time, and a comparison between the particle size distribution of the simulated particles and the actual ceramic beads to aid in the definition of the particle properties in the DEM domain, and to conduct the mesh independent study.

Acknowledgements We acknowledge Malaysia-Thailand Joint Authority (MTJA) for the grant awarded to carry out this research work.

References

1. Ismail NI, Kuang S, Yu A (2021) CFD-DEM study of particle-fluid flow and retention performance of sand screen. Powder Technol 378:410–420
2. Constien VG, Skidmore V (2006) Standalone screen selection using performance mastercurves. SPE International Symposium and Exhibition on Formation Damage Control
3. Ahad NA, Jami M, Tyson S (2020) A review of experimental studies on sand screen selection for unconsolidated sandstone reservoirs. J Petrol Explor Prod Techn 10(4):1675–1688
4. Wu B, Choi SK, Feng Y, Denke R, Barton T, Wong CY, Boulanger J, Yang W, Lim S, Zamberi M, Shaffee S (2016) Evaluating sand screen performance using improved sand retention test and numerical modelling. Offshore technology conference Asia
5. Chanpura RA, Hodge RM, Andrews JS, Toffanin EP, Moen T, Parlar M (2010) State of the art screen selection for standalone screen applications. SPE International Symposium and Exhibition on Formation Damage Control
6. Yang C, Zhou F, Feng W, Tian Z, Yuan L, Gao L (2019) Plugging mechanism of fibers and particulates in hydraulic fracture. J Petrol Sci Eng 1(176):396–402
7. Ul Haq I, Padmanabhan E, Khan JA, Uthayakumar T. Elastic properties and pore structure characterization of selected shales in the sungai Perlis beds, Terengganu, Peninsular Malaysia. in IOP Conference Series: Earth and Environmental Science. 2022. IOP Publishing.
8. Khan JA, Padmanabhan E, Ul-Haq I (2022) Hydraulic Fracturing to investigate impact of fracturing medium, bedding angle and perforation length on fracture growth in low and high Brittle Shale. in Offshore Technology Conference Asia. OnePetro
9. Guariguata A, Pascall MA, Gilmer MW, Sum AK, Sloan ED, Koh CA, Wu DT (2012) Jamming of particles in a two-dimensional fluid-driven flow. Phys Rev E 86(6):061311
10. Khan J, Pao WK (2013) Horizontal well cleanup operation using foam in different coiled tubing/annulus diameter ratios
11. Mahmoudi M et al. (2016) An experimental investigation of the effect of ph and salinity on sand control performance for heavy oil thermal production. in SPE Canada heavy oil technical conference. OnePetro
12. Khan JA, Pao WK (2014) Effect of different qualities of foam on fill particle transport in horizontal well cleanup operation using coiled tubing. In Advanced Materials Research. Trans Tech Publ
13. Alghurabi A, Mohyaldinn M, Jufar S, Younis O, Abduljabbar A, Azuwan M (2022) Erosion of sand screens by solid particles: a review of experimental investigations. J Petrol Explorat Prod Techn, 1–17

14. Pao W, Khan JA, Ofei TN, Irawan S (2015) Fill removal from horizontal wellbore using foam in different coiled tubing/annulus diameter ratios. Int J Oil, Gas Coal Techn 9(2):129–147
15. Xu S, Sun H, Cai Y, Geng X (2020) Studying the orifice jamming of a polydispersed particle system via coupled CFD–DEM simulations. Powder technology

Modelling of Density and Tensile Strength of Wollastonite-Filled Epoxy Composites

Anuja H. Karle, Namdev Ashok Patil, and Rachayya Arakerimath

Abstract This study examined how well wollastonite-filled epoxy composites might predict their density and tensile properties. Micro-composites of epoxy polymer with different loadings (1, 2, 3, 4, 5 wt.%) of wollastonite ($CaSiO_3$) filler were prepared by the magnetic stirring process. The properties were determined experimentally and modelled using a rule-of-mixture model for density and the Pukanzsky model for tensile strength. Then, the observed densities and tensile strengths were compared to values derived from theoretical models and to values computed using mathematical correlations created with Minitab 17. Over the range of $CaSiO_3$ levels examined, the predictions made by the models are in good agreement with the experimental data.

Keywords Micro-composites · Epoxy · $CaSiO_3$ · Mechanical properties · Theoretical modelling

1 Introduction

Polymers known for their versatility have the ability to modify the physical, mechanical, thermal, tribological and other properties by the incorporation of particulate fillers [1–3]. Ceramic fillers like Al_2O_3 [4–7], $CaSiO_3$ [6, 8–10], SiO_2 [11], SiC [12, 13], etc. are some of the newest particulate fillers widely used in the manufacturing of particulate-filled polymer composites. The use of wollastonite ($CaSiO_3$) microparticles can be an interesting topic in the study, considering the practical applications of the filler. In addition, the ceramic filler is inexpensive and widely available.

A. H. Karle · N. A. Patil (✉) · R. Arakerimath
Department of Mechanical Engineering, JSPM's Rajarshi Shahu College of Enfineerin, Tathwade, Pune, India
e-mail: napatil_mech@jspmrscoe.edu.in

A. H. Karle
e-mail: ahkarale_mech@jspmrscoe.edu.in

R. Arakerimath
e-mail: rrarakerimath_mech@jspmrscoe.edu.in

F. Ahmad et al. (eds.), *ICREEM 2022*, Lecture Notes in Mechanical Engineering,
https://doi.org/10.1007/978-981-99-5946-4_12

Several investigations in the literature have shown the addition of wollastonite particles to polymer matrices in small amounts. The microparticles are believed to improve the mechanical properties of composites as evidenced by their higher hardness, and tensile strengths to a certain extent when compared to those of neat polymers [14–16]. Among the thermoset polymers, epoxy resins are the extensively investigated commercial polymers because of their low cost, low density, good bonding strength with other materials and good mechanical properties [3, 17, 18]. Epoxy resin has been used for many applications such as paints and coatings, adhesives, industrial tools, aerospace industries, electronic industries, etc. However, its range of application is somewhat limited for further expansion due to poor toughness [19]. Various studies have reported an improvement in the mechanical and physical attributes of epoxy composites by lowering the amount of matrix materials [20–22].

Bernd Wetzel et al. [6] studied epoxy composites with Al_2O_3 and $CaSiO_3$ particles. They concluded that at low filler contents of 1–2 vol.% Al_2O_3, the addition of particles improved stiffness, impact energy and failure strain. Also, the introduction of $CaSiO_3$ in Al_2O_3 composite matrix resulted in an increase in flexural modulus to higher levels. Iskender Ozsoy et al. [23] revealed that the addition of micro-fillers Al_2O_3, TiO_2 and fly ash in the range of 10 wt.% to 30 wt.% and nano-fillers Al_2O_3, TiO_2 and clay in the range of 2.5 wt.% to 10 wt.% increased the tensile, flexural modulus and hardness of the composites with increasing filler content. Zakya Rubab et al. [24] investigated the mechanical properties of epoxy-titania composites. The study revealed 5 wt.% of titania reinforced epoxy composites significantly enhanced the tensile strength and Young's modulus of epoxy composites. Tayfun Uygunoglu et al. [3] observed that the inclusion of boron from 0 to 66 wt.% in polymer composite increases the compressive strength; however, it made the composites more brittle with low flexural strength. Wear resistance, water absorption and density are found to increase with increasing weight percentage of boron. Manoj Singla et al. [25] reported that the inclusion of fly ash particles in the matrix increased the compressive and impact strength of the material with the increase in fly ash content. Also, many researchers have investigated different models to predict the mechanical behaviour of particle-filled composites. A large number of parameters like the filler volume fraction, tensile strength and modulus, filler content, aspect ratio, shape factor of the filler and others were considered [2, 26–29]. Giulia Fredi et al. [27] assessed the effect of PCM microcapsules introduced in a polymer matrix on mechanical, thermal, rheological and microstructural properties experimentally. Microstructural characterization is also carried out to find the dispersion of MC phase in the matrix. The elastic modulus and tensile strength experimental data are fitted using the Nicolais-Narkis, Pukanszky and Lewis-Nielsen models, respectively, while the thermal conductivity of the composites is predicted using the Pal model. The study found that the experimental results and the values of theoretical models agree very well.

The aim of this work is to produce epoxy composites filled with wollastonite in various weight percentages and characterize them for mechanical properties. Mechanical properties viz. density and tensile strength of the wollastonite ($CaSiO_3$)-filled epoxy composites were evaluated. The experimental data of the properties

Fig. 1 Wollastonite powder

was then compared with the theoretical models considering parameters such as filler content, filler volume fraction and aspect ratio, to customize the suitability of the specific model. Finally, this work investigates how the introduction of wollastonite affects the mechanical performance of the epoxy resin not only experimentally, theoretically but also numerically using regression analysis.

2 Materials and Methods

2.1 Materials

Epoxy resin (AW106) and the corresponding hardener (HV953U) are chosen as the matrix material for the present research work. The matrix material is procured from Huntsman International (India) Pvt. Ltd, Mumbai. The wollastonite particles (Fig. 1) of average particle size 2 μm, purchased from Galaxy Corporation, Mumbai, are used as the filler material. All materials were used as received.

2.2 Preparation of epoxy/$CaSiO_3$ Composites

Curing agent and neat epoxy material were used to make a batch of samples. The other five batches of the samples were prepared using different fractions (1, 2, 3, 4 and 5 wt.%) of wollastonite ($CaSiO_3$) micro-particles into the epoxy matrix, in the presence of curing agent. Particles were mixed vigorously with the resin by a magnetic stirrer, prior to which particles were manually mixed. The obtained blend is manually mixed with the hardener for 5 min. The mixture is then applied to vacuum process to remove the entrapped air, after which it is moulded to a dimension of

Table 1 List of prepared composite samples

Sample	Composition ($CaSiO_3$ as filler)
1	Neat Epoxy (without filler)
2	Epoxy resin + 1 wt.% filler
3	Epoxy resin + 2 wt.% filler
4	Epoxy resin + 3 wt.% filler
5	Epoxy resin + 4 wt.% filler
6	Epoxy resin + 5 wt.% filler

150 mm × 25 mm × 5 mm and cured for 24 h at room temperature. Composites particulate of different compositions are shown in Table 1.

2.3 *Characterization*

2.3.1 Density Test

The experimental density of epoxy and epoxy/$CaSiO_3$ samples is determined by using ASTM D792 standard procedure. Density is measured by using simple water immersion technique. At least three specimens of 20 mm × 20 mm dimensions were measured for each sample.

2.3.2 Tensile Test

Universal testing machine was used to perform tensile tests of the sample at a constant cross-head speed of 50 mm/min. Room temperature at the time of testing being 24 ± 1 °C, the relative humidity was maintained at 50 ± 5%. The tests on composites were performed according to ASTM D638 with dog-bone shape specimens. The dimension of the specimen used were 125 mm × 25 mm × 5 mm considering a gauge length of 50 mm. Three specimens were tested for each composition until failure was declared. Typically fractured tensile test samples are shown in the Fig. 2.

2.4 *Modelling of Properties*

2.4.1 Density and Void Fraction

Theoretical Modelling

Rule of mixture model [30] was used to calculate theoretical density (ρ_{th}) of prepared samples which is expressed as

Fig. 2 Fractured test specimens

$$\rho th = \frac{1}{\frac{W_m}{\rho_m} + \frac{W_f}{\rho_f}} \tag{1}$$

W and ρ stand for the weight fraction and density, respectively. Matrix and particulate filler are indicated by the suffixes *m* and *f*, respectively. The density of the matrix was found to be 1.1 g/cm^3. The density of wollastonite particle is taken as 2.4 g/cm^3.

The volume fraction of voids (V_v) of the composite samples is determined by the equation:

$$V_v = \frac{\rho_{th} - \rho_e}{\rho_e} \tag{2}$$

where ρ_e is the experimental density of the composite. Void fraction of composites was calculated using comparison between theoretical and experimental density.

Proposed Polynomial Model

This model is dependent on the variant filler content (x). The nonlinear equation describes the relation between the response density and filler content. The general 3rd (Cubic) degree polynomial that best fits the experimental data is given as

$$\rho_c = a_1 + a_2 x + a_3 x^2 - a_4 x^3 \tag{3}$$

where the coefficients a_1, a_2, a_3 and a_4 are determined by regression analysis using Minitab 17 software. The other polynomial models of order 1 and order 2 were also developed using regression analysis and validated. The equation shows that density goes on increasing with increasing filler content conforming a cubic polynomial for use.

2.4.2 Tensile Strength

Tensile strengths of epoxy/$CaSiO_3$ composites were fitted by the model developed by Pukanzsky. The Pukanzsky model [31] is shown in Eq. (4) as

$$\sigma_C = \sigma_m\left(\frac{1-\varphi\prime_f}{1+2.5\varphi\prime_f}\right)e^{K\varphi\prime_f} \tag{4}$$

where σ_C is the tensile strength of composite and σ_m is the tensile strength of matrix, respectively, and $\varphi\prime_f$ is the corrected filler volume fraction. The parameter K is an adhesion parameter for the filler obtained by fitting the experimental data to the model.

The corrected filler volume fraction $\varphi\prime_f$ is represented in Eq. (5) as

$$\varphi\prime_f = \varphi_f.\lambda \tag{5}$$

where φ_f represents the volume fraction of wollastonite represented in Eq. (6) as

$$(\varphi_f)_i = \frac{w_i/\rho_i}{\sum(w_i/\rho_i)} \tag{6}$$

In the above equation, w_i and ρ_i represent the weight fraction and density of the ith composite, respectively. The densities of epoxy and $CaSiO_3$ are 1.1 g/cm^3 and 2.4 g/cm^3, respectively.

The λ in the equation represents the additive factor and is given as.

$$\lambda = a\left(A_f\right)^b \tag{7}$$

where A_f is the aspect ratio considered to be 3 [15]. The parameters a and b are the constants accounting non-linear behaviour of the filler.

Proposed Polynomial Model

For tensile strength obtained from the tests of wollastonite particulate-filled polymer composites, the recommended parametrization is the 4th (Quartic) degree polynomial ruling out the linear, quadratic and cubic polynomial models. The equation is given as

$$\sigma_c = a_1 + a_2x - a_3x^2 + a_4x^3 - a_5x^4 \tag{8}$$

where x is the variant viz. the filler content in the model. From the above equation, it is found that the tensile strength first increases and then decreases with the increasing filler content conforming a Quartic polynomial model.

2.5 *Quality Assessment*

In this study, a number of metrics, including the coefficient of determination (R2), the root mean squared error (RMSE) and the mean absolute error (MAE), are used to evaluate the findings provided by the regression models.

R^2, a key output in regression analysis, helps compare current model outputs with the actual outputs. The value of R^2 ranges from 0 to 1. The closest number to 1 implies a very high positive correlation between the predicted and actual values. Another measure for regression problems is the RMSE. It is the square root of the averaged squared difference between the predicted and actual values, whereas MAE calculates the absolute difference between them. In contrast to R^2, lower values of RMSE and MAE indicate a better performance of a regression model [32–35]. The quantities R^2, RMSE and MAE are given by the Eqs. (9–11):

$$R2 = 1 - \frac{\sum_{i=1}^{N} (x_i - y_i)^2}{\sum_{i=1}^{N} (x_i - \overline{y})^2} \tag{9}$$

$$RMSE = \sqrt{\frac{1}{N}\sum_{i=1}^{N} (x_i - y_i)^2} \tag{10}$$

$$MAE = \frac{1}{N}\sum_{i=1}^{N} (x_i - y_i) \tag{11}$$

where x_i and y_i are actual and predicted values respectively, N is the number of samples and $\overline{y}$ is the mean of predicted values in the above equations.

3 Result and Discussion

3.1 *Density and Void Fraction*

Results of density and void fraction of the prepared samples are tabulated in Table 2 and graphically depicted in Fig. 3.

The pure epoxy polymer has a measured density of 1.1 g/cm^3. The experimental densities disclose a steady increase from 0 to 5 wt.% of $CaSiO_3$ micro-particles in the epoxy matrix. Figure 3 depicts theoretical density of the epoxy/$CaSiO_3$ samples via the rule-of-mixture model.

Table 2 shows that the estimated densities of composite are not in complete agreement with the values obtained through experimentation. The disparity in densities serves as a gauge for the amount of voids in composite materials. It is seen from the graph (Fig. 3) the densities of the composites for lower filler content agree well with

Table 2 Densities (theoretical and experimental) of composite with per cent of void content

Filler content (wt.%)	Density (g/cc)		Void content (%)
	Theoretical	Experimental	
1	1.1068	1.1037	0.280
2	1.1149	1.1116	0.295
3	1.1209	1.1175	0.303
4	1.1280	1.1242	0.337
5	1.1353	1.1312	0.361

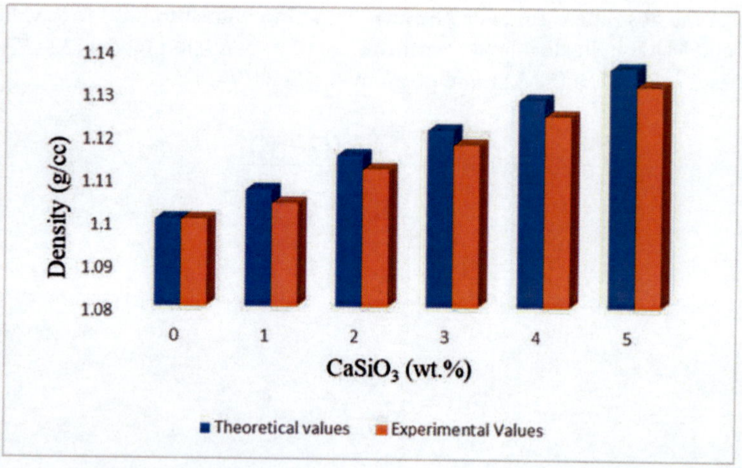

Fig. 3 Density of different composites

the theoretical values, but as the filler content increases the theoretical densities of the composites are on higher side as compared to the experimental densities. In order to confirm the growing difficulty of fabricating defect-free specimens for highly filled compositions, the void content was calculated. The void content showed an increase from 0.280 vol% for composites with 1 wt% CaSiO3 to 0.361 vol% for composites with 5 wt% filling. The reason for higher theoretical densities is the assumption of no voids and defects while calculating the theoretical values, while, in actual practice, fabrication of composites results in considerable number of voids which increases with filler loading as reflected in Table 2. The increase in the filler content beyond 5 wt.% $CaSiO_3$ may further increase the void content which may affect the mechanical properties of the composite. Hence, the particle loading has been restricted within 5 wt.%.

Moreover, the polynomial model with four parameters properly fits the experimental data as the model exhibits the most efficient values regarding all the criteria of evaluation. It had the best values of $R^2 = 0.998$, RMSE = 0.000594 and MAE = 0.000245. The response for density evaluated in terms of cubic model is given as

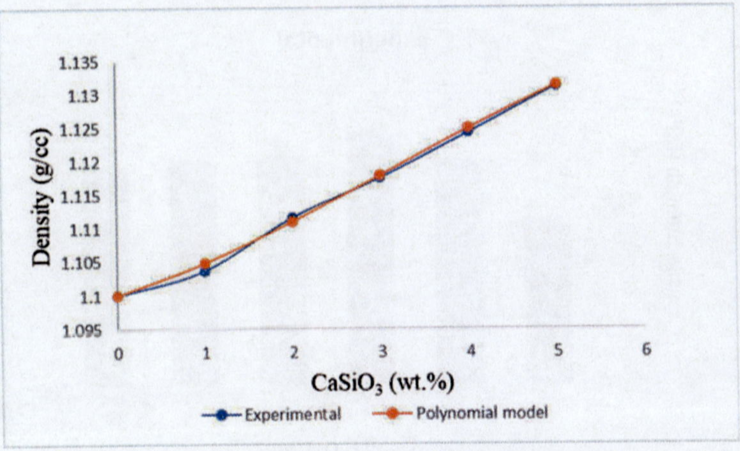

Fig. 4 Plot of density as a function of $CaSiO_3$ content of the composites

$$\rho_c = 1.1 + 0.00393x + 0.00098x^2 - 0.000103x^3 \tag{12}$$

Figure 4 shows the resulting data attained from the cubic model compared with the output data attained from the experiments for the increasing weight percentage of $CaSiO_3$ fillers.

3.2 *Tensile Strength*

Figure 5 reveals the plot of experimental tensile strength values for all the compositions of epoxy/$CaSiO_3$ composites. As shown in Fig. 5, the tensile strength of the developed composite is positively affected by an increase in the $CaSiO_3$ up to 2 wt.%. However, the tensile strength showed a dramatic decrease even at very low filler weight per cent of 3% to 5%. In spite of this, all the compositions showed a tensile strength higher than the neat epoxy. The change in the mechanical behaviour is evident by the incorporation of the filler in the resin. This can be attributed to the interaction between filler and matrix material. The maximum value obtained of 51.46 MPa for the specimen filled with 2 wt.% of the filler is approximately 1.5 times greater than the specimen with no fillers incorporated.

Figure 6 shows the experimental results compared with numerically computed values for increasing weight percentages of fillers. Among the several theories proposed to model the tensile strength of composite material, experimental data to some extent shows similar trend to the predicted values for the model developed by Pukanszky as plotted in Fig. 6. Quantitively, the accuracy offered by the Pukanzsky model is approximately 86% based on the experimental values obtained. The accuracy of the Pukanzsky model is also reported to be good in other composite systems

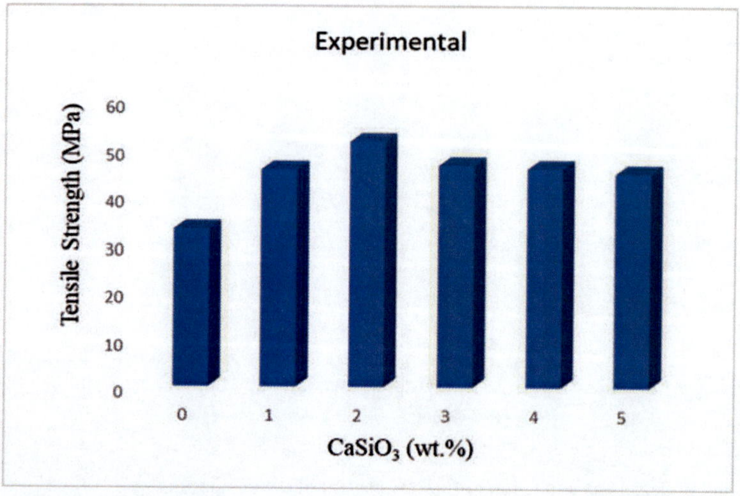

Fig. 5 Tensile Strength of different samples

[27, 31]. The observations of the model suggest that the optimum filler volume fraction and aspect ratio of filler play an important role in predicting the tensile strength of the composites.

Furthermore, the polynomial model showed a good match with the experimental values for the composites as shown in Fig. 6. The result indicates that five variables are required to accurately fit the tensile strength of the developed composite. Equation 13 shows the Quartic model for tensile strength that fits the experimental data:

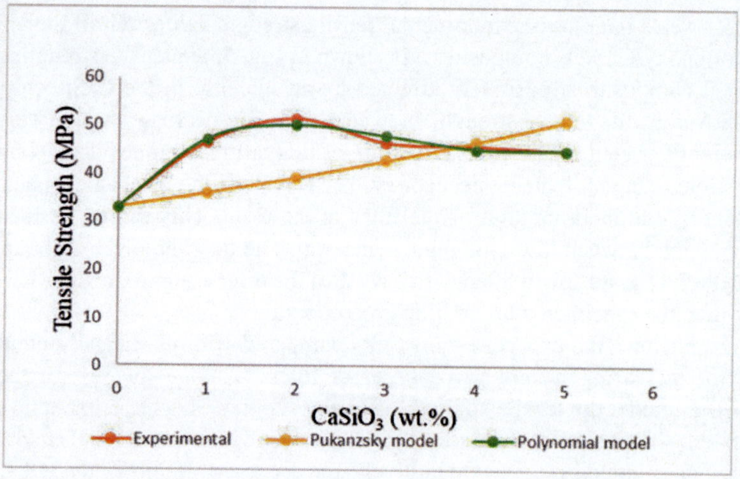

Fig. 6 Results of experimental, theoretical and polynomial model of Tensile strength

$$\sigma_c = 33 + 22.4487x - 9.61592x^2 + 1.47257x^3 - 0.0702706x^4 \quad (13)$$

This conclusion of Quartic model showing accurate prediction of experimental and interpolated values is made with respect to three statistical criteria R^2, RMSE and MAE. The values of R^2, RMSE and MAE are 0.997, 0.886 and 0.0230, respectively.

4 Conclusion

Based on the theoretical and experimental work, the following inferences can be made:

1. Different amounts of wollastonite particles (1, 2, 3, 4 and 5 wt.%) were used to fabricate epoxy/$CaSiO_3$ composite using the magnetic stirring process.
2. The inclusion of wollastonite up to 2 wt.% improved the tensile strength to 51.46 MPa with density of 1.1116 g/cm^3 which is a good strength-to-weight combination to be used in industrial applications. This clearly shows that low wollastonite content is effective for enhancements of mechanical properties.
3. The experimental data of the properties was fitted with the models of rule-of-mixture, Pukanszky and polynomial models. The modelling of density with rule-of-mixture and tensile strength with Pukanszky model evidenced similar trends when compared to experimental values.
4. The polynomial models in this work performed well in predicting the mechanical properties. The density and tensile strength of $CaSiO_3$ filled epoxy composites were best predicted by the proposed Cubic and Quartic models.

Acknowledgements The author takes this opportunity to convey their deepest gratitude to JSPM's Rajarshi Shahu College of Engineering, Mechanical Engineering Department, for funding publications.

References

1. Athimoolam M, Moorthy TV, Sureshbabu A (2016) Empirical modeling and experimental investigation of the amine treated nanoclay reinforced Epoxy-Polyurethane blended resin. Compos 24:539–546
2. Selvin TP, Kuruvilla J, Sabu T (2004) Mechanical properties of titanium dioxide-filled polystyrene microcomposites 58:281–289. https://doi.org/10.1016/S0167-577X(03)00470-1
3. Uygunoglu T, Gunes I, Brostow W, Faculty E (2015) Physical and mechanical properties of polymer composites with high content of wastes including boron 2. Exp Stud 18:1188–1196
4. Agrawal A, Satapathy A (2014) Experimental investigation of Micro-sized aluminium oxide reinforced epoxy composites for microelectronic applications, procedia. Mater Sci 5:517–526. https://doi.org/10.1016/j.mspro.2014.07.295

5. Cho J, Joshi MS, Sun CT (2006) Science and effect of inclusion size on mechanical properties of polymeric composites with micro and nano particles 66:1941–1952. https://doi.org/10.1016/j.compscitech.2005.12.028
6. Wetzel B, Haupert F, Zhang MQ (2003) Epoxy nanocomposites with high mechanical and tribological performance. Compos Sci Technol 63:2055–2067. https://doi.org/10.1016/S0266-3538(03)00115-5
7. Pawar JB, Tungikar VB (2022) Alumina—di water—based nanofluid process parameter optimization for stability. J Brazilian Soc Mech Sci Eng 8:1–13. https://doi.org/10.1007/s40430-022-03541-8
8. Lapčík L, Maňas D, Lapčíková B, Vašina M, Staněk M, Čépe K, Vlček J, Waters KE, Greenwood RW, Rowson NA (2018) Effect of filler particle shape on plastic-elastic mechanical behavior of high density poly(ethylene)/mica and poly(ethylene)/wollastonite composites. Compos Part B Eng 141:92–99. https://doi.org/10.1016/j.compositesb.2017.12.035
9. Karle AH, Nukulwar MR, Tungikar VB (2020) Materials Today : Proceedings Evaluation of mechanical and thermal properties of epoxy composites reinforced with CaSiO 3 particulate fillers. Mater Today Proc. https://doi.org/10.1016/j.matpr.2020.08.188
10. Karle AH, Tungikar VB (2021) Materials Today : Proceedings Optimization of wollastonite reinforced epoxy composites, Mater. Today Proc. 1–5. https://doi.org/10.1016/j.matpr.2021.01.688
11. Vaisakh SS, Azeez A, Mohammed P, Hassanzadeh M, Tortorici JF, Metz R (2016) Effect of nano-modified SiO 2/Al 2 O 3 mixed- matrix micro-composite fillers on thermal , mechanical , and tribological properties of epoxy polymers †, 2016. https://doi.org/10.1002/pat.3747
12. Madhusudhan T (2017) Investigation on wear resistance behavior of sic filled hybrid composites 8:82–92
13. Sharma SK, Kim Y (2016). Tribological behavior of silicon carbide Ceramics—A review tribological behavior of silicon carbide Ceramics—A review. https://doi.org/10.4191/kcers.2016.53.6.581
14. Hamzah AF, Alkawaz A, Bayan N (2018) Study of mechanical properties of wollastonite filled epoxy, (2018).
15. Chan JX, Wong JF, Hassan A (2019) Mechanical properties of wollastonite reinforced thermoplastic composites : A review, 1–35. https://doi.org/10.1002/pc.25403
16. Mahesh V (2022) Tribological characterization of sustainable Jute-Epoxy-Rubber crumb hybrid composite, Proc Inst Mech Eng Part C J Mech Eng Sci. 095440622211039. https://doi.org/10.1177/09544062221103941
17. Mahesh V, Harursampath D, Joladarashi S, Kulkarni SM (2022) Development of sustainable Jute/Epoxy composite and assessing the effect of rubber crumb on low velocity impact response, J Nat Fibers 0 1–12. https://doi.org/10.1080/15440478.2022.2054897
18. Mahesh V, Nagaraj SM, Subhashaya P, Singh G (2022) Physio-mechanical and thermal characterization of jute/rubber crumb hybrid composites and selection of optimal configuration using the MADM approach, Proc Inst Mech Eng Part C J Mech Eng Sci. 095440622210791. https://doi.org/10.1177/09544062221079166
19. Jin F, Li X, Park S (2015) Journal of industrial and engineering chemistry synthesis and application of epoxy resins : a review. J Ind Eng Chem 29:1–11. https://doi.org/10.1016/j.jiec.2015.03.026
20. Mahesh V, Mukunda S, Siddharth A, Joseph A, Harursampath D (2022) Effect of organic nanofillers on the free vibration and damping characteristics of Polymer-Based nanocomposites, in: pp 103–112. https://doi.org/10.1201/9781003173625-7
21. Mahesh V, Joseph A, Mahesh V, Harursampath D (2021) Investigation on the mechanical properties of additively manufactured PETG composites reinforced with OMMT nanoclay and carbon fibres, Polym Compos 42. https://doi.org/10.1002/pc.25985
22. Mahesh V (2019) Influence of areca nut nano filler on mechanical and tribological properties of coir fiber reinforced epoxy based polymer composite, Sci Iran. https://doi.org/10.24200/SCI.2019.52083.2527

23. Ozsoy I, Demirkol A, Mimaroglu A, Unal H, Demir Z (2015) The influence of micro- And nano-filler content on the mechanical properties of epoxy composites, Stroj. Vestnik/Journal. Mech Eng 61:601–609. https://doi.org/10.5545/sv-jme.2015.2632
24. Rubab Z, Afzal A, Siddiqi HM, Saeed S (2021) Preparation, characterization, and enhanced thermal and mechanical properties of Epoxy-Titania composites
25. Singla M, Chawla V (2010) Mechanical properties of Epoxy Resin—Fly ash composite, 9 199–210
26. Schlarb AK, Friedrich K (2007) Modelling of mechanical properties of nanoparticle filled polyethylene 945–964. https://doi.org/10.1080/00914030601139142
27. Dorigato A, Fambri L, Pegoretti A (2020) Detailed experimental and theoretical investigation of the thermomechanical properties of epoxy composites containing paraffin microcapsules for thermal management 1–19. https://doi.org/10.1002/pen.25374
28. Hazizan MD, Akil P, Firoozian AHPSAK (2010) Prediction of mechanical properties of Mica-filled epoxy composite using various mechanical models https://doi.org/10.1177/073168440 9351166
29. Sagar PDJS, Kashyap SJ, Madhu GM (2020) Investigation of mechanical , thermal and electrical parameters of gel combustion-derived cubic zirconia / epoxy resin composites for high-voltage insulation (Investigação de parâmetros mecânicos , térmicos e elétricos de, 66 186–196
30. Richard S, Selwin Rajadurai J, Manikandan V (2017) Effects of particle loading and particle size on tribological properties of biochar particulate reinforced polymer composites, J Tribol 139 1–10. https://doi.org/10.1115/1.4033131
31. Bouza R (2009) Effects of vinyltrimethoxy silane on mechanical properties and morphology of Polypropylene—Woodflour. Compos. https://doi.org/10.1002/pen
32. Nukulwar MR, Tungikar VB (2020) Evaluation of drying model and quality analysis of turmeric using solar thermal. System 56:233–241. https://doi.org/10.3103/S0003701X20040088
33. Nukulwar MR, Tungikar VB (2020) Thin-Layer mathematical modeling of turmeric in indirect natural conventional solar dryer 142:1–7. https://doi.org/10.1115/1.4045828
34. Shringi V, Kothari S (2014) Experimental investigation of drying of garlic clove in solar dryer using phase change material as energy storage, 533–539. https://doi.org/10.1007/s10973-014-3991-0
35. Validation methodologies (2017) 83–107. https://doi.org/10.1016/B978-0-12-805274-7.000 04-X

Thermal Analysis of Evacuated Tube Receiver for Solar Power Tower by Transient Simulation

Ahmad Zaimmul Adli Bin Abdullah, Mubarak Danladi Muhammad, Syed Ihtsham Ul-Haq Gilani, Muzaffar Ali, and Hussain H. Al-Kayiem

Abstract Solar thermal power is a promising and rapidly expanding source of carbon-free energy. Analysis and design techniques for solar thermal power generation for the Solar Power Tower (SPT) systems are currently mathematically difficult. We simulated a model of a SPT that advances the simulation of SPT performance by modelling them in Thermolib software. The model in this study simulates SPT with a direct steam generation system with design variables of 0.04 kg/s heat transfer fluid (HTF) flow rate at 6 bar operating pressure using water as HTF. The reaction to varied operating pressure, fluid flow rates, and insolation input has been analysed. The simulation results show that steam will be produced at 1–3 bar with maximum steam quality of 0.06, 0.02 and 0.0015, respectively. Lowering the HTF flow rates to 0.02 and 0.03 kg/s will produce steam with maximum quality of 0.15 and 0.028, respectively. Increasing the amount of heliostat from 15 to 20 in staggered formation results in a 14.5% increase in power received by the receiver, resulting in a maximum steam quality produced of 0.029. The analysis shows that high quality of steam can be achieved with a lower HTF flow rate and pressure, but it compromises the useful energy generated by turbine.

A. Z. A. Bin Abdullah (✉) · S. I. U.-H. Gilani · H. H. Al-Kayiem
Mechanical Engineering, Universiti Teknologi PETRONAS, Perak, Malaysia
e-mail: ahmad_21000813@utp.edu.my

S. I. U.-H. Gilani
e-mail: syedihtsham@utp.edu.my

H. H. Al-Kayiem
e-mail: hussain_kayiem@utp.edu.my

M. D. Muhammad
Department of Mechanical Engineering, Bayero University Kano, Gwarzo Road, P.M.B 3011, Kano, Nigeria
e-mail: mdmuhammad.mec@buk.edu.ng

M. Ali
Mechanical Engineering, University of Engineering and Technology, Taxila, Pakistan
e-mail: muzaffar.ali@uettaxila.edu.pk

F. Ahmad et al. (eds.), *ICREEM 2022*, Lecture Notes in Mechanical Engineering,
https://doi.org/10.1007/978-981-99-5946-4_13

Keywords Evacuated tube receiver · Steam quality · Direct steam generation · Solar power tower · Thermolib · Heliostats

1 Introduction

A renewable energy technology that is considered sustainable and economical is concentrating solar power (CSP). CSP technology is growing rapidly as it is to be one of the promising electricity generators from solar energy. Various CSP technologies are under development and researched in various parts of the world. Researchers are developing day by day the new and useful applications of renewable energy and contributing to the overall energy mix. All concentrating solar power (CSP) technologies use a mirror configuration to concentrate the sun's light energy onto a receiver and convert it into heat. The heat can then be used to create steam to drive a turbine to produce electrical power or used as industrial process heat. Figure 1 shows some CSP technologies already in practice and still in the development process at the same time for further improvements in their working or efficiencies.

The different type of CSP systems are as follows:

1. Parabolic Trough Collectors (PTC)
2. Linear Fresnel Reflectors (LFR)
3. Parabolic Dish Concentrators (PDC)
4. Solar Power Tower (SPT) or Solar Central Receiver.

These CSP systems possess the same energy conversion technique but differ in configuration and size. Faye et al. [1] reported that among the CSP systems, Solar Power Tower (SPT) technologies possess the most potential in producing alternative

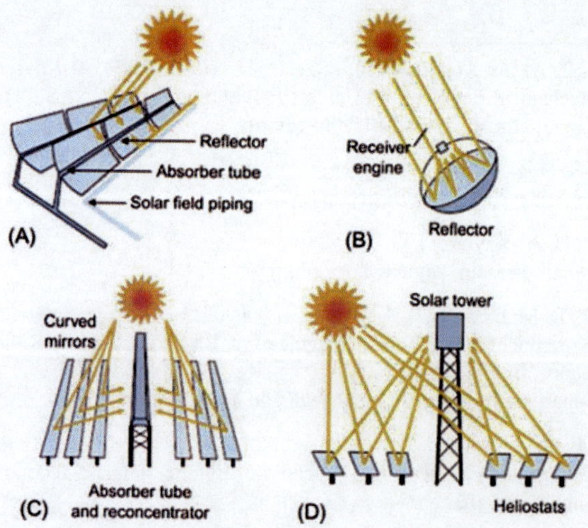

Fig. 1 Concentrated solar power technologies [1]

energy generation due to their nature to produce high temperatures and therefore, high efficiency. SPT is preferred in future development because it can provide dispatchable solar energy at a low cost [2]. An SPT system generally consists of a heliostat field, a steam turbine, and a thermal energy storage [3]. Typically, two promising types of solar power tower plants are discussed: the molten salt power tower plant and the direct steam generation (DSG) power tower plant. Molten salt is commonly used as both a heat transfer fluid and a thermal energy storage medium in molten salt power tower plants.

The use of mathematical modelling is essential since it is a way that can save money by extrapolating data from actual field experiences. This method can be utilized to acquire insight into the functioning principles of energy processes, and at the moment, a large number of models of DSG that call for numerical solutions have been published. [4], for instance, have developed a closed-form model of DSG that advances the simulation of DSG by defining their performance as a function of both time and axial position. Using the simulation tools Modelica, MATLAB, and ANSYS, [5–7] examined the DSG process and constructed a dynamic model of the parabolic trough collector. Silva et al. [8] established a one-dimensional thermal dynamic model to characterize the fluid temperature along the longitudinal axis of the absorber. Roldan et al. examined the temperature of absorber tubes in DSGs using the FLUENT software for the Finite Volume Method [9]. Yan developed a dynamic model for the DSG process of producing superheated steam [10]. Martinez and Almanza [11] developed a two-dimensional model using the finite difference numerical approach to examine the temperature profile in the absorber of a direct steam generator. For various compounds and mixtures, thermodynamic models have been developed. Rigid thermodynamic models, on the other hand, are computationally demanding and use cubic equations of state to compute vapour and liquid phase thermodynamic properties [12].

In this paper, we present a performance study of a solar power tower with an evacuated tube collector receiver using Thermolib, a dynamic simulation and optimization tool. It is written in MATLAB and C. Thermolib has a good set of routines for calculating analytical first and second derivatives with respect to temperature, pressure, and other variables. Symbolic differentiation is used to obtain the derivatives. The DIPPR database [13], the Peng-Robinson (PR) EOS [14], the Soave–Redlich–Kwong (SRK) EOS [15], and the van der Waals mixing rules [16] are used to build the library. There are already a few open-source libraries available, such as Chapman et al. [17] Simulink toolbox T-MATS and Martin et al. [18] MATLAB library. Furthermore, [19] developed the CALPHAD software Open-CALPHAD, which includes an open-source module for evaluating thermodynamic properties. None of these libraries, however, provides both first and second derivatives [12]. The findings of this study will be applied to the design and optimization evaluation of a system that will be installed at Universiti Teknologi PETRONAS (UTP), Malaysia, and will be powered by a solar thermal heat source.

Table 1 Design specification used in the SPT system

Parameters	Symbol	Value
Absorber diameter	D_a	0.04 m
Envelope diameter	D_e	0.07 m
Total length of main heater ETR	L_m	20 m
Total length of superheater ETR	L_s	7.5 m
Inlet pressure	P_{in}	5 bar
Feedwater temperature	T_0	25 °C
Air ambient temperature	T_{air}	25–30 °C
Saturation temperature @ 5 bar	T_{sat}	151.86 °C
Heliostat area	A_{heli}	3.75 m^2

A. *The SPT system configuration*

The system's essential components are the heliostat field, receiver, turbine, condenser, steam separator, and working fluid pump. The receiver is composed of a series of evacuated tube receivers (ETR) that are separated into two sections: the main heater and the superheater. The ETR is used to heat water utilized as a heat transfer fluid (HTF) in the solar circuit. The pump's pressured working fluid is preheated, evaporated, and superheated to a working temperature between 200 and 250 °C. The superheated steam then passes through the turbine, which generates useful electricity. The steam exiting the turbine is first cooled from superheated conditions to saturated vapour at variable temperatures in the de-superheater, and then condensed to saturated liquid using cooling water in the condenser. To continue the cycle, water is pumped again in the pump. Table 1 displays the operational parameters and specifications of the system under consideration.

The input data: direct normal irradiance (DNI) was extracted from the Solar Research Site (SRC) UTP weather station located at 4.37, 100.96 on 17 January 2020. The DNI data is then used as input in Tonatiuh software to simulate the heliostat field model performance in terms of power received on the receiver. Table 2 shows the result obtained by the simulation. The default amount of heliostat field used is 15 pieces arranged in staggered formation. Figure 2 presents the data obtained from weather station located at UTP, Perak.

The result as shown in Table 2 is used in Thermolib to simulate the thermal performance of the receiver main heater and eight evacuated tube receivers. The result shows that with 15 heliostats, the heliostat field can reflect at a condition of DNI of 869.21 W/m^2 up to 17 kW of power onto the 8 absorbers. Figure 3 shows the overview of modelled system in the Thermolib software which consists of a pump, heaters, steam separator, water tank, and sensors to assist in getting the required result.

B. Thermodynamic Analysis

Thermolib component blocks consist of input and outputs to reflect energy flows across the block's boundary [20]. Whenever such a signal has a value greater than

Table 2 Weather data on 17 January 2020 measured in UTP

Time	Direct normal irradiance (W/m^2)	Total power (W)	Average flux (W/m^2)	Maximum flux (W/m^2)	Estimated power received on 8 absorbers (W)
8:00 am	119.63	2544.34	678.49	1137.06	1705.23
9:00 am	341.61	8682.36	2315.30	3873.19	5818.98
10:00 am	571.01	15,526.80	4140.48	6722.24	10,406.16
11:00 am	745.62	20,886.30	5569.69	8656.70	13,998.16
12:00 pm	866.59	25,016.30	6671.02	10,175.80	16,766.10
1:00 pm	869.21	25,471.90	6792.50	10,181.80	17,071.41
2:00 pm	748.20	21,899.20	5839.78	8758.76	14,676.97
3:00 pm	574.53	16,528.00	4407.47	6722.18	11,077.18
4:00 pm	345.21	9661.03	2576.27	4037.28	6474.87
5:00 pm	122.83	3321.72	885.79	1429.67	2226.23

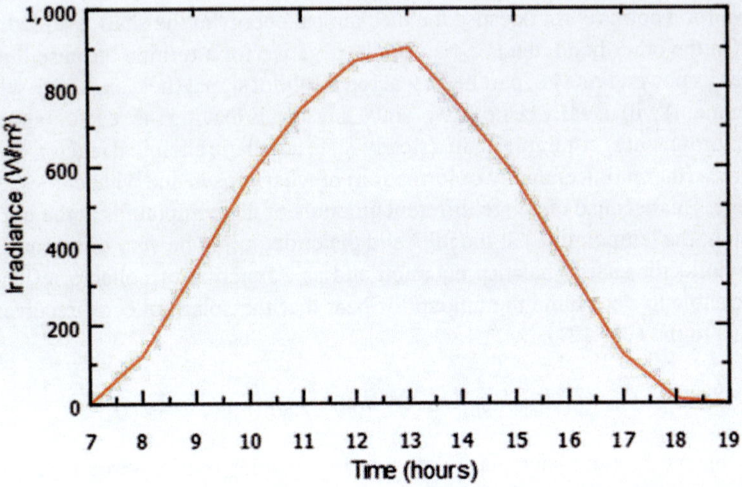

Fig. 2 DNI on 17 January 2020

zero, the energy flow is from the outside of the block to the interior. The energy/power/heat is subsequently transferred to the flow or stored inside. Whenever such a signal has a value greater than zero, the energy flow is from inside the block to the outside. The energy/power/heat is extracted from the flow or internal energy storage. In general, it holds [20]

$$\frac{dU}{dt} = \sum \dot{H}_{in,i} - \sum \dot{H}_{\text{out},j} + \sum \dot{Q}_k + \sum P_m \tag{1}$$

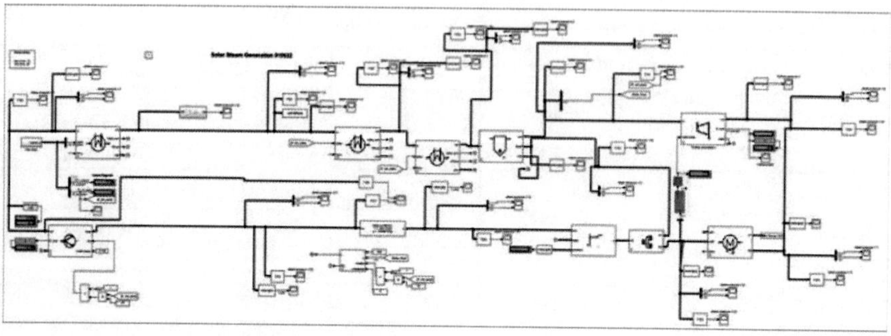

Fig. 3 Thermolib model used to simulate SPT

as the energy conservation law (1st law of thermodynamics). Kinetic and potential energies of the incoming and outgoing flows are neglected in Thermolib and in the equation above, where $H_{in,i}$ is the enthalpy flow of the input flow i, $H_{out,j}$ is the enthalpy flow of the output flow j, Q_k is the heat flow k, P_m is the mechanical power m, and U is the internal energy of the component. This means that the P_{mch} is usually positive for a compressor because the mechanical energy at the shaft is added to the fluid. On the other hand, the P_{mch} is usually negative for a turbine because the fluid releases its power. For a hot part like a reactor, the Qdot signal (heat exchange with the environment) will usually be negative, since this heat is leaving the reactor (suppose a cold environment is configured). In a steady-state adiabatic chemical reactor, the total enthalpies (thermal + enthalpy of formation) of what goes in and what comes out are the same. But because there are different amounts of the compounds at the inlet and the outlet, the temperatures at the inlet and the outlet might be very different. Based on the collector energy balance equation and as a function of collector efficiency, it is possible to determine the amount of heat that the solar collector receives and transfers to the HTF [21]:

$$\dot{Q}_{\text{in}} = G_t \cdot A_{\text{col}} \cdot \eta_{\text{col}} = \dot{m}_{htf} C_{p,htf} \left(T_{\text{col},,} - T_{\text{col,i}} \right) \tag{2}$$

where m_{htf} is the hot water mass flow rate in the solar heating circuit, A_{col} is the collector aperture area, η_{col} is the solar collector efficiency, and G_t is the total solar irradiation on the solar collector surface. Seri Iskandar, Perak, located on the west coast of the Malay Peninsula (4°22′25.0"N 100°58′08.3"E) was selected as a location to evaluate the SPT system. The solar collector's thermal efficiency can be calculated using the solar irradiation, average collector temperature, and air temperature in the following way [21]:

$$\eta_{col} = a_0 - a_1 \frac{\left(T_{col,m} - T_{amb} \right)}{G_t} - a_2 \frac{\left(T_{col,m} - T_{amb} \right)^2}{G_t} \tag{3}$$

Here a_0, a_1, and a_2 are efficiency equation constants for the solar collector. Table 2 presents the efficiency equation constants of the ETR for which calculations of the system performance have been carried out [21].

Constant	a_0[−]	a_1 [W/m²K²]	a_2 [W/m²K²]
Value	0.825	0.91	0.6×10^{-3}

The pump block raises the entering liquid's pressure to a specified target pressure. It determines the thermodynamic state of the outgoing flow as well as the pump's necessary mechanical power consumption. This block should only be utilized for liquid or incompressible flows. The pump is modelled using the isentropic efficiency and the following general relations: The first law of thermodynamics applies to pumps [20]:

$$\dot{m}\left(h_{in} + \frac{1}{2}v_{in}^2\right) = \dot{m}\left(h_{out} + \frac{1}{2}v_{\text{out}}^2\right) + \dot{W} \tag{4}$$

simplifying,

$$\dot{m}(h_{\text{out}} - h_{\text{in}}) = (p_{\text{out}} - p_{\text{in}})/\rho \tag{5}$$

where p_{out} and p_{in} are pressure of the outlet and inlet of the pump. The total heat transfer rate, Q_e from the HTF to the working fluid in the evaporator, is provided by [20]

$$\dot{Q}_e = \dot{m}_{wf}(h_3 - h_2) = \dot{m}_{htf}C_{p,htf}\left(T_{e,i} - T_{e,o}\right) \tag{6}$$

where h_2 and h_3 are the specific enthalpies of the working fluid at the inlet and outlet of the evaporator, and $T_{e,i}$ and $T_{e,o}$ are the temperatures of the hot water stream at the inlet and outlet, respectively. The vapour of the working fluid travels through the turbine to generate mechanical power. The turbine output capacity is determined by [20]

$$P_{mch} = \dot{m}(h_{\text{in}} - \text{h}_{\text{out}}) \tag{7}$$

and,

$$\eta_s = \frac{\Delta h}{\Delta h_s} \tag{8}$$

therefore,

$$P_{mch} = \dot{m} \times \eta_s \times \Delta h_s \tag{9}$$

Subscript s is for isentropic change. At the exit of the turbine, the exhaust vapour is sent to the condenser, where it is condensed by rejecting its heat to the cooling water. The condenser heat rejection rate, $\dot{Q}_c$, can be expressed as [20]

$$\dot{Q}_c = \dot{m}_{wf}(h_4 - h_1) = \dot{m}_w C_{p,w}\left(T_{c,o} - T_{c,i}\right) \tag{10}$$

where $\dot{m}_{wf}$ is the cooling water mass flow rate, and $T_{c,i}$ and $T_{c,o}$ are the cooling water temperature at the inlet and outlet of the condenser, respectively. The solar SPT system's net output power is described as [22]

$$\dot{W}_{net} = \dot{W}_t - \dot{W}_p \tag{11}$$

The ratio of net power output to heat input in the evaporator determines the SPT's thermal efficiency. It can be stated as follows:

$$\eta_{SPT} = \frac{\dot{W}_{\text{net}}}{\dot{Q}_e} \tag{12}$$

The overall efficiency of the SPT system can be defined as follows:

$$\eta_{sys} = \frac{\dot{W}_{\text{net}}}{G_t \cdot A_{col}} \tag{13}$$

To better understand the technical viability and commercial viability of the SPT, other characteristics must be determined in addition to the system efficiency, which is used to assess the system from a thermodynamic point of view. These variables include the working fluid pressure, flow rate of the HTF, as well as the solar field performance. The pressure, flow rate, and the HTF temperature are determined by the requirements of the procured Green Turbine to generate 1.5 kWe of electricity. Table 4 shows the parameters in order to achieve the target energy.

Thermolib 5.2 toolbox [23] from EUtech Scientific Engineering GmbH was used to create the simulation of the proposed solar ORC system in a MATLAB/Simulink environment. Based on the real gas behaviour of Pen Robinson Equation of State, the working fluid properties were assessed using Thermolib's built-in thermophysical properties database. Figure 3 presents the created Thermolib model for the solar SPT system. Figure 4 presents the result of water temperature at the conditions of 0.04 kg/s flow rate, pressure of 5 bar, and input data from Table 3. The first thing to notice is that the water did not reach the boiling point of 152 °C and thus no steam is produced: quality of steam (x = 0) throughout the day.

Thus, operational variables such as operating pressure, mass flow rate of HTF, and performance of solar field have been adjusted to find its effect on steam production.

C. *Effect of pressures*

The operating pressure value is critical in optimizing turbine efficiency. It also influences the boiling point of water, thus as the pressure rises, so will the boiling

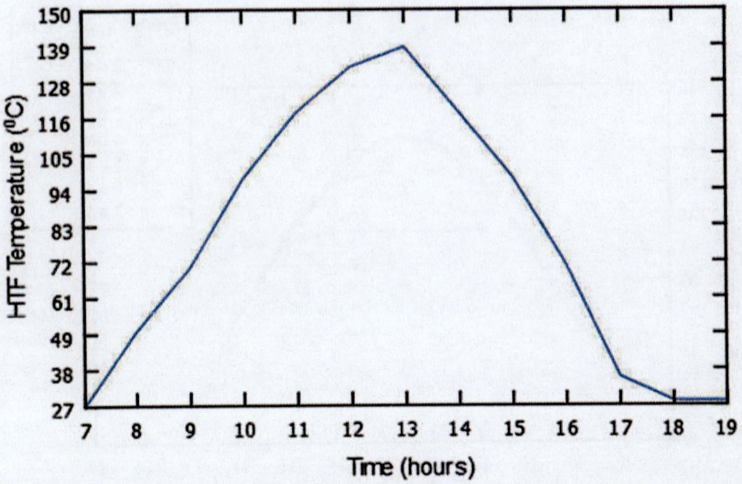

Fig. 4 Water temperature on 17 January 2020

Table 3 Green turbine requirements parameter

Pressure	5.2 bar
Flow rate	0.04 kg/s
Temperature of superheated steam	200–250 °C

point. Figure depicts direct comparison of thermal performance in terms of water temperature with a flow rate of 0.04 kg/s under various working pressures: 1–7 bar.

Varied pressures have different thermal properties on water temperature, as seen in Fig. 5. The results indicate that with the DNI input, only water with operating pressures of 1–3 bar will produce steam. The data is depicted in Fig. 6.

D. *Effect of flow rates*

The flow rate of water is critical in determining the thermal efficiency of the absorber as the primary heater. If it is too low, the turbine may not function properly; if it is too high, thermal efficiency may be reduced. Figure 7 compares the temperature of water with different flow rates of 0.02–0.06 kg/s.

The results demonstrate that water produces steam at flow rates of 0.02 and 0.03 kg/s as the water reaches the boiling point at 5 bar pressure. The steam produced is depicted in Fig. 8.

E. *Effect of Solar Field Performance*

The number of heliostats is increased from 15 to 20 since it is the appropriate amount for the project site sizes; if the number is increased, there will be a lot of obstruction. Figure 9 depicts the predicted data of the power received by the eight absorbers reflected by the 20 heliostats. The data reveals that employing only 20 heliostats

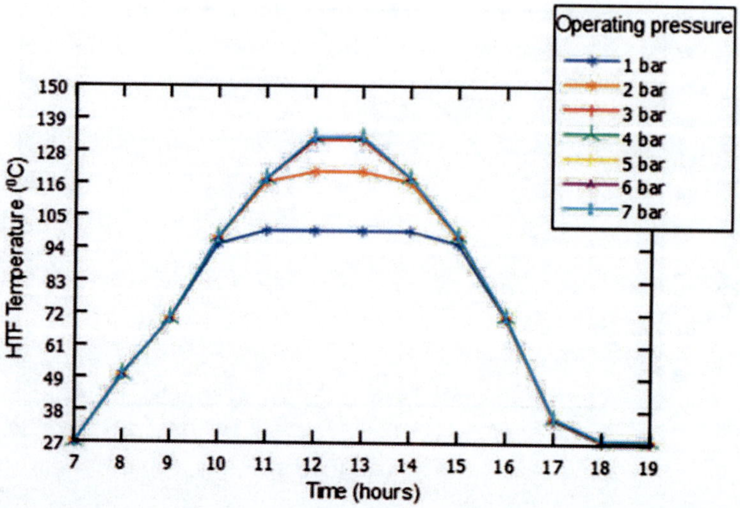

Fig. 5 Water temperature at different operating pressure

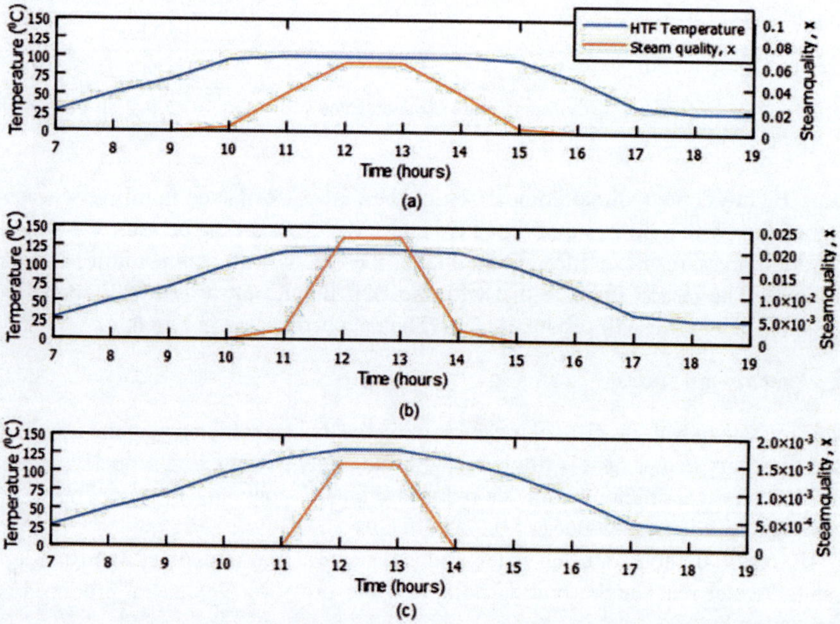

Fig. 6 Water temperature and steam quality with operating pressure. **a** 1 bar, **b** 2 bar, and **c** 3 bar

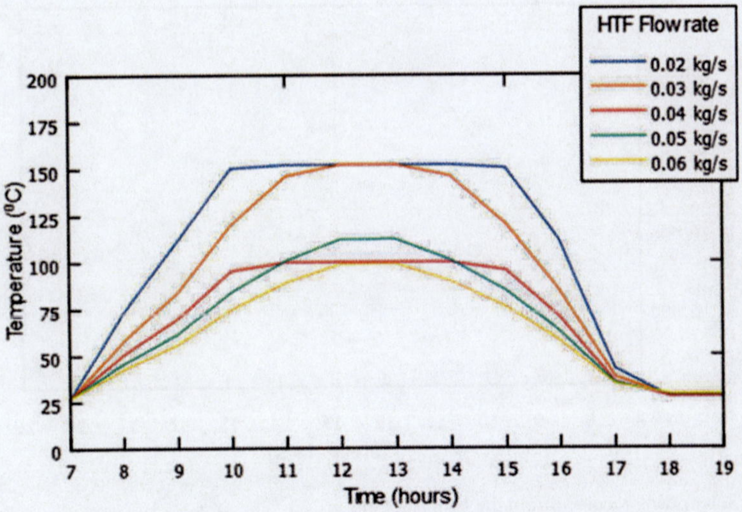

Fig. 7 Water temperature at different flow rate

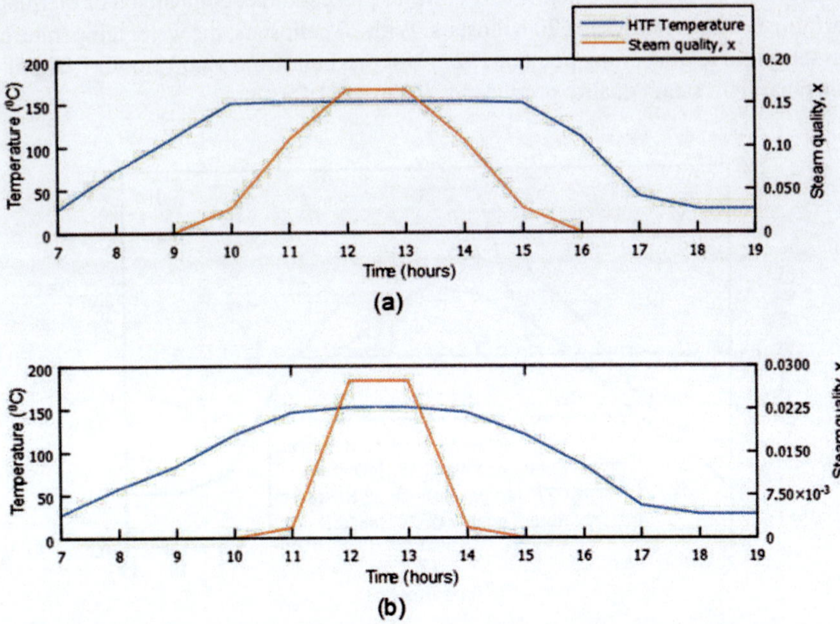

Fig. 8 Water temperature and steam quality with mass flow rate. **a** 0.02 kg/s and **b** 0.03 kg/s

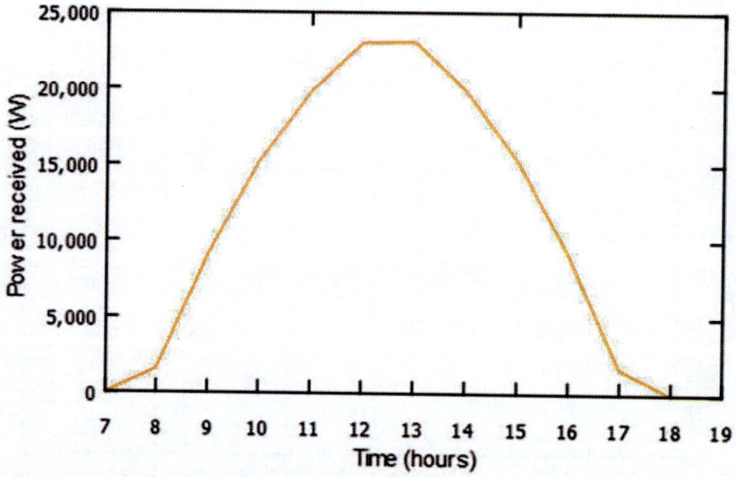

Fig. 9 Solar power received from the heliostat field

results in a 14.5% increase in power. Figure 10 present direct comparison of thermal performance between 15 and 20 heliostats. With 20 heliostats, the water temperature has increased to the boiling point of 152 °C for one hour from 12:00 pm till 1:00 pm. The maximum steam quality produced is 0.0269 at 1:00 pm.

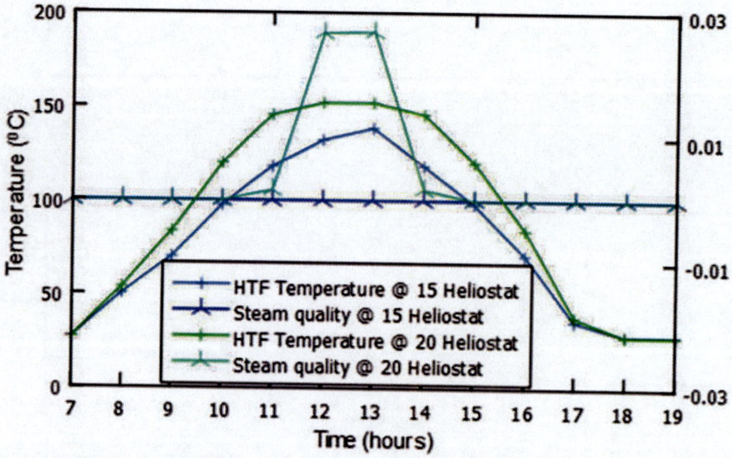

Fig. 10 Effects of number of heliostats on HTF temperatures

2 Conclusion

Thermolib was used to do a thermal simulation of the main heater and eight absorbers at SRC UTP on 17 January 2020. The steam turbine's operating conditions are 5 bar, 200 °C superheated steam, and 0.04 kg/s steam mass flow rate. As previously described in the SPT Configuration Section, we discovered that with 15 heliostats, the water did not reach boiling point, and so no steam was produced. Variables such as heliostat quantity, operating pressure, and mass flow rate have been varied to investigate their impact on the main heater's steam generation performance. For operating pressure, the simulation result presents that no steam will produce above 4 bar. However, for 1–3 bar steam is produced with maximum steam quality of 0.06, 0.02, and 0.0015, respectively. Increasing the HTF flow rate above 0.04 kg/s yields no steam. Yet, flow rate of 0.02 and 0.03 kg/s will produce steam with maximum quality of 0.15 and 0.028, respectively. Increasing the amount of heliostat from 15 to 20 shows 14.5% increase of power received on the ETR, therefore, steam is produced with maximum quality of 0.029. Though lower operating pressure and mass flow rate show promising steam generation results, this will not maximize the turbine electrical output because both variables are less than the turbine requirement. The most likely technique for ensuring steam generation without jeopardizing turbine requirements is to increase the number of heliostats. Though we can only offer up to 20 heliostats due to the limited field size of the SRC UTP; the simulation reveals that it is capable of generating a system for roughly 1 h. Implementing a recirculation system for steam generation instead of the one-through system executed in this study is recommended to achieve higher temperatures to produce more steam.

Acknowledgements The author would like to thank Ministry of Science, Technology and Innovation (MOSTI) IF102011308 (cost centre 015ME0-217) and Yayasan Universiti Teknologi PETRONAS (YUTP) FRGS (cost centre 0153AA-H28) for providing financial support for this project.

References

1. Faye K, Thiam A, Faye M (2021) Optimum height and tilt angle of the solar receiver for a 30 kWe solar tower power plant for the electricity production in the Sahelian Zone. Int J Photoenergy 2021:1–14. https://doi.org/10.1155/2021/1961134
2. Kolb GJ, Ho CK, Mancini TR, Gary JA (2011) Power tower technology roadmap and cost reduction plan, 38
3. Luo Y, Du X, Yang L, Xu C, Amjad M (2017) Impacts of solar multiple on the performance of direct steam generation solar power tower plant with integrated thermal storage. Front Energy 11(4):461–471. https://doi.org/10.1007/s11708-017-0503-5
4. Xu R, Wiesner TF (2015) Closed-form modeling of direct steam generation in a parabolic trough solar receiver. Energy 79:163–176. https://doi.org/10.1016/j.energy.2014.11.004
5. Eck M, Steinmann WD (2005) Modelling and design of direct solar steam generating collector fields. J Sol Energy Eng-Trans ASME 127(3):371e80

6. Hirsch T, Eck M, Steinmann W-D (2005) Simulation of transient two-phase flow in parabolic trough collectors using Modelica. In: Proceedings of the 4th international Modelica conference, Hamburg, Germany
7. Eck M, Hirsch T (2007) Dynamics and control of parabolic trough collector loops with direct steam generation. Sol Energy 81(2):268e79
8. Silva R, Perez M, Fernandez-Garcia A (2013) Modeling and co-simulation of a parabolic trough solar plant for industrial process heat. Appl Energy 106:287e300
9. Roldan MI, Valenzuela L, Zarza E (2013) Thermal analysis of solar receiver pipes with superheated steam. Appl Energy 103:73e84
10. Yan Q, Hu E, Yang YP, Zhai RR (2010) Dynamic modeling and simulation of a solar direct steam-generating system. Int J Energy Res 34(15):1341e55
11. Martínez I, Almanza R (2007) Experimental and theoretical analysis of annular twophase flow regimen in direct steam generation for a low-power system. Sol Energy 81(2):216e26
12. Ritschel TKS, Gaspar J, Jørgensen JB (2017) A thermodynamic library for simulation and optimization of dynamic processes. IFAC-Papers OnLine 50(1):3542–3547. https://doi.org/10.1016/j.ifacol.2017.08.951
13. Thomson GH (1996) The DIPPRR© databases. Int J Thermophys 17(1):223–232
14. Peng D-Y, Robinson DB (1976) A new two-constant equation of state. Ind Eng Chem Fundam 15(1):59–64
15. Soave G (1972) Equilibrium constants from a modified Redlich-Kwong equation of state. Chem Eng Sci 27(6):1197–1203
16. Shibata SK, Sandler SI (1989) Critical evaluation of equation of state mixing rules for the prediction of high-pressure phase equilibria. Ind Eng Chem Res 28(12):1893–1898
17. Chapman JW, Lavelle TM, May RD, Litt JS, Guo T-H (2014) Toolbox for the modeling and analysis of thermodynamic systems (T-MATS) user's guide. Technical report, NASA's Glenn Research Center
18. Martín A, Bermejo MD, Mato FA, Cocero MJ (2011) Teaching advanced equations of state in applied thermodynamics courses using open source programs. Educ Chem Eng 6(4):e114–e121
19. Sundman B, Kattner UR, Palumbo M, Fries SG (2015) OpenCalphad—a free thermodynamic software. Integr Mater Manuf Innov 4 (1)
20. Simulink, & User (nd) Simulation toolbox for the Design and Development of Thermodynamic Systems in MATLAB. https://www.thermolib.de/media/thermolib/downloads/Thermolib-UserManual.pdf
21. Tempesti D, Manfrida G, Fiaschi D (2012) Thermodynamic analysis of two micro CHP systems operating with geothermal and solar energy. Appl Energy 97:609–617
22. Hossin K, Mahkamov K (2015) Performance evaluation for a 10 kW solar organic rankine cycle power system to operate in the UK climate conditions. https://doi.org/10.13140/RG.2.1.4135.4721
23. Thermolib. http://www.eutech-scientific.de/productsservices/thermolib.html

Investigation of Range Extension of Personal Electric Mobility by Current-Limiting Mechanism

Ahmad Zaid Syakir Mohd Yazsid, Kundan Kumar, and Saiful Azrin Mohd Zulkifli

Abstract One of the major hurdles for adopting electric vehicles to the masses is their limited range per charge using their energy storage. This even applies to small personal electric transports such as electric scooters, where the relatively small battery size restricts the effective range of the vehicle. Multiple studies have been done to overcome this problem. While this research will attempt to implement a signal limiter from the user input to the control unit, this will reduce the current flow from the battery to the motor, thus in theory, reducing the overall power consumed by the motor, thereby extending the range of the vehicle. By using MATLAB-Simulink software, this study performs simulation of an electric scooter, along with a proposed current limiting mechanism. From the obtained results, it can be observed that proper selection of limit mechanism can increase the electric scooter's operating range, and this work is capable to serve as a benchmark for future conceptual and empirical electric mobility research.

Keywords Electric vehicles · Personal electric mobility · Range enhancement · Current limiting · MATLAB-Simulink

1 Introduction

Many believe that the future of mobility is in the hands of Electric Vehicles (EV) [1]. But the biggest hurdle EVs need to face before they can be approved by the mass consumers is the range anxiety and the relatively high price of the vehicle. While the most straightforward solution for range is to install a bigger battery to the vehicle, it would drastically increase the weight and the cost of the vehicle. Thus, other alternatives must be explored to tackle this problem.

A. Z. S. M. Yazsid · K. Kumar · S. A. M. Zulkifli (✉)
Electrical and Electronics Engineering Department, Universiti Teknologi PETRONAS, 32610 Seri Iskandar, Perak, Malaysia
e-mail: saifulazrin_mz@utp.edu.my

F. Ahmad et al. (eds.), *ICREEM 2022*, Lecture Notes in Mechanical Engineering,
https://doi.org/10.1007/978-981-99-5946-4_14

For a vehicle to be classified as an electric vehicle (EV), it needs to draw energy from electrical source or storage rather than a fuel tank, and the propulsion system must also run on electricity rather than an internal combustion engine [2]. EVs can also be in multiple types such as Battery Electric Vehicles (BEVs), Plug-in Hybrid Electric Vehicles (PHEVs), and Hybrid Electric Vehicles (HEVs). However, when someone mentions an electric car, they typically mean a BEV—a vehicle that runs entirely on electricity and has no combustion engine components and uses an electric powertrain, which is far different from the standard internal combustion engine powertrain.

An electric powertrain typically consists of four main parts [1]: a battery, an inverter that transforms direct current (DC) into multi-phase currents (AC) to drive the electric motor, the electric motor itself, which creates magnetic fields using electric energy to turn, and, in the majority of cases, a DC-DC converter that adjusts the battery's voltage to the electric motor's voltage and the power demand.

Electrically propelled vehicle appeared in the United States as early as 1890 [3], when a six-passenger electric vehicle with a top speed of 14 miles per hour was first developed. The first electric vehicle to be made available in Malaysia was the Mitsubishi i-MiEV, a pricey hatchback [4]. But Toyota had already introduced a hybrid electric car to the nation's market in 2009 [5]. As stated in [5], although there are already more than 31,000 registered EVs, of which 5% are BEVs, on Malaysian roads, the adoption rate is still rather low. However, EVs are becoming more and more popular every day because of their myriad of benefits and the urge for a greener alternative to internal combustion engines. Due to the widespread usage of gasoline-powered vehicles, emissions from the transportation industry significantly contribute to both climate change and health issues. To minimize pollution, nations around the world are trying to switch from gasoline-powered cars to electric vehicles (EVs).

EVs consume no fuel, while HPEVs and HEVs use less gasoline than a regular car would. The fuel or energy consumption of internal combustion engines was measured in kilometers per liter or l/100 km, but the energy consumption rate of an EV will be measured in kWh/km. The powertrain will be more efficient if the energy use figure is lower because it will require less energy to travel a given distance. The longer an EV can run before its battery runs out, the better. The efficiency of the powertrain will determine how far an EV can be driven without gaining weight [1, 6]. These days. EVs are more alluring because using an electric car for day-to-day use can be a much cheaper option to run on, given the recent surge in petrol prices. Not to mention the additional advantage of being greener than a traditional internal combustion engine. In addition, compared to conventional cars, electric vehicles require less maintenance.

Despite their increasing acceptance, EVs still have several disadvantages that keep some people from switching completely to full-time use. One of the reasons is that most places still have very few EV charging facilities. Furthermore, whether or not the user utilizes the specified electric car charging station, it takes a very long time to fully recharge an electric vehicle and may even take several hours. Not to mention that the restricted range of EVs is another significant reason why many people still do not want to fully transition to or adopt them. In addition, EVs are regrettably less dependable than traditional vehicles in terms of range.

Because an EV's range, even with a full charge, is significantly less than that of conventional cars, people sometimes experience range anxiety. Range anxiety, on the other hand, is the worry that an EV won't have enough energy in reserve to get the user where they want to go. Therefore, operating an EV necessitates careful planning on the part of the user. However, there is a range anxiety problem that prevents the adoption of EVs because there is a lack of suitable charging infrastructure, a lengthy charge time, a high cost for long-range EVs, and a limited range for affordable. Range-extended EVs (REEV) are viewed as a potential remedy for the short range and expensive price of EVs. By shrinking the battery, range extenders reduce the weight and initial investment of the vehicle while resolving the "range anxiety" problem that plagues battery-powered electric vehicles. Vehicles with extended range (REEV) are also referred to as series hybrid electric vehicles (SHEV).

In order to maximize EV autonomy, a range extender (RE) is a small electrical generator (APU) that only activates when necessary. The internal or external combustion engine and generator make up the range extender's major parts, and the engine is connected to the generator in a series configuration. The main purpose of the RE in an EV is to increase the vehicle's range. In the situation that the EV battery's state of charge (SOC) falls below a predetermined level, the range extender begins to operate. In this case, the engine keeps the car operating by supplying electricity via charging the battery or directly powering the EV while traveling. However, recent improvements in battery technology, such as faster charging times and larger batteries, have expanded the use of EVs, particularly in tiny vehicles and electric scooters for personal transportation, which has caused a recent boom in the market for personal electric mobility devices.

On the other hand, Personal Electric Mobility (PEM) is a term used to describe a class of electric vehicles that allow for single-person movement. PEM has a range of up to 30 km and a top speed of 40 km per hour depending on the vehicle. As a result, it became popular among college students who sought a practical mobility option but didn't want to bring a car to school. Even though there are still no restrictions on PEM use among the populace in Malaysia, more individuals are using bike lanes, pedestrian walkways, and occasionally main roadways.

Each PEM has a different size, form, and driving range capability. The number of battery electric vehicles on the road is currently small [7], despite efforts to reduce environmental pollution and overuse of fossil fuels. One may choose not to adopt EVs for a variety of reasons, including driving range, charging issues, and expense. Additionally, studies on how to help customers plan their trip time and distance sensibly, how to expand and finish the charging infrastructure to meet consumer demand, and how to teach consumers to calculate total cost based on their driving patterns can all help to overcome these obstacles [7]. Thus, keeping this all in mind, this paper focuses on the below-mentioned objectives:

- To explore the dynamics of an electric propulsion system (EPS) along with energy storage of an electric scooter using *MATLAB-Simulink*

- Implementing a limiting system to the control unit from the user input for range extension.

The paper is divided into 5 sections such that one may find the introduction related to personal electric mobility (PEM) under section I, whereas the literature review has been stated in section II. Section III provides the methodology and flow of our research work, and all results are discussed under section IV. Lastly, one may find the conclusion for this work under section V along with some future recommendations.

2 Literature Review

The efficiency of the electric powertrain is the ratio between the energy output of the battery and the energy output of the motor (Fig. 1a). A ratio of 100% would mean a perfect conversion of the electric energy into mechanical energy, but it is never the case. Many losses happen at different stages in the energy conversion process. The efficiency is not even a constant value. For example, the combined efficiency of an electric motor and inverter is ranging from 60 to 96%, depending on the drive profile, the speed and the torque of the motor, and its position on the drivetrain.

The efficiency is typically shown as a graph (efficiency map), where the y-axis and x-axis, which represent torque and speed, respectively, guarantee a certain efficiency for any given speed/torque combination. The ideal operating points, when efficiency is at its highest, are situated in a constrained region of the torque-speed map (Fig. 1b).

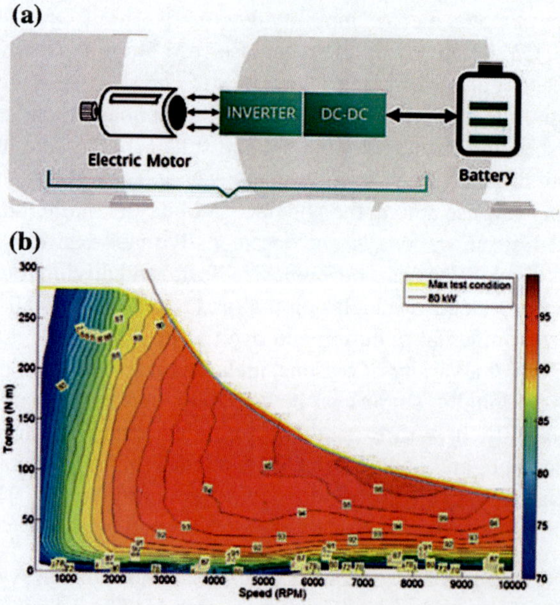

Fig. 1 **a** An electric powertrain efficiency [1]. **b** Combined efficiency map of an electric motor and inverter

Inverters do not only convert the DC from battery to AC and supply to the motors but also convert AC to DC during regenerative braking, acting as a rectifier, at which time the motor works as a generator. Energy efficiency is ensured by using the electric motor in this region [1, 7]. However, in actual driving situations, the inverter and electric motor are operated in a broader working range, making it difficult to maintain the system within its ideal operating range. Thus, to increase the operating range of the EV while maintaining high overall performance and low costs, new strategies are needed.

To predict the remaining range and help drivers locate a suitable charging station, several researchers have also researched real-time driving range prediction [8] and real-time charging station recommendation systems [9]. Additionally, several researchers investigated fast-changing technologies to shorten charging times and remove the impact of charging stations on power networks [10, 11]. Although the aforementioned electric car technologies may help make them more widely used, there are still several barriers to addressing the short-driving range issues. For instance, it will take some time before cities, small towns, and rural areas are entirely covered by charging stations. Another factor relates to the current limited battery technology which makes it challenging to significantly increase battery capacity. For these reasons, current research on extending the driving range of pure EVs focuses primarily on increasing energy use efficiency.

Another way to increase the driving ranges [12], driving stability, and safety [13, 14] for independent wheel drive electric vehicles is to distribute and control the braking and driving torques reasonably. The importance of torque/power distribution and control on independent drive electric vehicles has prompted extensive research in recent years, and the technology is now at a relatively advanced stage. According to the most recent study, the driving range of an electric car can be increased by taking energy expenditure into account when making steering and braking decisions. The electric motor's regenerative braking energy can be sent back to the battery of the car while it is braking. The driving range of an electric car can be increased in this way [15]. Additionally, the energy efficiency of the vehicle can be improved while it is being driven in accordance with the state of the vehicle, the status of the road, and the driver's actions. Driving after another vehicle is a typical event when traveling.

For SOC estimation, numerous innovative adaptive systems have been created. Finally, there are hybrid methods, which combine several distinct SOC estimation techniques to improve estimation accuracy [16]. The hybrid technique typically yields the best estimation since it integrates multiple methods rather than utilizing each one alone. Each technique has its own unique traits, therefore some are better suited for set discharging current conditions, while others are more suited for various conditions. Because numerous methods were being used with varied battery sizes and discharging conditions, it was difficult to compare the performance of different approaches. Therefore, it is challenging to choose only one technique that can be the best for SOC estimates globally.

Engineers and scientists have developed products that will change our planet using MATLAB to evaluate and design systems [17]. On the other hand, Simulink is a MATLAB environment that has been utilized to build systems with multidomain models. The proper level of detail for each task or project could be configured in the user's model. In order to create a control algorithm or limiting mechanism that will take into account the battery state-of-charge and then produce a new limited signal that controls the motor, this research therefore makes use of both platforms.

When these rules are applied, an updated list is produced, with direct feed-through blocks appearing after non-direct feed-through blocks in the sequence necessary to provide the blocks they drive with valid inputs. Simulink looks for and flags the existence of algebraic loops during the sorting process, which are signal loops in which an output of a direct feed through the block is connected to one of the block's inputs either directly or indirectly. Such loops appear to lead to a deadlock situation since Simulink requires a direct feed through block's input in order to compute its result. The name "algebraic loop" refers to a collection of concurrent algebraic equations where the input and output of the block are the unknowns. Furthermore, each time step in these equations may have a valid solution. *Simulink* therefore makes the assumption that loops containing direct feed-through blocks reflect a solved set of algebraic equations and make an effort to resolve them each time the block is updated throughout a simulation.

3 Methodology

This study starts with understanding the powertrain of the scooter, from the user input to the tires; the scooter will be built using a model in MATLAB-Simulink. Each parameter of the model will be based on a real working electric scooter that is available for purchase. The performance of the model will then be tested to see the similarities and the accuracy of the simulation, where redundancies will be ironed out to make sure the simulation is smooth and optimized.

The next step is to design a signal-limiting system where it will reduce the value of the signal from the user input and transmit the new value to the control unit. In MATLAB-Simulink, simple mathematical nodes will represent the limiting system in the simulation. But first, a test model will be programmed as a proof of concept before implementing it to the electric scooter simulation, where the results of the proof of concept will determine the success of the project. The overall *Simulink* model for the electric scooter is as shown in Fig. 2, where the entire powertrain is modeled with 4 subsystems, which are driver input subsystem, motor and controller subsystem, battery subsystem, and finally vehicle body and tire subsystem.

Figure 3 shows a *driver input subsystem* consisting of a longitudinal driver module having six ports that are preprogrammed in MATLAB to simulate the driver in a real vehicle, where we can insert a drive cycle and the values for breaking, acceleration, and actual velocity of the vehicle and it will give out values of the performance based on the motor rating and other vehicle parameters. Here, feedback velocity is taken as a reference from the velocity of the vehicle body as shown in Fig. 3. In other words, the output velocity of the vehicle is nothing other than feedback velocity. Here, gradient is selected as zero. The acceleration and brake commands have the

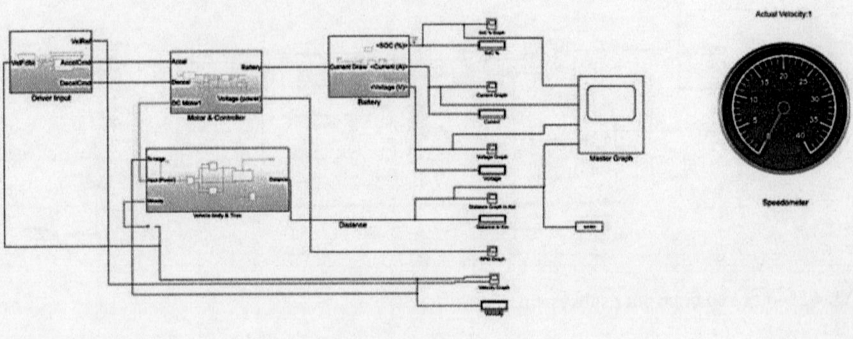

Fig. 2 Overall Simulink model for electric scooter

input voltage as that of the PWM controller. The PWM frequency chosen here is 1000 Hz.

The FTP75 drive cycle source is used to simulate moderate use of the electric scooter with ample acceleration, braking, and stoppage to simulate actual usage in real life.

The *motor and controller subsystem* consists of an electric motor that propels the electric scooter (Fig. 4). The rating and parameters of the motor are taken from manufacturers, and the motors are available for sale in the market. The output part 'R' of the shaft of the motor is connected to input 'B' of simple gear. The model parameterization of the motor is done here by rated load and speed having no-load speed 6000 rpm, rated speed is set 4000 rpm, motor weight is about 5000 kg, and rated DC supply voltage is 48-V. H-bridge controller is used here to control the motor speed with controlled PWM voltage.

The current sensor is connected to ' + ' terminal of the H-bridge and is used to measure the current of motor, which may be observed at scope by adding a converter in between scope and output pin of current sensor mentioned in the above diagram. The limiter is the added subsystem where the input from the user input will be altered depending on the battery's state of charge and change the value of the input to increase the range of the electric scooter. The *battery subsystem* is where the battery is located

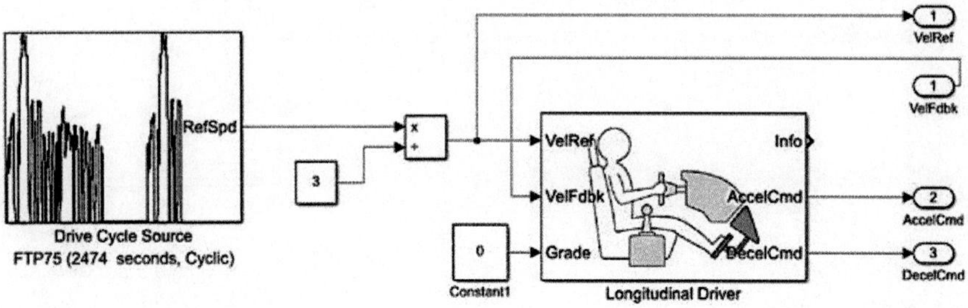

Fig. 3 Driver input subsystem

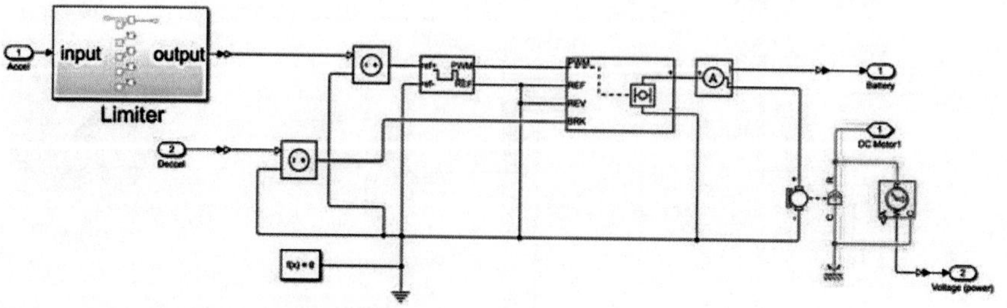

Fig. 4 Motor and controller subsystem

alongside the necessary connections for battery state of charge recording, where it will be readable through a scope (Fig. 5).

The *vehicle body and tire subsystem* will simulate where the power from the motor goes and propels the scooter (Fig. 6). Consisting of only two tires, the scooter will accurately simulate how an actual electric scooter behaves in the simulation. The output shaft rotates in the same direction as the input shaft. In the *vehicle body and tire subsystem*, '*B*' is input of simple gear and '*F*' is its output which is connected to input '*D*' of differential. The outputs (s_1 and s_2) of the differential must be connected to the axel of the vehicle body. The velocity may be observed at a scope mentioned in Fig. 5. The total force F_t exerted by the motor is given by the following equation:

$$F_t = m\frac{dv_x}{dt} + F_d + mgsin(\beta) \tag{1}$$

In Eq. 1, all three components are opposing the applied force exerted by the motor, where F_d is aerodynamic drag force, β is inclined angle with the surface, m is

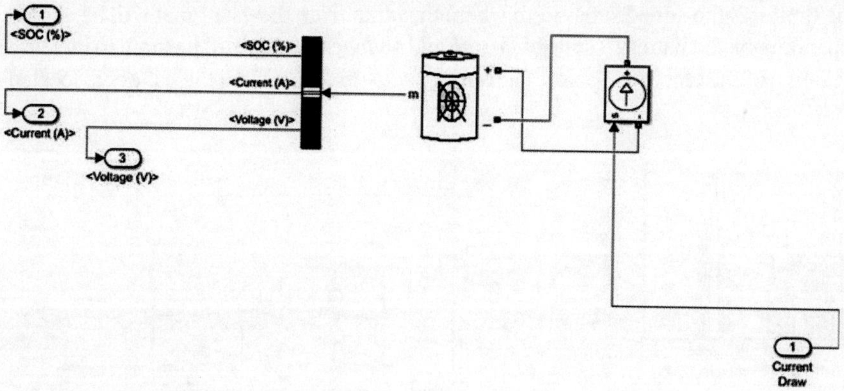

Fig. 5 Battery subsystem

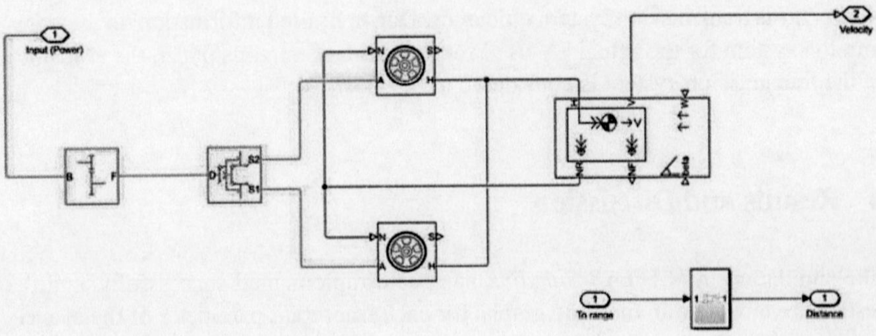

Fig. 6 Vehicle body and tire subsystem

mass of vehicle (kg), and $\frac{dv_x}{dt}$ is rate of change of velocity. The wheel traction power estimation is analytically evaluated as

$$P_t = \frac{\Delta E_p + \Delta E_k + W_{aero} + W_{rolling}}{t_{segment}} \quad (2)$$

where ΔE_p is the potential energy change, ΔE_k is the kinetic energy change, W_{aero} is the aerodynamic work, $W_{rolling}$ is the rolling resistance work, and $t_{segment}$ is the route segment duration. In the gearbox, the follower 'F' to teeth 'B' ratio is selected in simulation as '2' and analytically calculated as

$$G = \frac{\alpha_{smax}}{v_{EVmax}} X \frac{2\pi}{60} X \frac{r_w}{1.1} \quad (3)$$

where α_{smax} is the maximum motor speed (rpm), v_{EVmax} is the maximum vehicle speed (m/s), and r_w is the wheel radius (m). The traction torque T_t (Nm) and load torque T_L (Nm) of the motor are analytically evaluated as

$$T_t = \frac{P_t X r_w}{v_{EV}} \quad (4)$$

$$T_L = \begin{cases} \frac{T_t}{\theta_{TS} X G}, & P_t \geq 0 \\ \frac{\theta_{TS} X G}{T_t}, & P_t < 0 \end{cases} \quad (5)$$

where θ_{TS} is transmission system efficiency. Due to limited information on the transmission system for the actual EV used for the validation in this paper, the efficiency of the transmission system is considered to be 95%.

4 Results and Discussion

The simulation in *MATLAB-Simulink* has been implemented successfully, and the results are tabulated in multiple graphs for each important parameter of the electric scooter. Figure 7 shows the value of battery state of charge (SoC) over time in seconds after the simulation is started. The initial SoC level of the battery is set to 30% to reduce the amount of time needed to run the simulation until the battery's SoC is depleted.

In Fig. 7, one can observe the SoC values are changing because of braking and acceleration. At braking, battery is going to be charged because of regenerative braking, and during acceleration battery is used. The SoC may be calculated analytically as

$$SoC = \frac{\int_0^t I(t).dt}{C} \tag{6}$$

where c is battery capacity and I is current flowing from it. The vehicle speed W_t and shaft speed W_L in (rad/s) may be analytically evaluated as

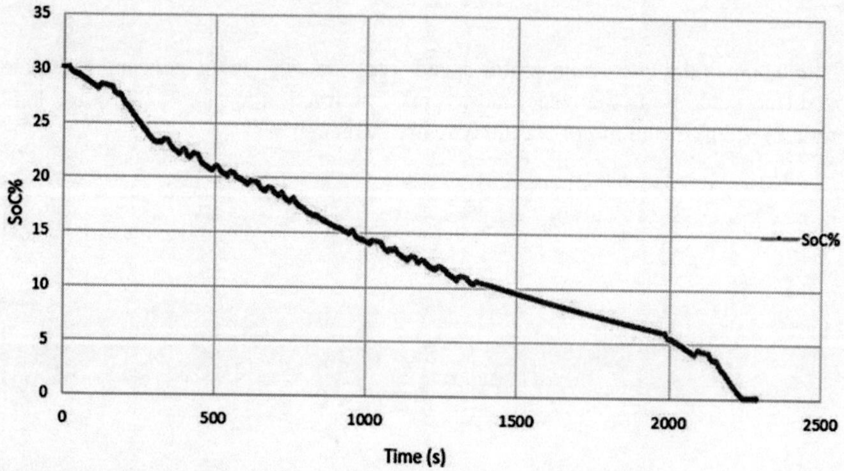

Fig. 7 Battery SoC versus time

$$W_t = \frac{V_{EV}}{r_w} \tag{7}$$

$$W_L = \frac{GXV_{EV}}{r_w} \tag{8}$$

Figure 8 shows the current over time graph from the electric scooter's battery drawn over the course of the simulation. These will be the control values that will be compared to when the limiter will be implemented, and the input values were changed. Figure 9 shows the distance traveled in km over time. Similar to the current values, the value for the traveled distance will set the benchmark for the upcoming limiter implementation test later in the paper.

Figure 10 represents the values of current and voltage over time in a single graph. It is noted that the when the value of voltage decreases, the value of current increases, thus the power from the battery that is stored is being used to power the electric motor of the scooter. The dot product of the two values will result in power that has been drawn from the battery. This explains the reduction of the battery's state of charge.

The limiter system implemented is a linear mathematical system that will reduce the input value from the user at a fixed percentage. The simulation tested various degrees of percentage such as 80, 60, 40, and 20%, respectively. The values will be changed depending on the required limiting factor. All the data has shown decreases in current drawn from the battery due to the limiting, but the most significant changes occur at 40 and 20% limiting. Figure 11 shows the comparison between the SoC values of all limiters, including without limit as a baseline. It shows that the difference

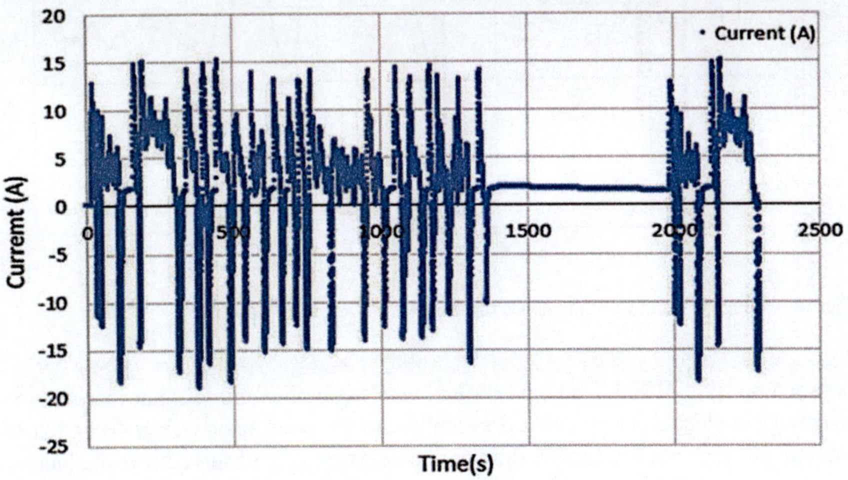

Fig. 8 Battery current versus time

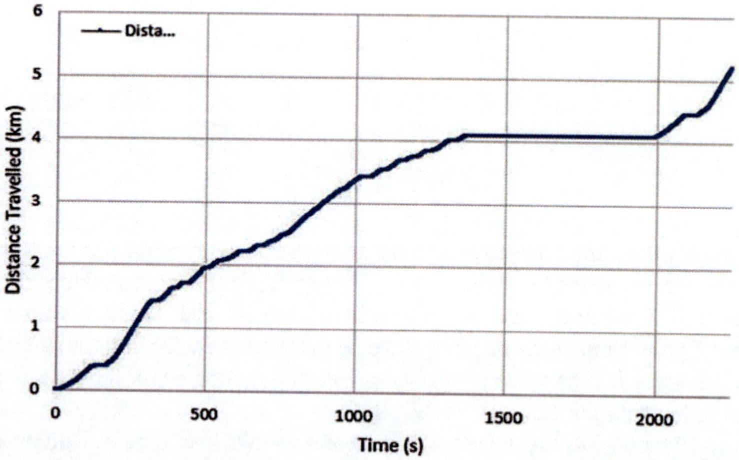

Fig. 9 Distance traveled versus time

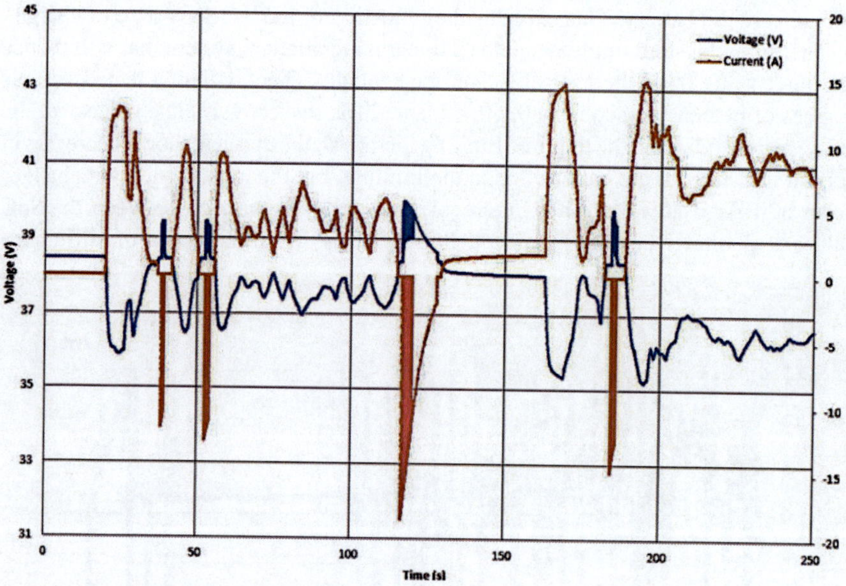

Fig. 10 Voltage (V) and Current (A) versus Time (s)

between no limit (100%), 80, and 60% is very negligible and will not increase the range of the electric scooter that drastically. On the other hand, when 40 and 20% limiters are applied, the current draw is much lower. It is observed that the battery SoC level for 20% limiting is not fully depleted while for other limiting settings, the battery SoC is fully depleted.

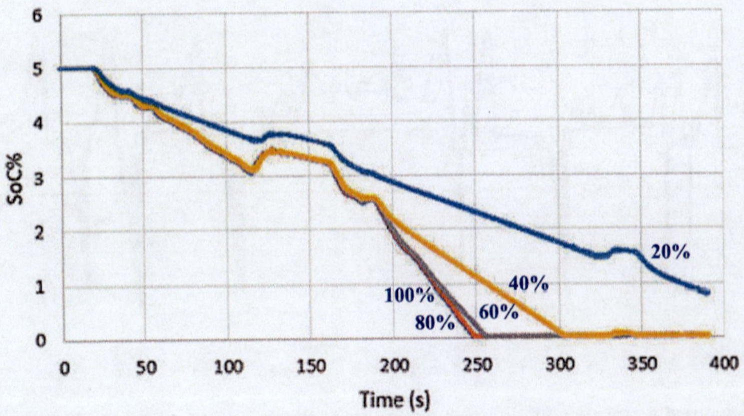

Fig. 11 Battery SoC versus time for different current limit settings

Figure 12 shows that the amplitude of the current drawn in the 20% limiting is far lower than the other settings. Due to the lower current drawn, the amount of power that will be used by the motor will be reduced as well. It is also observed that the current is no longer visible just before 310 secs when there is no limiter applied, while the current is still readable for the other limit settings. This tells us that the battery still has power left in it and, thus, is able to increase the overall range of the electric scooter. However, by limiting the motor current—by adjusting the value from the user input to the motor controller—the velocity and acceleration performance of the electric scooter are reduced, as can be seen in Fig. 13. This is by design one of the ways to improve the range which is to reduce the amount of energy drawn by the motor from the battery, sacrificing acceleration and velocity performance in the process. Yet, breaking performance will not be affected by the reduction of current drawn. It is important to note that, in Figs. 11, 12, and 13, velocity profiles for 80% and no-limit (100%) settings have negligible difference and thus are apparently indistinguishable.

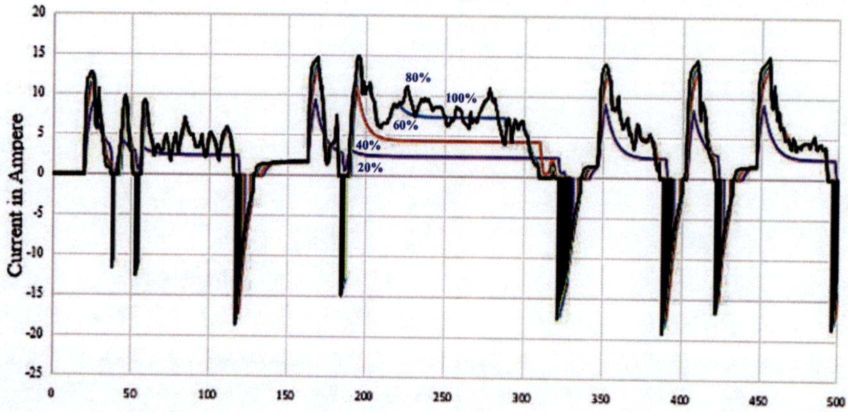

Fig. 12 Current versus time for different current limit settings

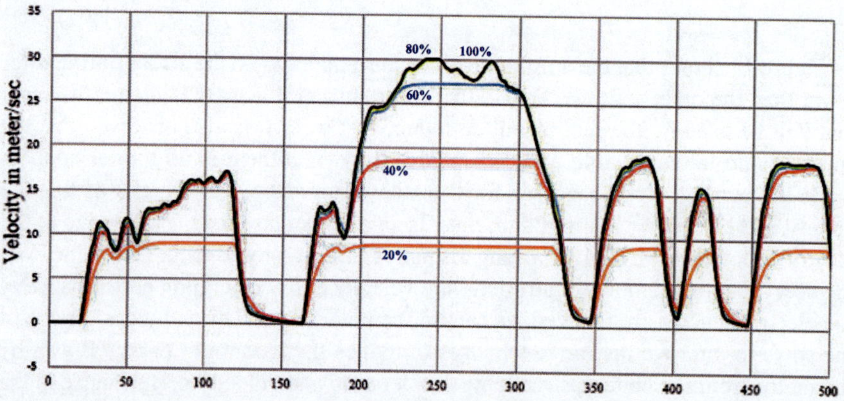

Fig. 13 Velocity versus time for different current limit settings

5 Conclusion

This paper proposes a method to reduce battery consumption in personal electric mobility equipment in order to extend its operating range. A simulation model in *MATLAB-Simulink* demonstrates the possibility to bypass and adjust input throttle request from the user to the motor controller. By reducing the input to a certain extent, the amount of reduced current can result in range extension as the amount of power drawn by the motor is reduced. The limiting mechanism used in this research is a straightforward mathematical subsystem, and it is not as dynamic as one might think. Current is limited using straightforward product blocks. When the restriction is set at 20%, the limit mechanism can increase the electric scooter's driving range by up to double that amount. From the obtained results of voltage and current profiles, the electric scooter range may be increased by adopting current limiting circuit. The goal is to implement the circuit in existing electric scooters and other micromobility or personal mobility options. There is much that can be carried out for future work, including design and investigation of dynamic control, whereby current limiting is dynamically adjusted and implemented based on battery SoC.

References

1. https://www.silicon-mobility.com/increasing-the-range-of-ev-with-the-same-battery-part-i-the-efficiency/
2. United States Environmental Protection Agency (EPA) (2021) . Available: https://www.epa.gov/greenvehicles/ explaining-electric-plug-hybrid-electric vehicles
3. Matulka R (2014) Energy.GOV," U.S Department of Energy. Available: https://www.energy.gov/articles/history- electric-car
4. "ExpatGo (2013). Available: http://www.expatgo.com/my/2013/08/17/electric-car-malaysia/
5. Scholarly Community Encyclopedia (2022). Available: https://encyclopedia.pub/entry/22506
6. Sadiq MM, Abro GEM, Yazsid AZSM, Zulkifli SA (2022) Development of range enhancement device for personal electric mobility. 7th International Conference on Electric Vehicular Technology (ICEVT), pp 144–152
7. Li W, Long R, Chen H, Geng J (2017) A review of factors influencing consumer intentions to adopt battery electric. Renew Sustain Energy Rev 78:318–328
8. Zulkifli SA, Mohd S, Saad N, Aziz ARA (2014) Impact of motor size & efficiency on acceleration, fuel consumption & emissions of split- axle through- the- road parallel hybrid electric vehicle. Appl Mech Mater 663:498–503
9. Sautermeister Stefan, et al. (2017) Influence of measurement and prediction uncertainties on range estimation for electric vehicles. IEEE Trans Control Syst Technol 19(8):2615–2626
10. Tian Zhiyong et al. (2016) Real-time charging station recommendation system for electric-vehicle taxis. IEEE Trans Control Syst Technol 17(11), 3098–3109
11. Rivera Sebastian et al. (2015) Electric vehicle charging station using a neutral point clamped converter with bipolar DC bus. IEEE Trans Ind Electron 62(4):1999–2009
12. Crosier Russell, Wang S (2013) DQ-frame modeling of an active power filter integrated with a grid-connected, multifunctional electric vehicle charging station. IEEE Trans Power Electron 28(12):5702–5716
13. Fujimoto Hiroshi, Harada S (2015) Model-based range extension control system for electric vehicles with front and rear driving– braking force distributions. IEEE Trans Ind Electron 62(5):3245–3254
14. Zhai Li, Sun T, Wang J (2016) Electronic Stability control based on motor driving and braking torque distribution for a four in-wheel motor drive electric vehicle. IEEE Trans Veh Technol 65(6):4726–4739
15. Kim DH et al (2010) Optimal brake torque distribution for a four-wheel drive hybrid electric vehicle stability enhancement. Drive Syst Techn 221(11):1357–1366
16. Yang Ming Ji et al (2009) A cost-effective method of electric brake with energy regeneration for electric vehicles. IEEE Trans Ind Electron 56(6):2203–2212
17. Chang W-Y (2013) The state of charge estimating methods for battery: a review. Intern Schol Res Not 2013:1–7
18. Mathworks. Available: https://www.mathworks.com/discovery/what-is-matlab.html

CFD Simulation of CO_2 Through a Converging–Diverging Nozzle

Mohamed Akram Azeerdeen and Tamiru Alemu Lemma

Abstract In the process of CO_2 separation from natural gas, research has been done by various parties. Even though they were successful to a certain extent, a far more clear and optimized process is being searched for. In this study, simulation software is used to replicate experimental data. Supersonic nozzles are used, as there is a reduction in the pressure and temperature of the gas when it flows through a converging–diverging nozzle, and the condensation process can take place. The aim of this study is to find a set of parameters that can be used to achieve a higher efficiency in CO_2 condensation. At the conclusion of this study, a sharp pressure drop, of more than 50%, is observed when using lamina flows for both Ideal and Peng-Robinson CO_2. When the K-Epsilon turbulence model is used, the pressure drop is only 25%. CFD simulations were conducted on ANSYS CFX. Past literature has shown how achievable the partial condensation of both natural gas mixtures containing CO_2 and pure CO_2 is.

Keywords Converging–diverging nozzle · CFD · Turbulence modeling

1 Introduction

There has been a great deal of study in the area of water vapor condensation. Nevertheless, new findings are always being made. Condensation is a phenomenon that occurs naturally. However, there is less information on the condensation properties of the gases found in nature. When we dive into the petroleum industry, there are unnecessary byproducts in the natural gas extraction process. Carbon dioxide is one of the main byproducts. The entry of these unwanted gases causes serious damage to

M. A. Azeerdeen · T. A. Lemma (✉)
Department of Mechanical Engineering, Universiti Teknologi PETRONAS, Seri Iskandar, Perak, Malaysia
e-mail: tamiru.lemma@utp.edu.my

M. A. Azeerdeen
e-mail: mohamed_20001003@utp.edu.my

F. Ahmad et al. (eds.), *ICREEM 2022*, Lecture Notes in Mechanical Engineering,
https://doi.org/10.1007/978-981-99-5946-4_15

the pipelines. A key research area by various oil companies and universities revolves around the removal of carbon dioxide from the stream before it enters the pipelines. The entry of CO_2 into the pipelines results in the formation of hydrates and causes carbonic acid corrosion. Different techniques are applied in order to remove the hydrates and remove the corrosion. Unfortunately, these measures are temporary and cost petroleum companies a large amount of money, as production is halted for maintenance to take place. Taking all of this into consideration, the present study aims to remove CO_2 before it enters the pipeline in an efficient manner.

2 Literature Review

2.1 Condensation Inside Supersonic Nozzles

The issue of condensation occurrences inside supersonic nozzles has been approached many times. The primary fluid in these nozzles is water. Wegener and Mack [1] mentions, at supersonic velocities, the condensation of water can be clearly seen. Frank [2] states the condensation of water vapor in a Laval nozzle results in an increased temperature. Wegener and Pouring [3] conducted a study to observe temperature increase due to condensation inside supersonic nozzles. They concluded by stating that at temperatures between 200–220 K ice condensates were formed from water vapor. Karimi and Abdi [4] was able to separate water from high-pressure natural gas, successfully, using supersonic nozzles and altering the design parameters. For offshore applications related to gas separations, Jassim et al. [5] commends the use of supersonic separators, also by adding on that gas flow at high speeds through nozzles will prevent any fouling or solid depositions.

Vijayakumaran and Lemma [6] using simulations was able to prove that the condensation efficiency of water vapor increased by 26% when swirling flow was being used inside the supersonic nozzle. A higher number which is being defined by swirl intensity shows an increase in the Liquid Mass Fraction of water at the exit of the nozzle. The swirl numbers used in the simulation are shown in Table 1. The results acquired when comparing swirling flows and normal flows are shown in Figs. 2 and 3.

Table 1 Values of Swirl number

Swirl ID	Radial velocity	Tangential velocity	Axial velocity	Swirl number
Units	m/s	m/s	m/s	–
1	180	100	240	0.417
2	27	48	86	0.558
3	43	30	42	0.714
4	19	64	75	0.853

2.2 Carbon Dioxide Condensation

Duff [7] studied the condensation taking place inside supersonic nozzles, and it was concluded that prior to condensation there is a definitive area CO_2 supersaturation. Further studies were carried out by Duff [8] where it was shown the formation of particles during the condensation process of CO_2 inside the supersonic nozzles was gradual unlike the results obtained with water vapor, which were abrupt.

According to Ramos et al. [9], there is a formation of two types of CO_2 condensate clusters. This depends on the size of the cluster. In the initial stage small molecules form. These then grow when they collide with other molecules and clusters. The three-body collision process was reported as the main collision mechanism. In this mechanism, excess collision energy is absorbed which prevents the formed cluster from breaking. After the cluster reaches the critical radius, cluster–cluster collision occurs and is the primary process which leads to the formation of large clusters and condensation. Cluster-mediated collisions are known to be far greater than molecule-mediated collisions.

According to Duff [8], the CO_2 condenses into ice crystals instead of liquid droplets provided the pressure is below the CO_2 ice points, which is 75.1 psia (2.18 bar). Therefore, it is unclear whether the solid density or the liquid density should be taken as the droplet density. In the case of water, it can be seen that as the droplet size decreases, the spontaneous freezing temperature decreases. This made it quite difficult to separate the liquid and solid CO_2 condensates of the clusters, at the critical radius. Duff [8] also conducted an experimental study, where it showed the low temperature behavior of CO_2 was adequately portrayed by the Plank equation of state. The rational prediction conditions of the condensation of CO_2 are provided by the classical nucleation theory and drop growth theory. For this experiment, the droplet density was assumed to be 1.18 gm/cm^3, which is equal to the triple point of CO_2. In order to reduce uncertainties in future works, Duff [8] has recommended to contain the CO_2 inside large amounts of a different gas which does not condense in the CO_2 temperature and pressure range. The other gas should have properties close to a perfect gas.

According to the simulation conducted by Hammer et al. [10], whereby CO_2 was captured using a Laval nozzle, the expansion inside the supersonic nozzle results in the freezing of CO_2 and condensate to dry ice. With the help of swirling action created at the entrance of the nozzle, the heavier dry ice is separated from the main gas. It is argued that due to the supersonic velocities, there will be no deposition of CO_2 in the nozzle. From the subcooled metastable state, the formation of solids in a fast expansion is said to occur. According to this study, a large volume of gaseous CO_2 to dry ice condensates if the inlet pressure is increased and the inlet temperature is decreased.

In recent times, Sun et al. [11] carried out numerical simulations in order to simulate the condensation of CO_2 from a methane and CO_2 gas mixture. It was found in this study that prior to homogeneous nucleation, phase transition of CO_2 from gas to liquid takes place. From a variety of inlet conditions, CO_2 was successfully reduced from the main gas stream.

2.3 Experimental Study

Ahmad Samawe et al. [12] reports that PETRONAS proved, by carrying out experiments, the concept of CO_2 separation using a TWISTER supersonic separator. Before proceeding with the experimental study, a simulation was carried out using ANSYS FLUENT. Static guide vanes were fixed in the supersonic separator, in order to generate swirling of the natural gas mixture. Only when the temperature is drastically reduced does condensation of CO_2 take place inside the Laval nozzle. The natural gas flowing at supersonic velocities results in temperature reduction. The gas mixture expands through the nozzle while flowing into the diffuser. The centrifugal forces generated due to the swirling action separate the CO_2 condensate and the dry gas. Then the liquid CO_2 flows into a secondary outlet. Between the supersonic separator inlet and the primary outlet, in the experimental setup, it was recorded there was a 2.1 mol% average reduction. The collection of liquid from the secondary outlet confirms the presence of CO_2 condensates. As compared with the inlet, there was a higher percentage of methane, CH_4, found in the primary outlet. When compared with the prediction of the simulation, the outcome was supposed to have a more conservative CO_2 reduction of 1.8 mol%. It was noticed that, as soon as the gas mixture entered the nozzle, CO_2 began to condense. Due to the swirl, CO_2 in liquid form was pushed to the wall of the nozzle, expelling itself through the secondary outlet.

In the process of increasing the reliability of the supersonic separator, for use on-site, the authors have advised conducting further studies to design a tool which is more accurate.

McGuire [13] filled for a patent, explaining a method to separate carbon dioxide from methane gas with the help of a convergent-divergent nozzle, at the United States Patent and Trademark Office (USPTO). The method used is the same as [11], whereby CO_2 condenses to a solid dry ice form. The pure methane gas moves along the middle of the nozzle and the CO_2 moves adjacent to the walls. The solid CO_2 goes through a phase transfer to fluid, after separation. This is because of the high expansion temperatures.

2.4 Properties of Carbon Dioxide (CO_2) and Equation of State

The physical properties of CO_2 are presented in Table 2.

Table 2 Properties of CO_2

Properties	Unit	Value
Critical temperature	°C	31.1
Critical pressure	MPa	7.4
Boiling point	°C	−78.4
Melting point	°C	−56.6
Saturation pressure, *Psat* @ 70 °F	MPa	852.8
Liquid density @ 70 °F	kg/m^3	763.12
Gas density@ 70 °F	kg/m^3	1.8
Specific volume @ 70 °F	m^3/kg	0.55
Specific gravity	–	1.555
Specific heat @ 70 °F	J/kgK	37,346.2

The properties of CO_2 for modeling and simulation purposes are based on the National Institute of Standards and Technology (NIST) real gas models which is available in ANSYS [10].

Two equations of state for CO_2 were found in the literature, namely Plank's equation of state as illustrated in [4] and the new equation of state for CO_2 as mentioned in [11]. Plank's equation of state is as follows:

$$v = \frac{RT}{p} + \frac{(0.0825 + 0.0012657p) \times 10^3}{(0.01T)^3} \tag{1}$$

whereas the new fundamental equation of state suggested by Span and Wagner [11] expressed in the form of Helmholtz energy is

$$A(\rho, T)/(RT) = \varnothing(\delta, \tau) = \varnothing^{\circ}(\delta, \tau) + \varnothing^{r}(\delta, \tau) \tag{2}$$

$$\delta = \frac{\rho}{\rho_c}$$

$$\tau = \frac{T_c}{T}$$

Span and Wagner's [11] fundamental equation is divided into two parts, an Ideal Gas part $\varnothing^{\circ}(\delta, \tau)$ and Residual part $\varnothing^{r}(\delta, \tau)$. The final expression for the Ideal Gas part is

$$\varnothing^{o}(\delta, \tau) = \ln(\delta) + a_1^o + a_2^o\tau + a_3^o\ln(\tau) + \sum_{i=4}^{8} a_i^o\ln\left[1 - e^{(-\tau\theta_i^o)}\right] \tag{3}$$

Coefficients for a_i^o, θ_i^o and all derivatives required for the calculation of the thermodynamic properties for the Ideal Gas part $\emptyset^o(\delta, \tau)$ are explicitly specified in the literature. The formula for the Residual part of the equation is

$$\emptyset^r(\delta,\tau) = \sum_{i=1}^{7} n_i\delta^{d_i}\tau^{t_i} + \sum_{i=8}^{34} n_i\delta^{d_i}\tau^{t_i}e^{-\delta c_i} + \sum_{i=35}^{39} n_i\delta^{d_i}\tau^{t_i}e^{-\alpha_i(\delta-\epsilon_i)^2-\beta_i(\tau-\gamma_i)^2} + \sum_{i=40}^{42} n_i\Delta^{b_i}\delta e^{-C_i(\delta-1)^2-D_i(\tau-1)^2} \quad (4)$$

where $\Delta = \left\{(1-\tau) + A_i\left[(\delta-1)^2\right]^{1/(2\beta)_i}\right\}^2 + B_i\left[(\delta-1)^2\right]^{a_i}$.

Similar to the Ideal Gas part, the final values for the coefficients of n_i are explicitly stated in the literature. These values were determined by a direct nonlinear fit to the linear and nonlinear data.

For this study, it is viable to use Plank's equation of state as it has been tested experimentally.

A mathematical expression of the change of states within CO_2, using the equation of state, is suggested by Duff [7]. The equation has successfully been tested in experimental studies and is therefore applicable.

2.5 *Isentropic Equations*

The isentropic equation for ideal gas is used to plot an isentropic curve. The symbols k and Ma stand for heat capacity ratio and Mach number, respectively:

$$\frac{P}{P_o} = \left(1 + \frac{k-1}{2} * Ma^2\right)^{\frac{-k}{k-1}} \quad (5)$$

3 Simulation Setup

The geometry for the simulation was taken from [14]. The research performed in [14] was experimental for different Inlet conditions. The main change is in the inlet and outlet pressure. Figures 4 and 5 show the solid geometry and the meshing.

There are 5 different cases in [14] that were presented. Each case had 2 sets of data so 10 different simulations were performed. Only case 4 simulation results were used to compare with the experimental data. The inlet boundary conditions were set to 80 bar and 313 K. Outlet pressure was set at 26.92 bar (Fig. 3).

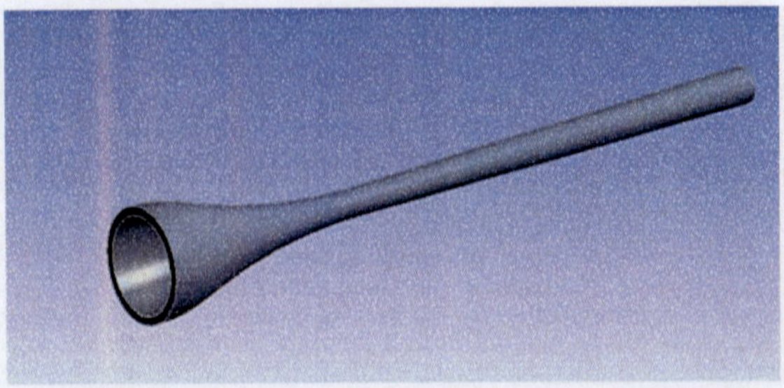

Fig. 1 Model geometry using ANSYS Designmodeler

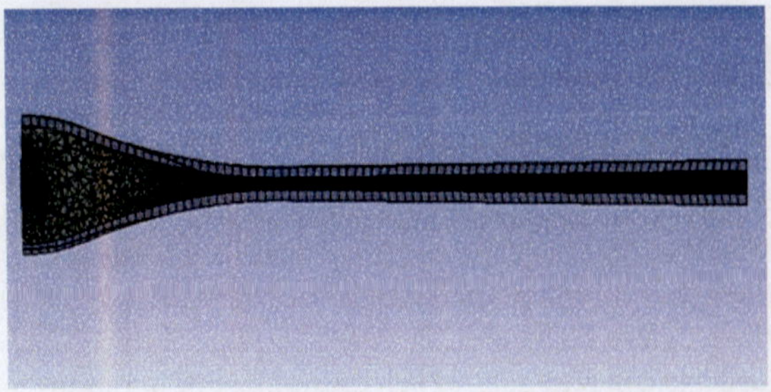

Fig. 2 Mesh

3.1 Mesh

The mesh is very fine and the default setting is shown in Fig. 2.

3.2 Modeling

The software used for simulation is ANSYS CFX. The simulation is a steady-state simulation. The materials and equations of states used are found in it. The K-Epsilon turbulence model was used, and Lamina was also used. The number of iterations was varied, and the working fluid used was CO_2 both ideal and Peng-Robinson. In Fig. 3, the experimental data was compared to the isentropic flow which is created using the isentropic compressible flow equations which is displayed as Eq. (5). According to Fig. 4, the simulation of Ideal gas CO_2 with K-Epsilon produced results that were closest to the experimental data. One of the objectives of this study is to replicate the experimental results using ANSYS CFX. The isentropic conditions were created by setting the heat transfer setting to isothermal and the wall condition to no slip with the heat transfer setting of adiabatic. The turbulence setting is set to Lamina. The fluid used is CO_2 under ideal conditions. The inlet conditions are 80 bar and 313 K. The outlet condition for the simulation was set to a very low number which is 26.92 Pa. The reason this was done was for the flow to mimic the experimental values so a similar graph could be plotted. The same was done for the flow with CO_2

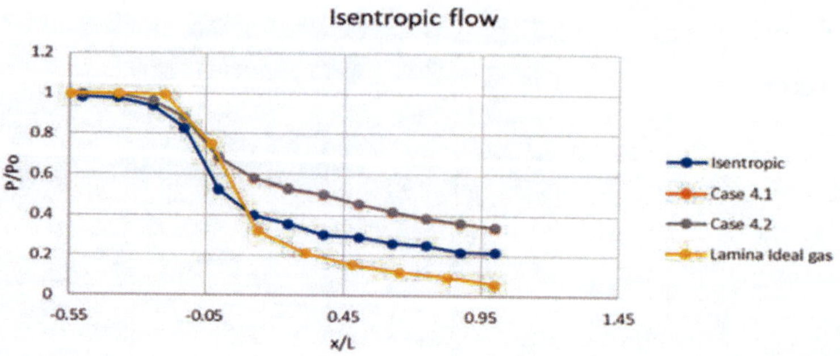

Fig. 3 Experimental case compared with isentropic flow based on equations and simulated lamina flow

under the Peng-Robinson equation of state. However, when the turbulence model was set to K-Epsilon and the outlet condition was changed to the experimental data, which is 26.92 bar the curve was almost identical to the experimental curve. When the outlet condition of 26.92 bar was applied to the lamina flows there was a spike in the pressure at the end, as shown in Fig. 5. It was observed the number of iterations also had a part to play. When the number of iterations increased, the spike in pressure at the end increased as well. Further simulation needs to be run in order to understand this phenomenon more.

All 5 cases were simulated using the K-Epsilon turbulence model. The inlet and outlet conditions were that from the experimental data. Pressure against temperature was plotted. All the results are close to the critical point which means it exists in vapor and liquid forms, which is favorable for separation in the future.

Figures 7 and 8 show the Mach number and pressure through the center of the nozzle. It can be observed that the flow does not reach supersonic at the center. However, when compared with Fig. 6 the fluid exists in liquid form which is the desired outcome.

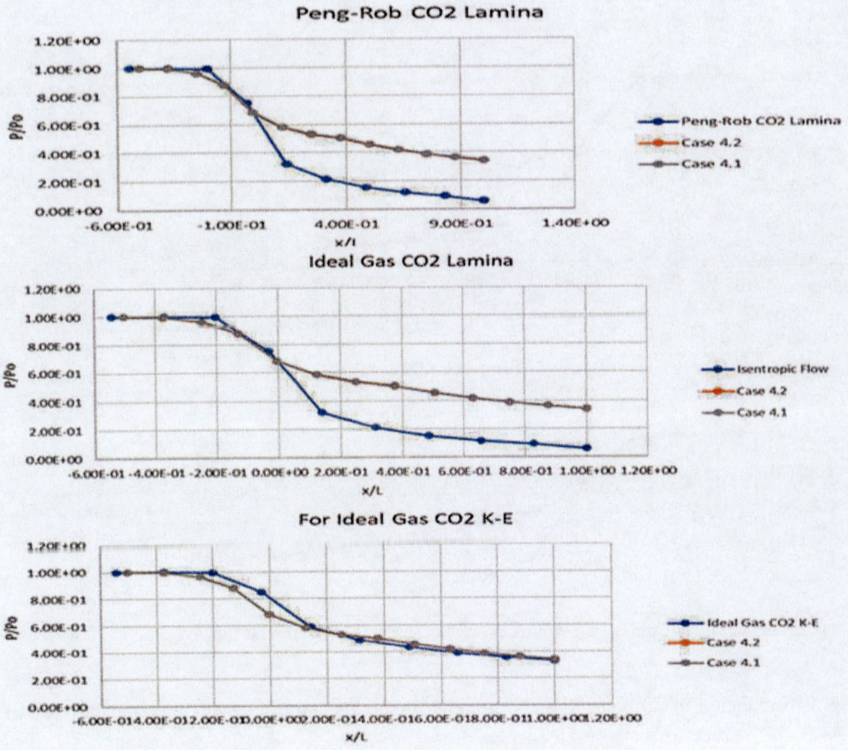

Fig. 4 Simulation of different flows compared with experimental data of Case 4

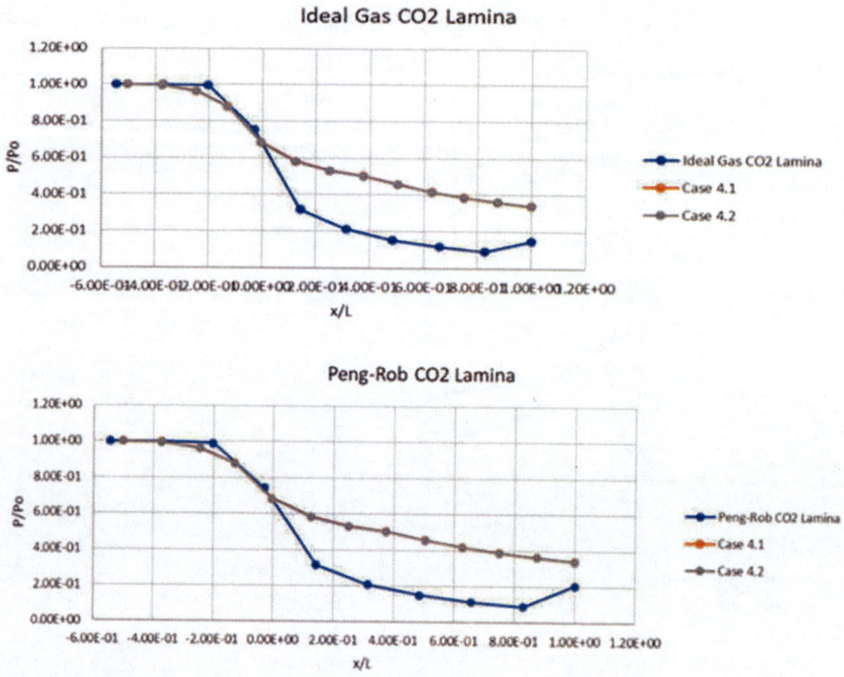

Fig. 5 Simulation of different flows with higher iterations compared with experimental data of Case 4

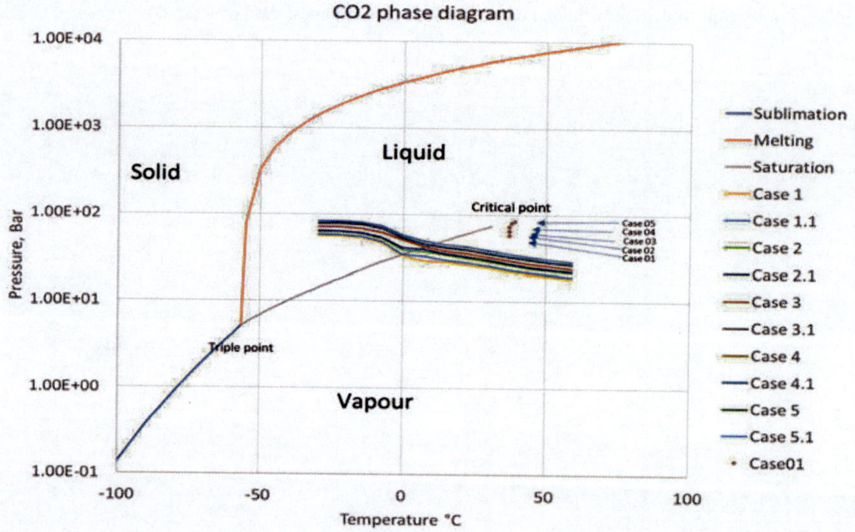

Fig. 6 CO_2 phase diagram

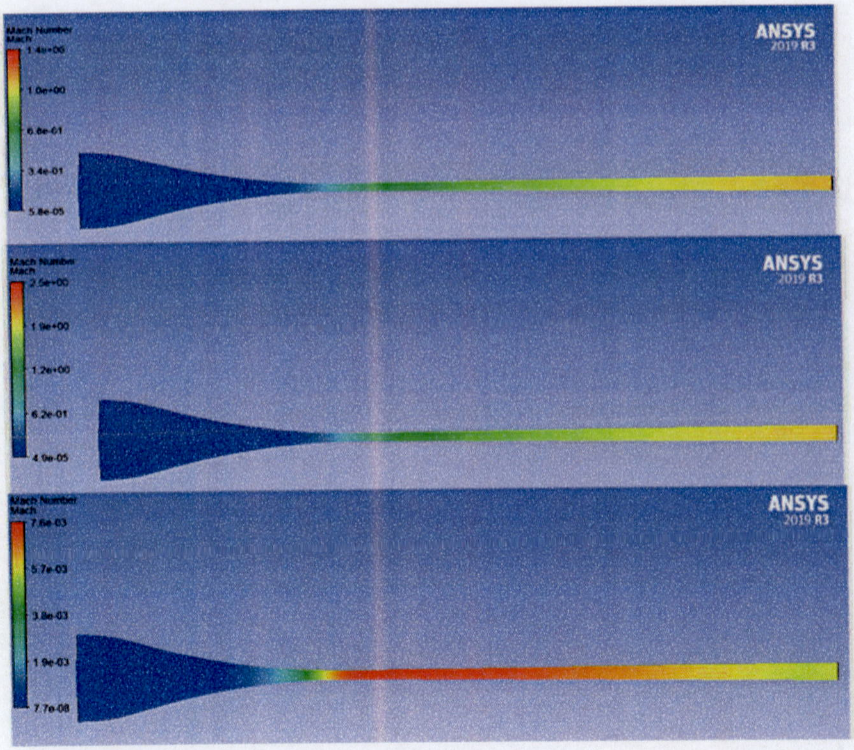

Fig. 7 Mach number of flow CO_2 through nozzle **a** isentropic Peng-Robinson, **b** isentropic ideal gas, and **c** K-Epsilon ideal

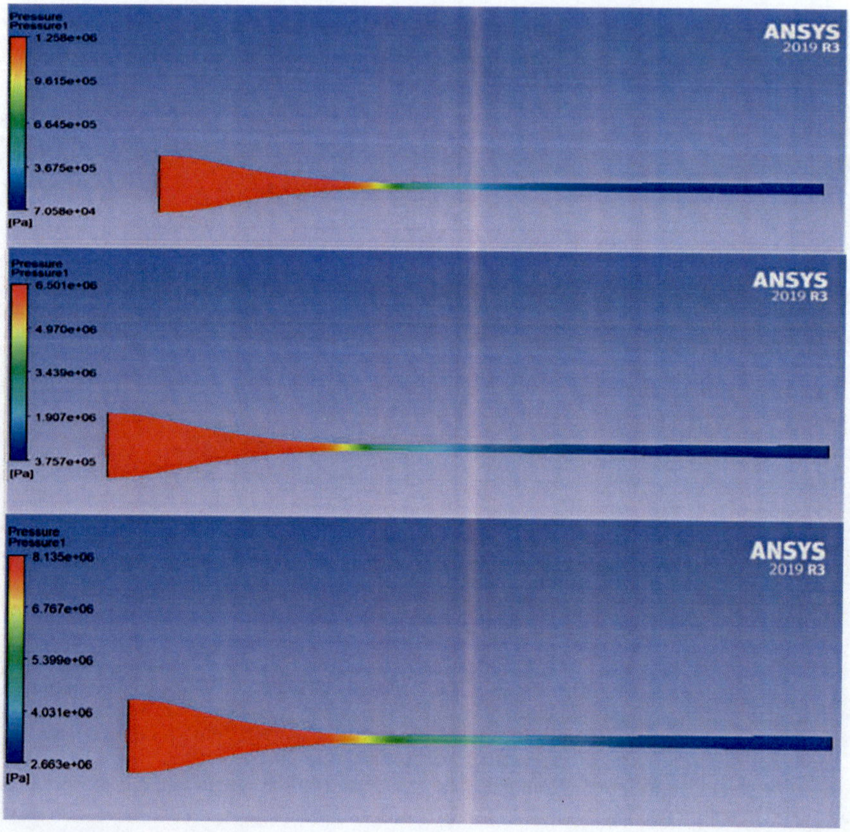

Fig. 8 Pressure of flow CO_2 through nozzle **a** isentropic Peng-Robinson, **b** isentropic ideal gas, and **c** K-Epsilon ideal

4 Conclusion

Overall, the attempt to replicate experimental data is successful under turbulent conditions. It is observed that when the pressure drop from inlet to outlet is greater, the flow creates a curve similar to the experimental data. All the results show the fluid exists in critical condition at the throat which is necessary if a converging–diverging nozzle is used for gas separation. The results show that there is more than a 50% pressure drop after the throat area in the first two simulations. This drop in pressure in real-life conditions might lead to the freezing of the fluid which is not favorable. In the conditions which mimic the experimental data, i.e. the third simulation, the pressure drop is only 25%. There is more that can be done in this area. Instead of using only CO_2, various components of natural gas can be added in their respective volume fractions found in the literature. This can further improve accuracy and more detailed observations can be made when other fluids are involved. In the future, a transient simulation can be done using the same experimental data.

Acknowledgements The authors are pleased to recognize the support of Universiti Teknologi PETRONAS and supervisors in terms of providing a research environment. The research is supported by YUTP-FRG grants with cost centers: 015LC0-400. The authors would also like to recognize the feedback provided by the reviewers.

References

1. Wegener PP, Mack LM (1958) Condensation in supersonic and hypersonic wind tunnels. In: Dryden HL, von Kármán T (eds) Advances in applied mechanics. Elsevier, pp 307–447
2. Frank W (1985) Condensation phenomena in supersonic nozzles. Acta Mech 54(3–4):135–156
3. Wegener PP, Pouring AA (1964) Experiments on condensation of water vapor by homogeneous nucleation in nozzles. Phys Fluids 7(3):352–361
4. Karimi A, Abdi MA (2009) Selective dehydration of high-pressure natural gas using supersonic nozzles. Chem Eng Process 48(1):560–568
5. Jassim E, Abdi MA, Muzychka Y (2008) Computational fluid dynamics study for flow of natural gas through high-pressure supersonic nozzles: Part 1. Real gas effects and shockwave. Petroleum Sci Technol 26(15):1757–1772
6. Vijayakumaran H, Lemma TA (2018) Simulation of swirling wet steam flow through a supersonic nozzle. In: AIP conference proceedings. AIP Publishing LLC
7. Duff KM (1964) Condensation of carbon dioxide in supersonic nozzles. Massachusetts Inst of Tech Cambridge Gas Turbine Lab
8. Duff KM (1966) Non-equilibrium condensation of carbon dioxide in supersonic nozzles. Massachusetts Institute of Technology
9. Ramos A et al (2012) Cluster growth in supersonic jets of CO_2 through a slit nozzle. In: AIP conference proceedings. American Institute of Physics
10. Hammer M et al (2014) CO_2 capture from off-shore gas turbines using supersonic gas separation. Energy Procedia 63:243–252
11. Sun W et al (2017) Numerical simulation of CO_2 condensation process from CH_4-CO_2 binary gas mixture in supersonic nozzles. Sep Purif Technol 188:238–249
12. Ahmad Samawe R et al (2014) Concept proofing of supersonic nozzle separator for CO_2 separation from natural gas using a flow loop. In: Offshore technology conference-Asia. 2014. Offshore technology conference
13. McGuire P (2002) Method of removing carbon dioxide or hydrogen sulfide from a gas. Google Patents
14. Lettieri C et al (2018) Characterization of nonequilibrium condensation of supercritical carbon dioxide in a de Laval nozzle. J Eng Gas Turbines Power 140(4):041701

Variable Geometry Industrial Gas Turbine Part-Load Performance

Waleligne Molla Salilew, Zainal Ambri Abdul Karim, and Tamiru Alemu Lemma

Abstract This paper provides an investigation on the effect of variable inlet guide vane and physical faults on the performance of a three-shaft gas turbine engine at part-load operation. Using the engine manufacturer's data as input, both design-point and off-design engine performance models have been developed. The model was validated with manufacturer data. Upon validation of the model, the clean engine model was subjected to variable inlet guide vane drift, fouling, and erosion conditions to simulate a degradation state. Two scenarios that cause gas turbine degradation have been considered and simulated: First, how would the combined effect of fouling and variable inlet guide vane drift cause the degradation of the engine performance. Second, how would the combined effect of erosion and variable inlet guide vane drift cause the degradation of the engine performance? The result prevailed that the up-VIGV drift, which is combined fouling and erosion, shows a small deviation because of offsetting the isentropic efficiency drop caused by fouling and erosion. Furthermore, the deviation of performance and output parameters due to the combined faults is discussed following the plots.

Keywords Gas turbine · Performance model · Variable inlet guide vane · Physical faults

W. M. Salilew · T. A. Lemma (✉)
Mechanical Engineering Department, Universiti Teknologi PETRONAS, Bandar, 32610 Seri Iskandar, Perak, Malaysia
e-mail: tamiru.lemma@utp.edu.my

Z. A. A. Karim
Centre for Automotive Research and Electric Mobility (CAREM), Universiti Teknologi PETRONAS, 32610 Seri Iskandar, Perak, Malaysia

F. Ahmad et al. (eds.), *ICREEM 2022*, Lecture Notes in Mechanical Engineering,
https://doi.org/10.1007/978-981-99-5946-4_16

1 Introduction

The gas turbine is the primary source of power for industrial systems and aircraft. It is a thermal machine that works in conditions of severe stress, high pressure, high temperature, and high speed. It is a complex machine that has mechanical, electrical, and hydraulic systems [1]. The tip clearance variations in the rotating components, actuator wear, seal wear, blade erosion, blade fouling, blade warping, blocked fuel nozzles, sensor issues, and domestic object damage and foreign object damage are the factors that contribute to a decline in the gas turbine performance over time [2–4]. Due to harsh operating conditions, the performance of the gas turbine steadily declines over time. Anomalous working circumstances must be avoided as soon as possible to reduce the associated increase in energy consumption and environmental damage. Even though the engine is working normally because of having a decent inlet filtration system to use clean air, the parts of the engine flow stream will become eroded, corroded, covered in rust scale, and damaged [5]. The use of an efficient condition monitoring and diagnostic technique that can identify, analyse, and predict impending or incipient failure conditions and recommends the necessary maintenance measures at the appropriate time is most sought for. Energy usage and environmental pollution will be significantly reduced by improving the gas turbine's operational performance. To evaluate the gas turbine's overall performance and satisfy operational and maintenance requirements, knowledge of the gas turbine's fundamental state is crucial [6]. A number of commercial simulation software have shown to be a helpful tool for creating models of certain engine performance models [7]. The software helps to develop both steady-state and transient models using component characteristics and produces the matching results. To simulate the performance of both clean and deteriorated gas turbine engines, the thermodynamic interactive generalized turbine engine model must be created [8].

Modern gas turbines include advanced features to achieve improved reliability, availability, and sustainability [9]. Variable geometry in compressors and turbines, which comprises variable inlet guide vane (VIGV), variable stator vanes (VSVs), variable bleed valve (VBV), and variable area nozzle (VAN) are some of the features which increase the performance of the gas turbine [10]. Since variable guide vanes (VIGVs and VSVs) are scheduled as a function of spool speed, they are designed to operate when the engine experiences a speed fluctuation during startup, shutdown, and load changes [11]. Surging and stalling that frequently occur at low speeds during startup and shutdown can always be injurious to the engine's health and performance [12]. Therefore, stability is ensured, and performance is improved by removing compressor surge phenomena using the VIGVs and bleed control system. Additionally, nonlinear and complicated dynamics have considerable negative impact on the performance of the latest generation of gas turbines. Other problems, such as variable geometry mechanism faults, axial compressor fouling, and increasing ambient temperature, may also contribute to degradation [13]. Different connections attached to guiding vanes are used as part of an actuation system to regulate the VIGVs [14]. The VIVGs could have various issues, like drifting beyond the standard functioning range. These issues and faults finally cause an unexpected shutdown due to surge and choking in the compressor stages.

Salar et al. [15] identified that low-speed stop, high-speed stop, hydraulic ram leakage, and rotary variable displacement transducer (RVDT) misalignment are the main causes of VIGV drift. According to Tsalavoutas et al. [16], other factors, such as the wearing of the actuation mechanism linkages and the misalignment or stacking of the vanes because of loosened bolts, caused VIGVs to deviate from their typical schedule. Continuous monitoring is required to make sure that the movement and position of the actuation mechanism are in sync with the actual design schedule to avoid the occurrence of these issues. The difficulty of monitoring the position of an actuation mechanism for a large number of vanes makes this solution to be impractical. To overcome this, an adaptive performance model to identify VIGV problems was proposed by Stamatis et al. [17]. By artificially introducing various VIGV drifts, some researchers have attempted to observe the impact of VIGV drift on gas turbine performance. To investigate the impact of VIGV drift on the performance of the compressor, Cruz-Manzo et al. [18] recently created a MATLAB Simulink-based performance model. A comparison of the difference between the position that the VIGV was intended to be in and the drift position was investigated, and a significant variation has been observed. A soft sensor was created by Salar et al. [15] to forecast VIGV drift. The defect was located by comparing the deviation between the actual guide vane position from the control panel and the VIGV position provided by the soft sensor. Additionally, Razak and Dosanjh [19] covered the significance of the condition monitoring system for identifying VIGV drift.

Compressor fouling has an impact on the performance of gas turbines. When sticky particles are accumulated on the annulus passage of the compressor, including the rotors and stators, compressor fouling occurs [20]. Ajoko and Tolumoye [21] studied the monitoring of industrial gas turbine performance which can be affected by fouling, erosion, and other physical faults. The main goal of the study is to forecast an industrial gas turbine's performance utilizing a data analysis process to check, balance, and monitor the behaviour of the gas turbine while it is in operation. According to established facts, this low performance is caused by a drastic drop in the measurable performance parameters, which can be rectified and controlled with ambient temperature by regulating the intake supply fluid with air which is dense and close to the intended specification. Finally, Ajoko and Tolumoye advised that, in order to avoid and minimize total gas turbine downtime, gas turbine operators must be familiar with such information for early prognosis. Mishra [4] studied on aero gas turbine compressor fouling and corrosion. The conclusion showed that the physical faults have been demonstrated to be well-addressed by compressor washing. The frequency of the compressor wash must be carefully calculated from the perspectives of corrosion rate and performance degradation. The research work [22, 23] provide more information on the impact of fouling and erosion on the gas turbine performance. To prevent a severe performance decline, adaptive performance models and simulations were conducted to estimate the fouling rate during operation.

As discussed above, researchers have worked on the simulation of gas path faults and gas turbine performance degradation. However, only a few number of researchers

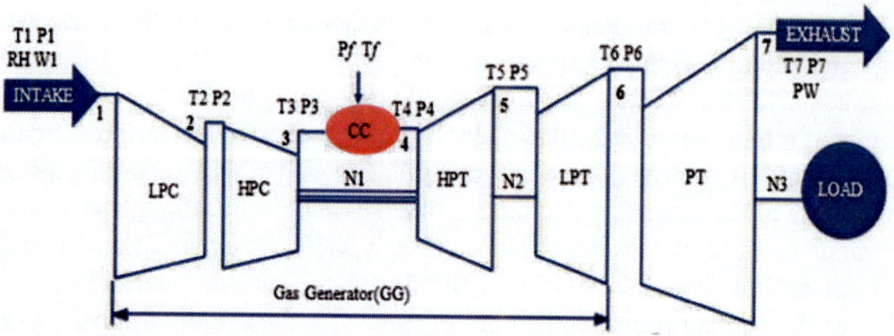

Fig. 1 Configuration of a three-shaft gas turbine

have studied the performance degradation of all significant gas turbine components. Variable inlet guide vane drift and its combined effect with other physical faults remained unexplored. Therefore, the purpose of this study is to study the effect of variable inlet guide vane drift and its combined effect with fouling and erosion on the three-shaft gas turbine performance at part-load operation. The effects on component isentropic efficiency and output parameters are studied and discussed. During developing the engine performance model, secondary air or cooling system and variable inlet guide vane scheduling were considered. With the help of engine manufacturer data, the gas turbine design-point and off-design performance models have been built using GasTurb 13 and validated. Using the correlation between physical faults and performance parameters, fouling and erosion physical faults were introduced to the healthy engine model. Investigations on the effect of combined effect of variable inlet guide vane drift with fouling and variable inlet guide vane drift with erosion on performance and output parameters are conducted.

A three-shaft gas turbine engine was used as the study's case engine. It consists of six primary gas-path components, which include a low-pressure compressor and turbine, high-pressure compressor and turbine, power turbine, and combustion chamber. It has a single-shaft power turbine and a two-shaft gas generator. The gas generator includes the low- and high-pressure compressor and turbine. The low-pressure turbine and high-pressure turbine are driven by the low-pressure compressor and the high-pressure compressor, respectively. The power turbine has two stages and is a free axial turbine. The middle and last stages of the low- and high-pressure compressors are used to extract the cooling air. Figure 1 depicts a configuration for the three-shaft gas turbine.

2 Gas Turbine Performance Model Development

Many researchers used different methods to develop the gas turbine performance model. Using the method of determining enthalpies and entropies at several key cycle points to assess the gas turbine performance can be the most accurate method

[24]. The performance model for gas turbines requires the developing of design-point and off-design models [25]. The design-point model simulates the engine operating at design load while computing all parameters at each component station. In off-design performance simulation, the design point is treated as one operating point. Combining the performance of each component of the gas turbine yields the overall performance of the actual gas turbine cycle [24, 26]. The input data for the developing of the design-point model was gathered from the technical manuals or catalogue of the manufacturer. For assessing the design-point and off-design model, the energy balance approach and the iteration and optimization method were used to make the model accurate and verify the produced results [27]. If the model result matches with the catalogue data, then, the model is considered as an accurate model to predict the engine performance [28].

2.1 Design-Point Performance Modelling

The parameters at each station point of the engine were determined at design-point calculation. The ambient conditions, component efficiencies, compressor pressure ratio, inlet mass flow rate, air–fuel ratio in the combustion chamber, and turbine inlet temperature are among the input data. A design-point performance model is developed at a single operating point by employing thermodynamic equations to calculate all unknown parameters. It is essential to examine the compatibility or energy equilibrium between components that use a common shaft. Programming in Matlab or other programming language can be used to create these thermodynamic equations. The model will continue to be improved until the percentage error becomes smaller, and the energy is balanced. In this study, the performance model was developed using GasTurb 13 commercial software. The software works by employing the most accurate and promising method called determining enthalpy-entropy values at each gas path point. The input data for modelling the design and off-design performance as well as model validation are gathered from several sources. The majority of the engine information, including thermal efficiency, power output, exhaust temperature, spool speeds, heat rate, inlet mass flow rate, pressure ratio, compressor, and turbine stages, was gathered from engine manufacturer datasheets. The remaining input parameters were gathered from scientific journal papers, and during model optimization, engineering judgments and estimated values were used. The engine catalogue/data datasheet was utilized to evaluate the off-design model over the entire operating range, including data on thermal efficiency versus ambient temperature and power output versus ambient temperature. The ambient conditions are international standard organization conditions. Thermodynamic parameters such as P and T at various gas path points, fuel flow rate, compressor inlet air flow rate, component isentropic efficiencies, gas generator speed, net power output, and power turbine speed were among the data to be calculated in design-point simulation. Some of these parameters were available in the catalogue and in technical documents from the manufacturer.

Table 1 RB211-24G engine design-point model input data [29]

Parameter	Unit	Value	Source
Pressure ratio	–	20:1	Catalogue
Power output	MW	26.025	Catalogue
Thermal efficiency	%	35.8	Catalogue
Heat rate	kJ/kWh	10,043	Catalogue
Exhaust mass flow rate	kg/s	92.2	Catalogue
Exhaust temperature	°C	488	[29]
Turbine inlet temp	°C	1193	Catalogue
HPC rotational speed	RPM	9445	Catalogue
LPC rotational speed	RPM	6643	Catalogue
FPT rotational speed	RPM	4950	Catalogue
HPC stages	–	6	Catalogue
LPC stages	–	7	Catalogue
LPT stages	–	1	Catalogue
HPT stages	–	1	Catalogue
FPT stages	–	2	Catalogue

The parameters needed to be determined are low-pressure exit pressure and temperature, high-pressure compressor exit temperature, high-pressure turbine inlet pressure, high-pressure turbine exit pressure and temperature, low-pressure turbine exit pressure and temperature, power turbine exhaust pressure, each component's isentropic efficiency, and fuel flow rate. The input data shown in Table 1 were used to develop the engine performance model and to simulate all these parameters. Using the optimization constraints and parameters listed in Tables 2 and 3, the model has been optimized. Table 4 shows the objective function of the figure of merit.

Design-point performance was developed using the known input parameters collected from the catalogue and available literature, along with some engineering judgments [7, 30]. The secondary air system was taken into consideration; the percentage of the bleed air is depicted in Table 5.

The combustion chamber performance model was created using the energy balance equation, taking pressure loss into account:

$$\dot{m}_a h_3 + \dot{m}_f \times LHV \times \eta_{cc} = (\dot{m}_a + \dot{m}_f) h_4 \tag{1}$$

Table 2 Parameters used as constraint

Parameters	Minimum value	Optimized values	Maximum value
Heat rate	8852	10,040.2	10,043.5
Thermal efficiency	38	35.85	41
Exhaust temperature (T5)	733	752.076	769.7

Table 3 Parameters to be optimized

Variables	Min value	Optimized value	Max value
HPT Rotor 1 cooling air	0.03	0.051	0.054
HPT NGV 1 cooling air	0.04	0.061	0.065
IPT Rotor 1 cooling air	0.008	0.021	0.025
IPT NGV 1 cooling air	0.008	0.021	0.025
IPC isentropic efficiency	0.9	0.9	0.95
Exhaust pressure ratio	1	1.1620	1.2
HPT isentropic efficiency	0.89	0.8977	0.93
HPC isentropic efficiency	0.9	0.85	0.95
PT isentropic efficiency	0.89	0.8963	0.92
LPT isentropic efficiency	0.91	0.9125	0.94

Table 4 Objective function

Parameter	Value
Power output (kW)	26,025

Table 5 Secondary air fraction used during simulation

No.	Components	Unit	Amount
1	Overboard air	%	0.5
2	HPT NGV 1 cooling air	%	6.1
3	HPT Rotor 1 cooling air	%	5.1
4	IPT NGV 1 cooling air	%	2.1
5	IPT Rotor 1 cooling air	%	2.1

$$\dot{m}_f = \frac{\dot{m}_a(h_4 + h_3)}{\eta_{cc} \times \text{LHV} - h_4} \tag{2}$$

Compatibility of the components work:

$$W_{\text{HPC}} = W_{\text{HPT}}, W_{\text{LPC}} = W_{\text{LPT}}, W_{\text{PT}} = W_{\text{Load}} \tag{3}$$

where LHV is lower heating value, η_{cc} is combustion efficiency, $\dot{m}_a$ is inlet air mass flow rate, $\dot{m}_f$ is fuel mass flow rate, and h_3 is high-pressure compressor exit enthalpy. W_{LPC} is low-pressure compressor work, W_{HPC} is high-pressure compressor work, W_{IPT} is high-pressure turbine work, W_{PT} is power turbine work, and W_{Load} is the load.

The model was optimized till the validation was matched. Finally, the output of the design-point model was compared to the design parameters from the gas turbine product datasheet, and as shown in Table 6; it was determined that the model has minimal discrepancies from the engine manufacturer data. Due to the agreement to

Table 6 The design-point model output

Parameter	Units	Catalogue	GasTurb13 model	% Error
Power output	kW	26,025	26,025.5	0.0019
Thermal efficiency	%	35.8	35.8	0
Pressure ratio	–	20:1	20:1	0
Fuel flow rate	kg/s	–	1.53281	–
Lower heating value	MJ/kg	–	47.16	–
Turbine inlet temperature	°C	1193	1193	0
Exhaust temperature	°C	488	479.5	1.74
Heat rate	kJ/(kW*h)	10,043	10,040.2	0.027

the actual design values and very low variance, the design-point values produced by the GasTurb 13 simulation were found to be acceptable.

2.2 *Off-Design Performance Modelling*

Following a successful development of the cycle design-point calculations, an off-design model was developed. Off-design is the potential of the engine to run for the entire operating conditions including the design point. It is all about the state change in ambient conditions and engine load. For instance, the ambient temperature may change drastically from winter to summer which has a considerable impact on engine performance. As a result, the engine must perform efficiently in both design-point and off-design operating points. The initial step in off-design simulations is the adaptation of the target engine's design point using the scaling approach and already existing compressor and turbine mappings. The design point was scaled uniformly for each of the five component maps. Component matching is the second step in off design, and it is done by employing the Newton–Raphson iterative process to ensure mass flow and work compatibility [25, 31]. A suitable characteristics map comprising design-point data was selected for the alteration of the compressor and turbine maps. An additional coordinate was added to the component map digitization process, as suggested by Kurzke [32], to maintain an accurate component matching. Component maps illustrates the interrelationship between corrected mass flow rate, corrected rotational speed, isentropic efficiency, and pressure ratio. The primary variables in the compressor and turbine map were the intake mass flow rate (kg/s), rotational speed (rpm), inlet temperature (K), and inlet pressure to the compressor or turbine (kPa). The performance maps corrected speed, corrected mass flow, pressure ratio and isentropic efficiency were standardized using design-point values. In order to scale the map, for example, for a high-pressure compressor, the values read from the map table must be corrected for Reynolds number effects, using the Reynolds Number Index (RNI), to ensure alignment with the design-point efficiency and corrected flow.

The auxiliary point beta for the map scaling point in GasTurb was set by default to be equal to 0.5 and the corrected speed to be equal to 1.0. The design point (subscript *dp*) is matched with the map scaling point (subscript *R*, map), which is utilized as a reference point:

$$\eta_{\text{dp,map}} = \eta_{R,\text{map}} \cdot f_{\eta,RNI} \tag{4}$$

$$\left(W\sqrt{\Theta_R}/\delta\right)_{dp,\text{ map}} = \left(W\sqrt{\Theta_R}/\delta\right)_{R,\text{map}} \cdot f_{W,\text{RNI}} \tag{5}$$

where f_{η}, RNI is the Reynolds number index to correct the reference point efficiency read from the map, $\eta_{dp,\text{ map}}$ is the scaled map design-point efficiency, $(W\sqrt{(\Theta R)}/\delta)_{dp,\text{ map}}$ is the scaled map design point corrected mass flow, $\eta_{R,\text{ map}}$ is the reference point efficiency from unscaled map, $(W\sqrt{(\Theta R)}/\delta)_{dp,\text{ map}}$ is the reference point corrected mass flow from unscaled map. The corrected temperature (Θ) and corrected pressure (δ) are expressed below. $f_{W,\text{ RNI}}$ is the Reynolds number index to correct the reference point corrected mass flow read from the map.

$$\theta = \frac{T_o}{288.15\,K'} \tag{6}$$

$$\delta = \frac{P_o}{101.325\text{ KPa}} \tag{7}$$

where P_o and T_o is the inlet pressure and temperature. To be comparable with the design-point efficiency *dp*, the value taken from the map needs to be corrected for Reynolds number effects using the terms $f_{\eta,\text{ RNI}}$ and $f_{W,\text{ RNI}}$. Then, the map scaling factors will be calculated as follows by assuming $f_{\eta,\text{ RNI}}$ and $f_{W,\text{ RNI}}$, and commonly assumed as $f_{\eta,\text{ RNI}} = 0.99$ and $f_{\eta,\text{ RNI}} = 0.995$ [7]:

$$f_{\text{Mass}} = \frac{\left(\frac{W\sqrt{\Theta_R}}{\delta}\right)_{dp}}{\left(\frac{W\sqrt{\Theta_R}}{\delta}\right)_{R,\text{map}} \cdot f_{W,RNI}} \tag{8}$$

$$f_{Eff} = \frac{\eta_{dp}}{\eta_{R,\text{ map}} \cdot f_{\eta,\ RNI}} \tag{9}$$

$$f_{P_3/P_2} = \frac{(P_3/P_2)_{dp} - 1}{\left(\frac{P_3}{P_2}\right)_{R,\text{ map}} - 1} \tag{10}$$

$$f_{\text{Speed}} = \frac{1}{N_{R,\text{map}}} \tag{11}$$

where $f_{P3/P2}$, f_{Mass}, f_{Speed}, and f_{Eff} are pressure ratio, mass flow rate, speed scaling factors, and efficiency, respectively. $N_{R,\text{ map}}$ is compressor speed in the map. P_3 and

P_2 are compressor exit and inlet pressure. Once all these scaling factors have been established, the efficiency, corrected mass flow, pressure ratio, and corrected speed were multiplied by the scaling factors to scale the map.

At this point, the suitable component maps were chosen, and the cycle design point was matched with the selected map for the accurate off-design simulations. In order to ensure that the flow and operation of the connected individual components are compatible, GasTurb 13 matches the component by the interaction of components. A steady-state off-design operating line is typically built using the Newton-Raphson iterative methodology because of its simplicity in nonlinear systems [32]. The off-design model was created and validated using manufacturer data after the relevant component maps were chosen, and the cycle design point was correlated. Commercial software, GasTurb 13, was used to develop the off-design model. Catalogue data were used to validate the off-design model, including efficiency versus ambient temperature and power output versus ambient temperature. The VIGV scheduling and bleed air were taken into consideration in this study. The VIGV scheduling graph depicted in Fig. 2 was imported into the engine simulation software. This scheduling data was gathered from the engine manufacturers catalogue.

The off-design model was used to generate data after incorporating the power output versus ambient temperature and VIGV scheduling into the software. The generated data was then compared with the validation data. The validation data and the model that produces the efficiency versus ambient temperature and power output versus ambient temperature were matched closely. At each operational point, the power output versus ambient temperature error from the validation data was 0.02%, while 4.5% error was for the efficiency versus ambient temperature output data as shown in Figs. 3 and 4, respectively.

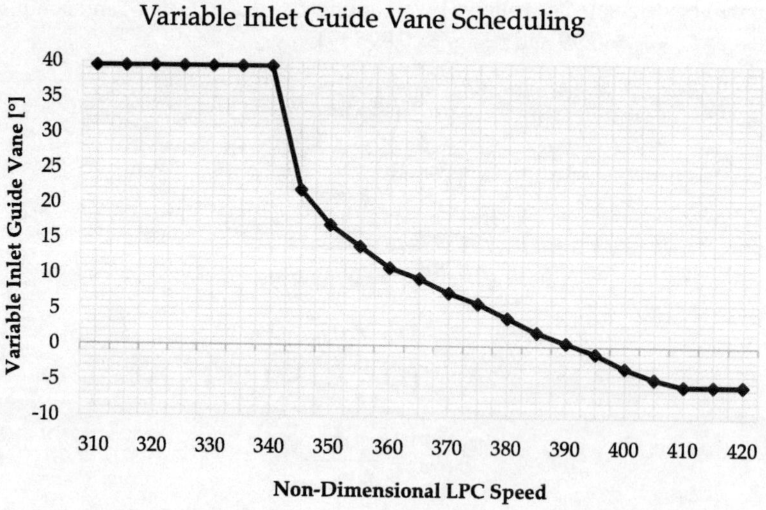

Fig. 2 RB211-24G VIGVs schedule ($N_1/\sqrt{T_0}$)

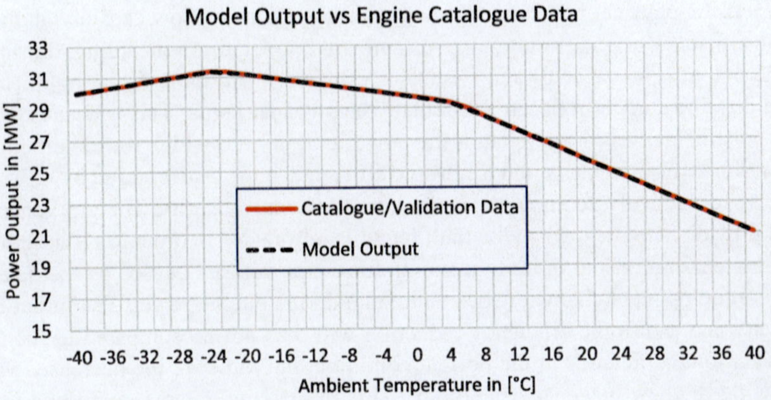

Fig. 3 Validation of off-design model with power output

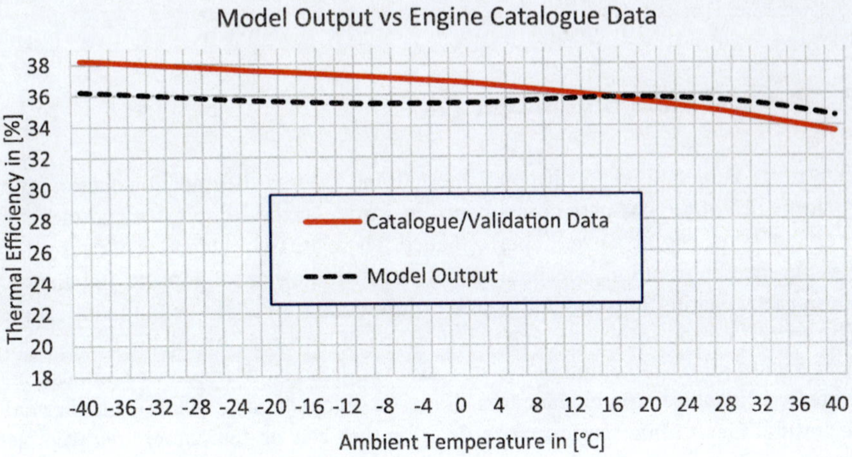

Fig. 4 Validation of the off-design model with thermal efficiency

Observation of Figs. 3 and 4 exhibited that the off-design model results closely matched the validation data. This indicates that the model is trustworthy and acceptable in predicting the gas turbine's performance. Another method of optimizing the off-design model was to directly schedule validation data into the GasTurb 13, such as efficiency versus ambient temperature and power output versus ambient temperature. The graphs of power output and efficiency, as well as the exhaust temperature in relation to ambient temperature, are identical in both approaches. However, the second method's shortcoming is that it is impossible to simulate gas turbine performance when the engine is operating at part-load. Due to the scheduled power output in the GasTurb 13 programme, the limiting value option for power out will not be active. If the physical faults are injected with manipulating flow capacity and isentropic efficiency at power output versus ambient temperature scheduling, the values of the output flow capacity and isentropic efficiency will be reduced significantly more than the amount reduced to simulate the physical faults. This is because more heat must be delivered to maintain the power output unit, which lowers efficiency. For instance, flow capacity and isentropic efficiency must be decreased by −7.5 and −2.5%, respectively, to simulate fouling at 100% fault severity. The gas generator must provide more heat when the fault factor is introduced in order to produce more high-temperature gas in order to make up for the power loss caused by the deterioration of component characteristics and maintain the output power. The initial flow capacity and isentropic efficiency reduction will thus not be compared to the flow and isentropic efficiency in the normal condition, but rather to the decreased value following the new power balance point. This means, two factors contribute to the change in component performance: the first is a fault, such as a decrease in flow capacity of 7.5% and an isentropic efficiency of 2.5%; the second is the addition of more heat to maintain the power output production.

3 Physical Fault Simulation

Due to its vulnerability to different internal and external degradation factors, the performance of the gas-path components, in particular the compressor and turbine, is crucial to the engine's overall performance. There are two categories of causes for gas turbine deterioration. Mechanical failures including misalignment, imbalance, loose parts, bearing failures, and a lack of lubricating oil could be the main cause. The second reason for gas turbine deterioration is performance-related issues such debris and fouling in compressors, corrosion, edge erosion, improper combustion, increased clearance around blade tips, domestic object damage (DOD), and thermal distortion. Gas turbine performance deterioration can be temporary (recoverable by washing) or permanent (non-recoverable by washing) deterioration. Permanent degradation requires replacement, whereas temporary degradation can be recovered during engine overhaul and operation. Fouling, erosion, corrosion, and blade tip clearance cause temporary deterioration, whereas airfoil distortion, untwist, and base distortions are permanent deterioration. The most frequent cause of gas turbine performance decline is the existence of fouling and erosion. Hence, fouling and erosion physical faults combined with variable inlet guide vane drift were simulated in this study. One of the most common causes for component deterioration is fouling, which is responsible for more than 70% of the overall loss of engine performance during the period of operation [5, 33]. It is known that during gas turbine operation, a volume of air containing air pollutant enters the engine. Using health parameters, it is possible to monitor the decline in gas turbine engine performance [34]. The performance changes caused by various defects are depicted by health parameters. The ratio of degraded component isentropic efficiency and mass flow and those that are in good condition component's isentropic efficiency and mass flow are called component health parameters as shown in Eqs. (12) and (13). It defines that gas turbine performance degradation occurred when the health parameter is less than 1, but not when it is greater than 1 [35]. $H_{\Gamma,i}$, $H_{\eta,i}$ are health parameters:

$$H_{\Gamma,i} = \frac{\dot{m}_i\sqrt{T_i}}{P_i} / \left(\frac{\dot{m}_i\sqrt{T_i}}{P_i}\right)_{\text{ref}} \tag{12}$$

$$H_{\eta,i} = \eta_i / \eta_i \text{ ref} \tag{13}$$

where P_i is total pressure $H_{\Gamma,i}$, $H_{\eta,i}$ are health parameters, $\dot{m}_i$ is flow rate, and η_i is isentropic efficiency.

To study of how physical faults affect gas turbine performance, physical faults were introduced into the model. The relationship between physical faults and performance parameter deviation was used to simulate physical faults. The method used to simulate physical faults required a deliberate deviation in flow capacity and isentropic efficiency from the normal condition by maintaining the relationship of physical faults and performance parameters are shown in Table 7. Prior to simulation and data generation from the model, the deviating values were scheduled in the software using the modifier option. As shown in Table 7, to simulate compressor fouling from 0 to 100% fouling severity level, mass flow must be purposefully decreased at intervals of 0 to −7.5%, and isentropic efficiency was also reduced at intervals of 0 to −2.5%. The flow capacity and isentropic efficiency relationship is 3:1 [35, 36]. For instance, the flow capacity and efficiency will drop by −0.75 and −0.25%, respectively, at a 10% low- and high-pressure compressor fouling severity level scenario. This means 10% of −7.5 and −2.5%. To simulate compressor erosion and turbine fouling simulation at 10% fault severity, the efficiency and mass flow must be reduced by −0.2 and −0.4% respectively. However, it is important to note that erosion in the turbine leads to an increase in flow capacity. Thus, to simulate turbine erosion at 10% fault severity, the isentropic efficiency was decreased by −0.2%, while mass flow was increased by 0.4%.

Table 7 Health parameters and physical faults relationships [35–37]

Physical fault	Flow capacity change (A)	Isentropic efficiency change (B)	Ratio A:B	Range
Compressor fouling	$\Gamma_C\downarrow$	$\eta_C\downarrow$	3:1	(0, −7.5%) (0, −2.5%)
Compressor erosion	$\Gamma_C\downarrow$	$\eta_C\downarrow$	2:1	(0, −4%) (0, −2%)
Turbine fouling	$\Gamma_T\downarrow$	$\eta_T\downarrow$	2:1	(0, −4%) (0, −2%)
Turbine erosion	$\Gamma_T\uparrow$	$\eta_T\downarrow$	2:1	(0, +4%) (0, −2%)

4 Results and Discussion

4.1 *The Combined Effect of Fouling with VIGV Drift and Erosion with VIGV Drift on the Component Isentropic Efficiency*

After developing and validating of the design-point and off-design performance model, fouling and erosion were simulated using the relationship between physical faults and health parameter as shown in Table 7. The GasTurb 13's Modifier option was used to implant the physical faults using isentropic efficiency and flow capacity. To simulate the engine model at a deteriorated condition, the compressor maps and turbine maps mut be updated using the following scaling factors [35]:

$$SF_{\Gamma,C} = 1 + \frac{\Delta H_{\Gamma,C}}{100} \tag{14}$$

$$SF_{\eta,C} = 1 + \frac{\Delta H_{\eta,C}}{100} \tag{15}$$

$$SF_{\Gamma,T} = 1 + \frac{\Delta H_{\Gamma,T}}{100} \tag{16}$$

$$SF_{\eta,T} = 1 + \frac{\Delta H_{\eta,T}}{100} \tag{17}$$

where $SF_{\eta,C}$ and $SF_{\Gamma,C}$ are the scaling factors of the compressor efficiency and flow capacity, $SF_{\Gamma,T}$and$SF_{\eta,T}$ describe the scaling factors of the turbine and compressor flow capacity and efficiency, while $\Delta H_{\Gamma,C}$ and $\Delta H_{\eta,C}$ are the change of compressor flow capacity and efficiency, $\Delta H_{\Gamma,T}$ and $\Delta H_{\eta,T}$ are the change of turbine flow capacity and efficiency. After implanting physical faults using the relations, the combined effect of fouling and variable inlet guide vane drift and erosion and variable inlet guide vane drift on component isentropic efficiency was investigated. Five different drift angles at 100% fouling and erosion fault severity level were simulated. The VIGV drift angles were −6.5°, −3.5°, 0°, 3.5°, and 6.5°. The part load ranges were from 100 to 60% incremented by 10%. The combined effect of fouling and VIGV drift and the combined effect of erosion and VIGV drift on the main components isentropic efficiency are shown in Fig. 5a–j.

Figure 5a–j shows the combined effect of variable inlet guide vane and fouling and the combined effect of variable inlet guide vane and erosion on the component isentropic efficiency at part-load operation, 100% load, 90% load, 80% load, 70% load, and 60% load. The fouling and erosion severity was at 100% and the VIGV drift angles were −6.5°, −3.5°, 0°, 3.5°, and 6.5°.

Figure 5a shows the deviation of LPC's isentropic efficiency when LPC fouling and VIGV drift occur at the same time. The results show that the highest deviation of LPC isentropic efficiency was recorded when VIGV drift was at −6.5° and 100%

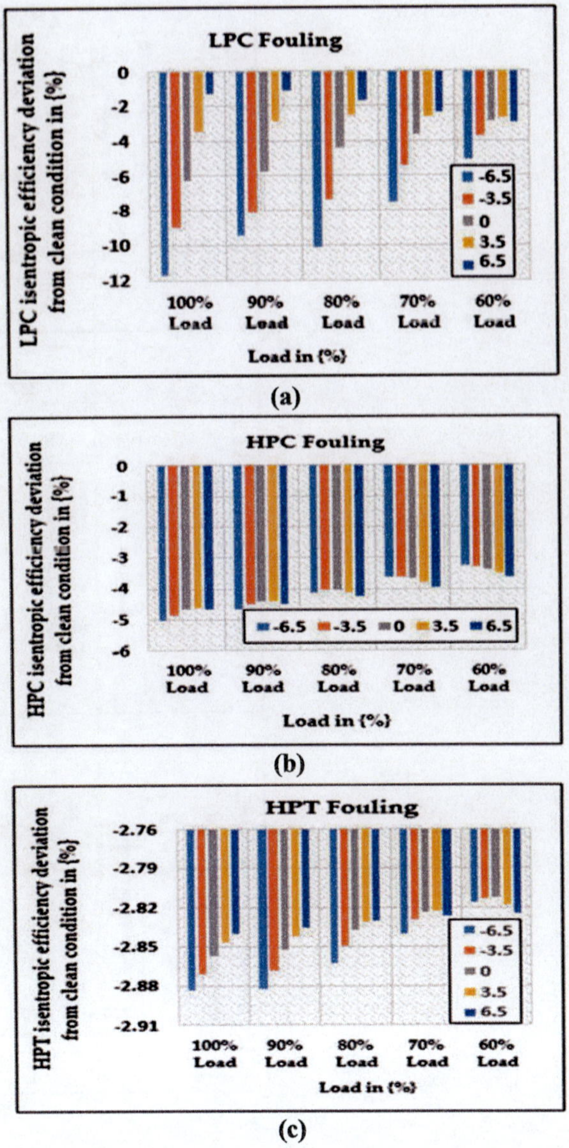

Fig. 5 Deviation of component isentropic efficiency: **a** LPC fouling with VIGV drift, **b** HPC fouling with VIGV drift, **c** HPT fouling with VIGV drift, **d** LPT fouling with VIGV drift, **e** PT fouling with VIGV drift, **f** LPC erosion with VIGV drift, **g** HPC erosion with VIGV drift, **h** HPT erosion with VIGV drift, **i** LPT erosion with VIGV drift, and **j** PT erosion with VIGV drift

fouling severity existed at the same time, while the lowest deviation was shown when VIGV drift was at 6.5° and 100% fouling severity existed at the same time. The result also shows that the deviation of lower drift is greater than the deviation of up-VIGV drift for each load. Figure 5b shows the deviation of HPC's isentropic efficiency when HPC fouling and VIGV drift occur at the same time. The results show that, although the deviation magnitude is different, the trend is similar to that of LPC's isentropic deviation. The highest and lowest deviation occurred at VIGV angle −6.5°

Fig. 5 (continued)

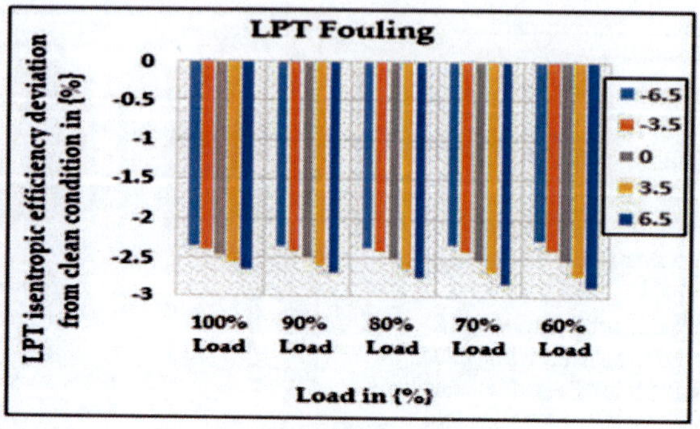

(d)

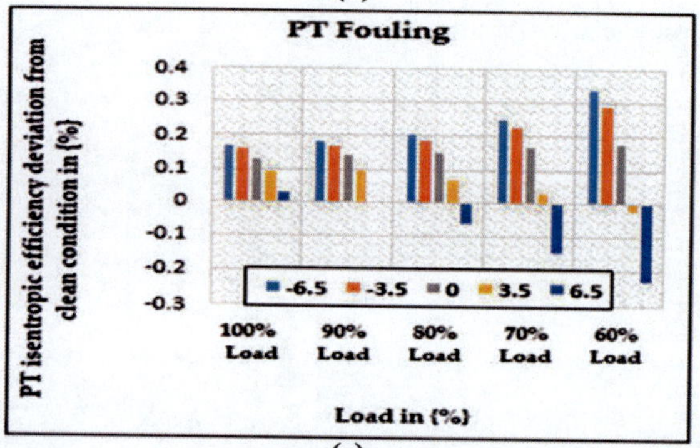

(e)

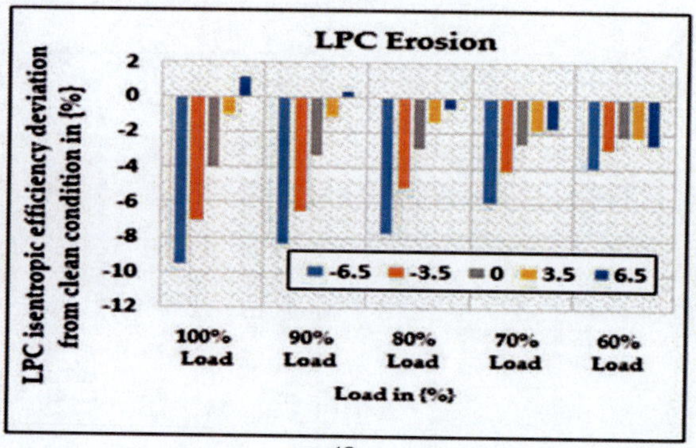

(f)

Fig. 5 (continued)

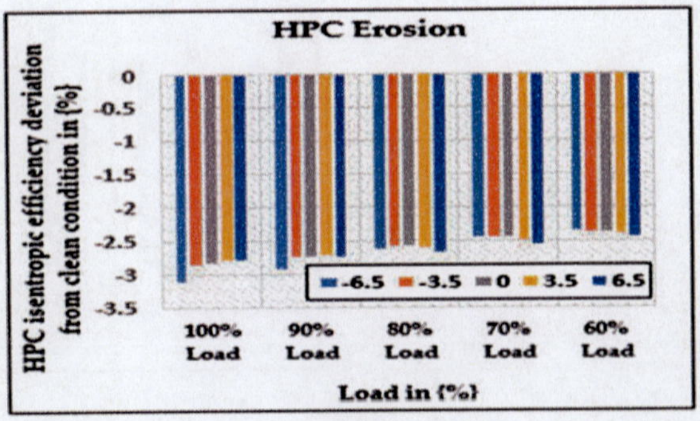

(g)

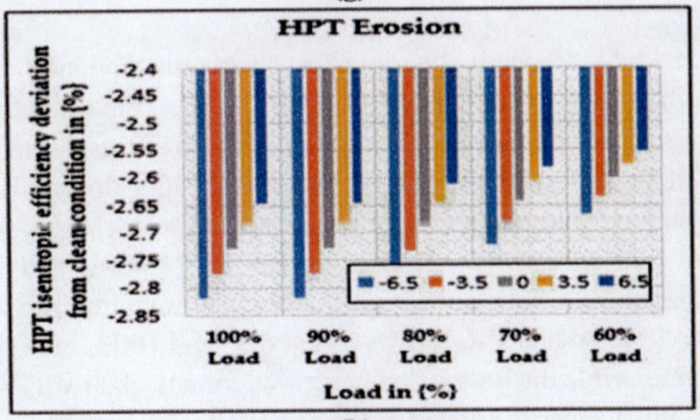

(h)

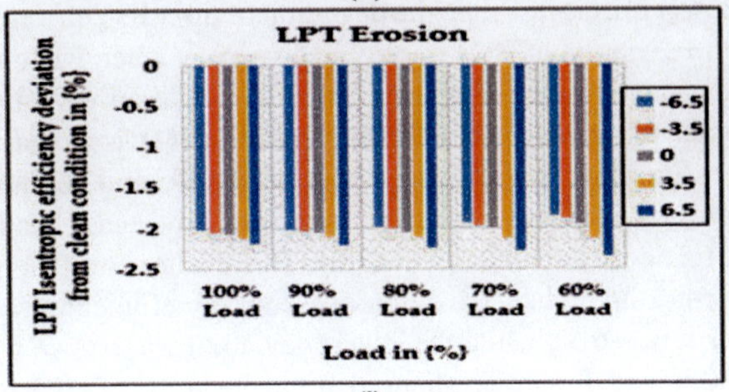

(i)

Fig. 5 (continued)

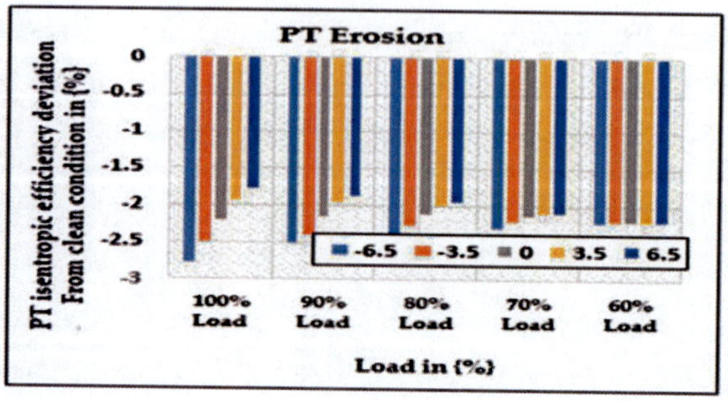

(j)

and 6.5°, respectively. Figure 5c shows the deviation of HPT's isentropic efficiency when HPT fouling and VIGV drift occur at the same time. The results show that the isentropic efficiency deviation at each VIGV drift and fouling is between −2.8 and 2.88%. This shows that the variation of VIGV drift on HPT isentropic efficiency does not have that much effect on increasing the deviation. Figure 5d shows the deviation of LPT's isentropic efficiency when LPT fouling and VIGV drift occur at the same time. The results show that the highest deviation of LPT isentropic efficiency was recorded when VIGV drift was at 6.5° and 100% fouling severity existed at the same time, while the lowest deviation was shown when VIGV drift was at −6.5° and 100% fouling severity existed at the same time. The result also shows that the deviation of lower drift is less than the deviation of up-VIGV drift for each load. Figure 5e shows the deviation of PT's isentropic efficiency when PT fouling and VIGV drift occur at the same time. The results show the deviation is minimal. At 100% load and 90% load, the deviation of PT isentropic efficiency is increasing at each VIGV drift angle.

Figure 5f, g shows the deviation of LPC and HPC isentropic efficacy when LPC erosion and VIGV drift occur concurrently, and when HPC erosion and VIGV drift occur concurrently, respectively. The results show that in both LPC and HPC components, the highest deviation of isentropic efficiency was recorded at a down VIGV drift of −6.5°, while the lowest deviation was shown when VIGV drift was at 6.5 at each load. Figure 5h shows that the deviation of HPT isentropic efficacy when HPT erosion and VIGV drift occur concurrently, and the result shows the same trend as the trend shown in the LPC and HPC isentropic deviation due to the combined effect of erosion and VIGV drift. Figure 5i shows the deviation of LPT's isentropic efficiency when LPT erosion and VIGV drift occur at the same time. The results show that the highest deviation of LPT isentropic efficiency was recorded when VIGV drift was at 6.5° and 100% erosion severity existed at the same time, while the lowest deviation was shown when VIGV drift was at −6.5° and 100% erosion severity existed at the same time. The result also shows that the deviation of lower drift is less than the deviation of up-VIGV drift for each load. Figure 5j shows that the deviation of PT isentropic efficacy when HPT erosion and VIGV drift occur concurrently, and the result shows the same trend as the trend shown in the LPC, HPC, and HPT isentropic deviation due to the combined effect of erosion and VIGV drift. But the deviation magnitude is different and minimal in PT.

After simulating the combined effect of variable inlet guide vane and fouling and the combined effect of variable inlet guide vane and erosion on the component isentropic efficiency at part-load operation, the combined effect of variable inlet guide vane and fouling and the combined effect of variable inlet guide vane and erosion on gas turbine output parameters such as low-pressure compressors exit temperature, low-pressure compressor exit pressure, high-pressure compressor exit pressure, high-pressure turbine exit pressure, high-pressure compressor exit temperature, low-pressure turbine exit pressure, fuel flow, power turbine exit temperature, low-pressure speed, high-pressure speed at part-load operation, 100% load, 90% load, 80% load, 70% load, and 60% load has been simulated. The fouling and erosion severity was at 100% and the VIGV drift angles were –6.5°, –3.5°, 0°, 3.5°, and 6.5°. The results are shown in Fig. 6a–j.

Figure 6a–j shows how output parameters change at different loads, 100% load, 90% load, 80% load, and 70% load, when fouling and VIGV drift are concurrently existed. The fouling and erosion severity were applied at 100% and the VIGV drift angles, –6.5°, –3.5°, 0°, 3.5°, and 6.5° were considered and the fouling and erosion severity is at 100%.

Figure 6a shows the deviation of output parameters when LPC fouling and VIGV drift occur at the same time. The results show that the deviation of T5, SFC, TIT, and HR was increasing while the deviation of ηth and PR was decreasing at each load and VIGV drifts. The highest deviation of the output parameters observed when VIGV drift was at –6.5° and 100% fouling severity existed at the same time, while the lowest deviation was shown when VIGV drift was at 6.5 and 100% fouling severity existed at the same time for all parameters, at each load and VIGV drift. Figure 6b shows the deviation of output parameters when HPC fouling and VIGV drift occur at the same time. The results show the same trend as the output parameter deviation shown in the LPC is due to the combined effect of fouling and VIGV drift. But the magnitude of output parameter deviation was lower than the deviation shown due to LPC faults. Figure 6c shows the deviation of output parameters when HPT fouling and VIGV drift occur at the same time. The results show that, except for the up-VIGV drift angle of 6.5°, the deviation of T5, SFC, PR, TIT, and HR was increasing while the deviation of ηth was decreasing at each load and VIGV drift. The highest deviation of the pressure ratio was observed when VIGV drift was at 6.5 and 100% fouling severity existed at the same time, while the highest deviation for other parameters was shown when VIGV drift was at –6.5° at each load and VIGV drifts and 100% fouling severity existed at the same time. Figure 6d depicts the output parameter deviation when LPT fouling and VIGV drift occur concurrently, and Fig. 6e depicts the output parameter deviation when PT fouling and VIGV drift occur concurrently. Both the LPT fouling and PT fouling graphs show the same trend as the output parameter deviation shown in the LPC and HPC due to the combined effect of fouling and VIGV drift. But the magnitude of exhaust temperature deviation is increased due to PT faults while it is minimal due to LPT faults.

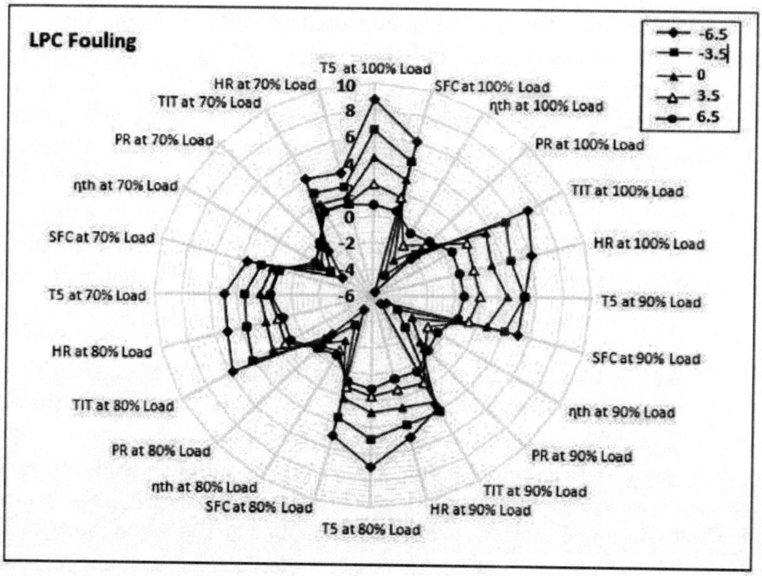

(a)

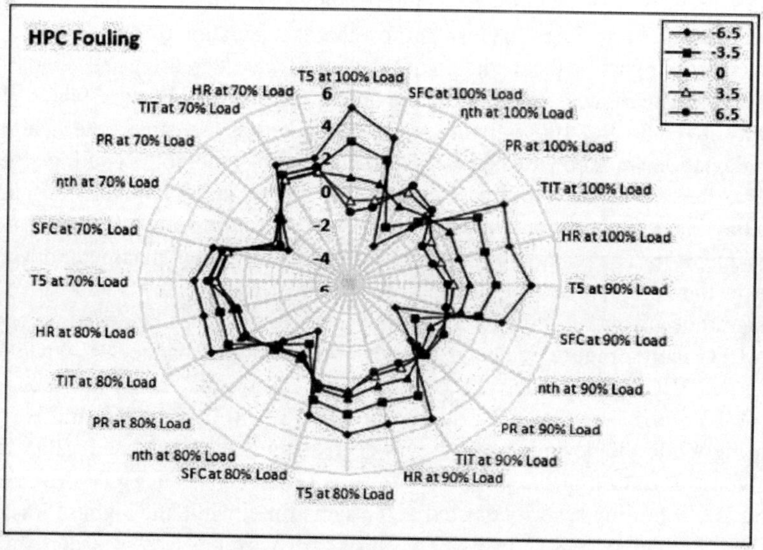

(b)

Fig. 6 Deviation of gas turbine output parameters: **a** LPC fouling with VIGV drift, **b** HPC fouling with VIGV drift, **c** HPT fouling with VIGV drift, **d** LPT fouling with VIGV drift, **e** PT fouling with VIGV drift, **f** LPC erosion with VIGV drift, **g** HPC erosion with VIGV drift, **h** HPT erosion with VIGV drift, **i** LPT erosion with VIGV drift, and **j** PT erosion with VIGV drift

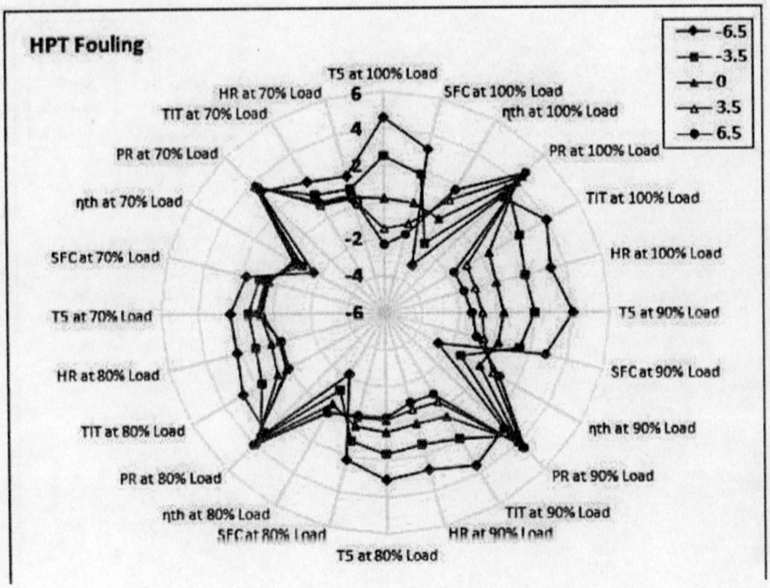

(c)

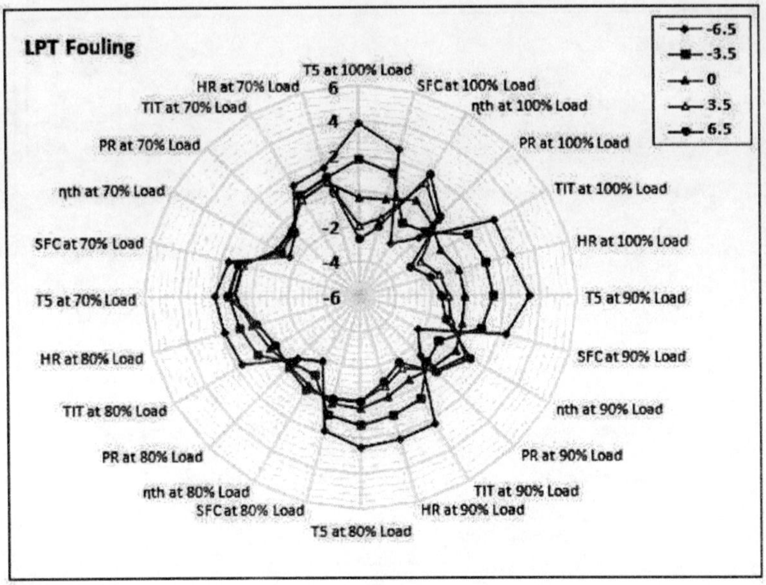

(d)

Fig. 6 (continued)

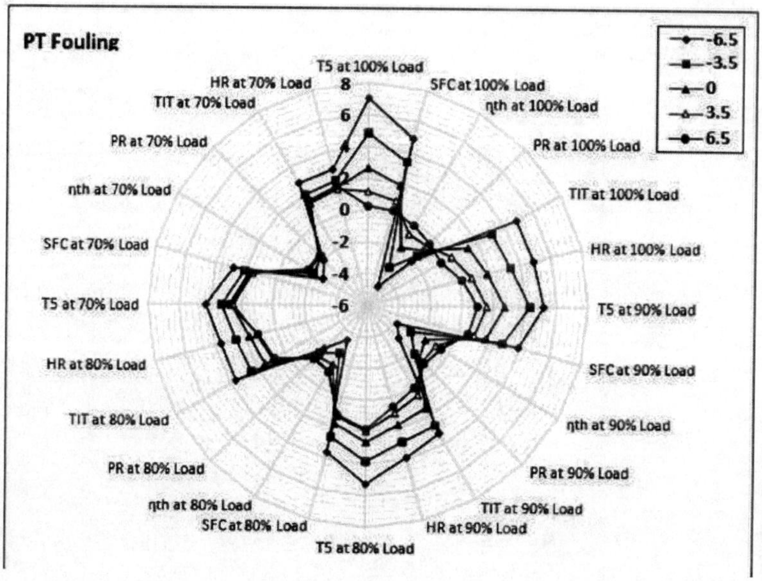

(e)

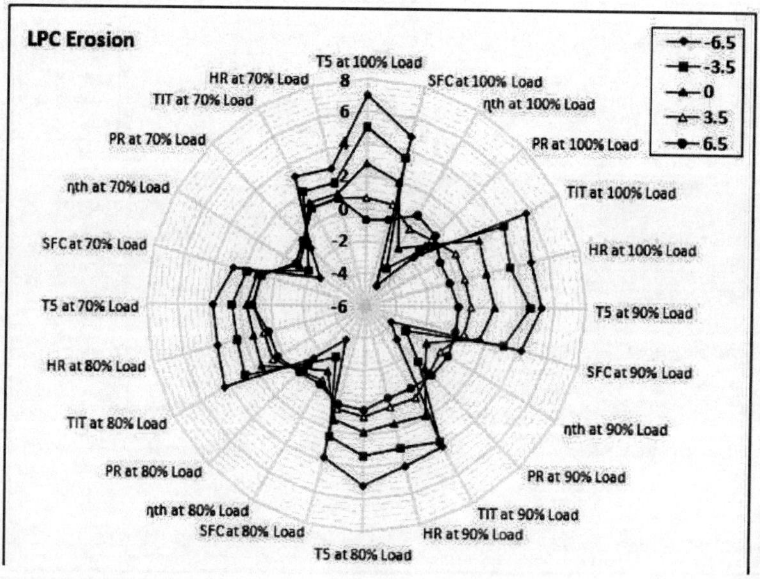

(f)

Fig. 6 (continued)

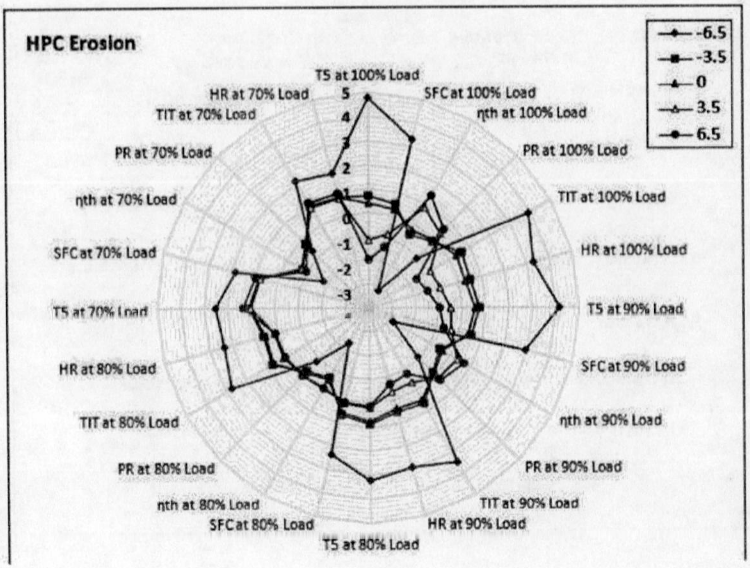

(g)

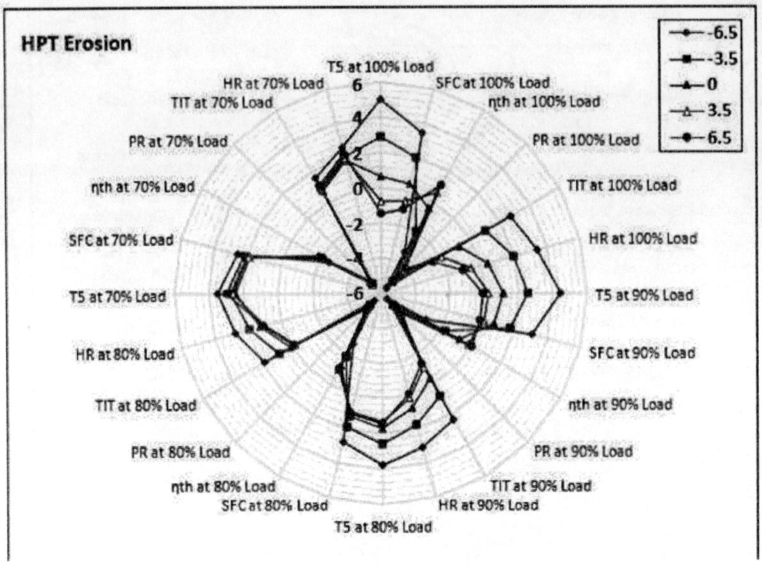

(h)

Fig. 6 (continued)

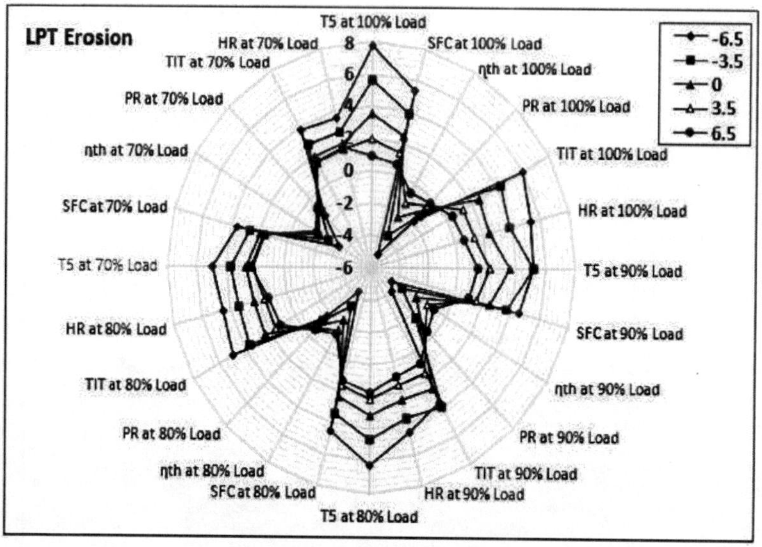

(i)

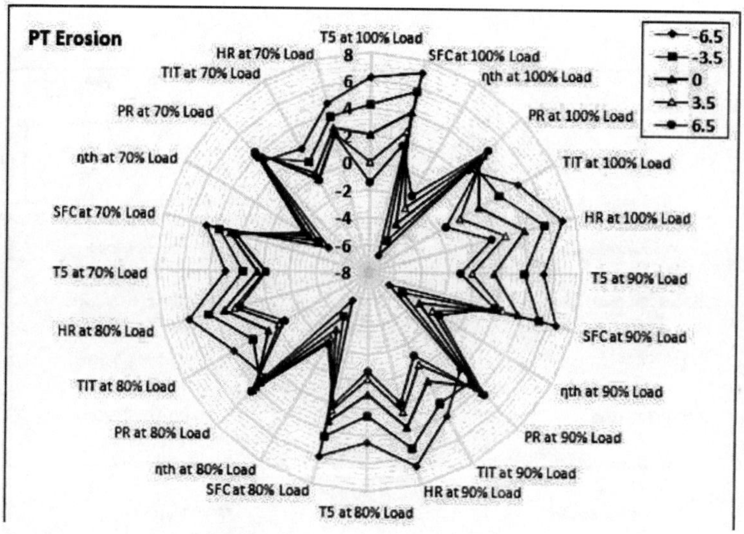

(j)

Fig. 6 (continued)

Figure 6f shows the deviation of output parameters when LPC erosion and VIGV drift occur at the same time. The results show that, except for the VIGV drift of 6.5°, the deviation of T5, SFC, TIT, and HR was increasing while the deviation of ηth and PR was decreasing at each load and VIGV drifts. It is also observed that the highest deviation of the output parameters occurred when VIGV drift was at −6.5 and 100% erosion severity existed at the same time, while the lowest deviation was shown when VIGV drift was at 6.5 and 100% erosion severity existed at the same time for all parameters, at each load and VIGV drift. Figure 6g shows the deviation of output parameters when HPC erosion and VIGV drift occur at the same time. The results show the same trend as the output parameter deviation shown in the LPC due to the combined effect of erosion and VIGV drift. But the magnitude of output parameter deviation was lower than the deviation shown due to LPC faults. Figure 6h shows the deviation of output parameters when HPT erosion and VIGV drift occur at the same time. The results show that the ηth and PR deviation were decreasing at VIGV drift angles of −6.5°, −3.5°, and 0°, where they increased at 3.5° and 6.5° VIGV drifts. The deviation of T5, SFC, and TIT HR was increasing at VIGV drift angles of −6.5°, −3.5°, and 0°, where they decreased at 3.5° and 6.5° VIGV drifts. Figure 6i depicts the output parameter deviation when LPT erosion and VIGV drift occur concurrently, and the result follows the same trend as the output parameter deviation shown when HPC erosion and VIGV drift are combined and existed. But the output parameter deviation due to LPT erosion and VIGV drift is greater than the output parameter deviation due to the combined effect HPC erosion and VIGV drift, at each load and VIGV drift. Figure 6j depicts the output parameter deviation when PT erosion and VIGV drift occur concurrently. The result shows that the deviation of ηth is decreasing at each load and VIGV drift angles while the deviation of other parameters was increasing, except T5 and TIT at a VIGV drift angle of 6.5°.

5 Conclusion

The results showed the upstream components are more affected by fouling, while the downstream components are affected by erosion. It is also seen that VIGV drift has a significant impact on the gas turbine performance. When component fouling and VIGV drift are combined and exist, and when component erosion and VIGV drift are combined and exist, the highest isentropic efficiency deviation is shown in the LPC, while the lowest deviation is shown in the LPT. The combined effect of fouling and up-VIGV drift and the combined effect of erosion and up-VIGV drift show a small deviation due to offsetting the isentropic efficiency decline caused by fouling and the isentropic efficiency decline caused by erosion. Furthermore, the combination of fouling and down-VIGV drift, as well as erosion and down-VIGV drift, reduced the component's isentropic efficiency. It was clearly observed that the deviation of component isentropic efficiency and output parameters increases as the load increases. The combined effect of fouling and down-VIGV drift, and the combined effect of erosion and down-VIGV drift caused the exhaust temperature to have the highest deviation at each load and VIGV drift angles. In general, in this paper, the deviation of isentropic efficiency and output parameters due to the combined effects of fouling and VIGV drift, and the combined effects of erosion and VIGV drift at part-load operation are studied and discussed. The results will be very helpful in developing a fault detection, isolation, and identification model. The authors suggest that future research could use the results presented in this paper to develop a diagnosis model using machine learning.

Acknowledgements For providing the resources, the authors are thankful to Universiti Teknologi PETRONAS.

Author Contributions Z. A. A. K., T. A. L., K. G. K., and A. D. F. reviewed the manuscript, made critical revisions, and analysed the data. W. M. S. created the model, ran the simulation, produced the graphs, and wrote the draught paper. Additionally, Z. A. A. K. and T. A. L. provided the funds necessary to purchasing the commercial software, GasTurb 13. The manuscript's current form was approved by all the authors after they read it.

Funding Universiti Teknologi PETRONAS provided funding for this study under grant number YUTP-FRG, Cost Center 015LC0-347.

Conflicts of Interest The authors of the current work do not have conflict of interest.

Data Availability Statement Not applicable.

Nomenclature

HPC	High-pressure compressor
LPC	Low-pressure compressor
LPT	Low-pressure turbine
HPT	High-pressure turbine
PT	Power turbine
T24	Low-pressure compressor exit temperature
P24	Low-pressure compressor exit pressure
P3	High-pressure compressor exit pressure
P43	High-pressure turbine exit pressure
T3	High-pressure compressor exit temperature
P47	Low-pressure turbine exit pressure
FF	Fuel flow
T5	Power turbine exit temperature
N1	Low-pressure speed
GT	Gas Turbine
N2	High-pressure speed
CC	Combustion chamber
VIGV	Variable inlet guide vane

NGV	Nozzle guide vane
VAN	Variable area nozzle
VSV	Variable stator vane
VBV	Variable bleed valve
RVDT	Rotary variable displacement transducer
RNI	Reynolds number index
DOD	Domestic object damage
dp	Design point

References

1. Li Z, Zhong SS, Lin L (2017) Novel gas turbine fault diagnosis method based on performance deviation model. J Propul Power 33(3):730–739. https://doi.org/10.2514/1.B36267
2. Merrington G, Kwon OK, Goodwin G, Carlsson B (1990) Fault detection and diagnosis in gas turbines. In: Proceedings of the ASME turbo expo, vol 5. https://doi.org/10.1115/90-GT-339
3. Saravanamuttoo HIH, Lakshminarasimha AN (1985) A preliminary assessment of compressor fouling. Am Soc Mech Eng
4. Mishra RK (2015) Fouling and corrosion in an aero gas turbine compressor. J Fail Anal Prev 15(6):837–845. https://doi.org/10.1007/s11668-015-0023-8
5. Diakunchak IS (1991) Performance deterioration in industrial gas turbines. In: Proceedings of the ASME Turbo Expo, vol 4. https://doi.org/10.1115/91-gt-228
6. Singh Grewal N (1988) Gas turbine engine performance deterioration modelling and analysis supervisor
7. G. Gmbh (2018) GasTurb 13 design and off-design performance of gas turbines, 85221 Dachau, Max Feldbauer Weg 5, Germany, 2018. [Online]. www.gasturb.de. Accessed 2 Aug 2022
8. North Atlantic Treaty Organisation (2007) Performance prediction and simulation of gas turbine engine operation for aircraft, marine, vehicular, and power generation (Estimation et simulation des performances du fonctionnement), vol 323
9. Tahan M, Tsoutsanis E, Muhammad M, Abdul Karim ZA (2017) Performance-based health monitoring, diagnostics and prognostics for condition-based maintenance of gas turbines: a review. Appl Energy Elsevier Ltd 198:122–144. https://doi.org/10.1016/j.apenergy.2017.04.048
10. Haglind F (2010) Variable geometry gas turbines for improving the part-load performance of marine combined cycles—gas turbine performance. Energy 35(2):562–570. https://doi.org/10.1016/j.energy.2009.10.026
11. Kim JH, Song TW, Kim TS, Ro ST (2002) Dynamic simulation of full start-up procedure of heavy-duty gas turbines. J Eng Gas Turbines Power 124(3):510–516. https://doi.org/10.1115/1.1473150
12. Bringhenti C, Takachi Tomita J, de Souza Júnior F, Barbosa JR (2006) Gas turbine performance simulation using an optimized axial flow compressor. In: Power for land, sea, and air, Barcelona, Spain. [Online]. https://doi.org/10.1115/GT2006-91225. Accessed 12 Aug 2022
13. Salilew WM, Abdul Karim ZAA, Lemma TA, Fentaye AD, Kyprianidis KG (2022) Predicting the performance deterioration of a three-shaft industrial gas turbine. Entropy 24(8):1052. https://doi.org/10.3390/e24081052
14. Song TW, Kim TS, Kim JH, Ro ST (2001) Performance prediction of axial flow compressors using stage characteristics and simultaneous calculation of interstage parameters. Proc Inst Mech Eng Part A J Power Energy

15. Salar A, Hosseini SM, Sedigh AK, Zangmolk BR (2010) Improving VIGV predictive monitoring by developing a failure mode virtual sensor. In: Proceedings—UKSim 4th European modelling symposium on computer modelling and simulation, EMS2010, pp 240–247. https://doi.org/10.1109/EMS.2010.47
16. Tsalavoutas A, Mathioudakis K, Stamatis A, Smith M (2022) Identifying faults in the variable geometry system of a gas turbine compressor, Munich, Germany. [Online]. https://doi.org/10.1115/1.1330267. Accessed 12 Aug 2022
17. Stamatis A, Mathioudakis K, Papailiou KD (1989) Adaptive simulation of Gas Turbine performance. In: Proceedings of the ASME Turbo Expo, vol 1. https://doi.org/10.1115/89GT205
18. Cruz-Manzo S, Maleki S, Panov V, Zhang YE (2018) Performance analysis of a twin-shaft gas turbine with fault in the variable stator guide vane system of the axial compressor. Belgium, Brussels
19. Razak AMY, Dosanjh MS (2002) Application of an advanced performance monitoring system to detect an implanted fault on a twin spool aero derived gas turbine. Amsterdam, The Netherlands. https://doi.org/10.1115/GT2002-30022
20. Enyia J, Thank-God I, Igbong D, Diwa J (2015) Industrial gas turbine on-line compressor washing for power generation. Int J Eng Res Technol (IJERT)
21. Ajoko, Tolumoye J (2014) Performance monitoring of industrial gas turbine. Int J Eng Sci Invent 3(5):62–68
22. Salilew WM, Abdul Karim ZAA, Lemma TA, Fentaye AD, Kyprianidis KG (2023) Three shaft industrial gas turbine transient performance analysis. Sensors 23(4):1767. https://doi.org/10.3390/s23041767
23. Molla Salilew W, Ambri Abdul Karim Z, Alemu Lemma T (2022) Investigation of fault detection and isolation accuracy of different Machine learning techniques with different data processing methods for gas turbine. Alexandria Eng J 61(12):12635–12651. https://doi.org/10.1016/j.aej.2022.06.026
24. Razak AMY (2007) Industrial gas turbines: performance and operability. Woodhead Pub
25. Hashmi MB, Lemma TA, Karim ZAA (2019) Investigation of the combined effect of variable inlet guide vane drift, fouling, and inlet air cooling on gas turbine performance. Entropy 21(12). https://doi.org/10.3390/e21121186
26. Razak AMY (2013) Gas turbine performance modelling, analysis and optimisation. Woodhead Publishing Limited. https://doi.org/10.1533/9780857096067.3.423
27. kurkz (2020) Design-point calculations of industrial gas turbines. ASME, pp 376–397
28. Ao SI, Gelman L, Hukins DWL, Hunter A, Korsunsky A, International Association of Engineers (2018) Design and off-design operation and performance analysis of a gas turbine. In: Proceedings of the world congress on engineering, vol II. London, U.K. ISSN: 2078-0966 (Online) WCE 2018
29. Jasmani MS, Li YG, Ariffin Z (2011) Measurement selections for multicomponent gas path diagnostics using analytical approach and measurement subset concept. J Eng Gas Turbines Power 133(11). https://doi.org/10.1115/1.4002348
30. Gao JH, Huang YY (2011) Modeling and simulation of a aero turbojet engine with GasTurb. In: Proceedings—2011 international conference on intelligence science and information engineering, ISIE 2011, pp 295–298. https://doi.org/10.1109/ISIE.2011.149
31. Chen YZ, Zhao XD, Xiang HC, Tsoutsanis E (2021) A sequential model-based approach for gas turbine performance diagnostics. Energy 220. https://doi.org/10.1016/j.energy.2020.119657
32. Kurzke J (2007) Simplifications in gas turbine performance calculations. In: ASME turbo expo 2007: power for land, sea and air, Montreal, Canada
33. Boyce MP, Meher-Homji CB, Lakshminarasimha AN (1994) Modeling and analysis of gas turbine performance deterioration. ASME, Boyce Engineering International Inc.
34. Ntantis E (2008) Capability expansion of non-linear gas path analysis. PhD thesis, Cranfield University
35. Qingcai Y, Li S, Cao Y, Zhao N (2016) Full and part-load performance deterioration analysis of industrial three-shaft gas turbine based on genetic algorithm. In: Proceedings of ASME turbo expo 2016: turbomachinery technical conference and exposition, South Korea

36. Mohammadi E, Montazeri-Gh M (2014) Simulation of full and part-load performance deterioration of industrial two-shaft gas turbine. J Eng Gas Turbines Power 136(9). https://doi.org/10.1115/1.4027187
37. Salilew WM, Abdul Karim ZAA, Lemma TA, Fentaye AD, Kyprianidis KG (2022) The effect of physical faults on a three-shaft gas turbine performance at full- and part-load operation. Sensors 22(19):7150. https://doi.org/10.3390/s22197150

Assessment of Heat Exchanger Tube Bundle Using STELLAR: Industrial Case Study

M. Adnan Ayob, Sia Hua Jiuh, A. R. Othman, and M. Mohammad

Abstract A step change is inevitable in maintenance and equipment management strategy to mitigate risk for failures. This paper presents the assessment tool developed based on the statistical approach by the probability analysis to access the reliability and remaining life of heat exchanger tube bundles. The assessment includes data sanitization involving screening, classification and validation of tube testing data. Subsequently, statistical analysis of the probability of failure, remaining life and the risk matrix on the heat exchanger tube bundle are determined. The case study for this paper includes 57 heat exchangers operated in petrochemical plant, with the healthiness of the equipment was assessed and discussed. The findings indicated that 22 (38%) out of 57 heat exchangers required high focus and priority in maintenance plan, cleaning and inspection during next opportunity. The results could be used for inputs in planning schedule for heat exchanger management to ensure effective maintenance strategies to prevent equipment failure.

Keywords Statistic · STELLAR · Heat exchanger · Tube bundle

M. A. Ayob (✉)
Petronas Chemicals MTBE Sdn Bhd, Kawasan Industrial Gebeng, Lot 111, 26080 Kuantan, Pahang, Malaysia
e-mail: adnan_ayob@petronas.com

S. H. Jiuh
Technical Services Department, ASEAN Bintulu Fertilizer Sdn Bhd, 18th Km Tanjung Kidurong, 97008 Bintulu, Sarawak, Malaysia
e-mail: sia_huajiuh@petronas.com

A. R. Othman · M. Mohammad
Faculty of Engineering, Mechanical Department, Universiti Teknologi PETRONAS, Seri Iskandar, Malaysia
e-mail: rahim.othman@utp.edu.my

M. Mohammad
e-mail: mazian.mohammad@utp.edu.my

F. Ahmad et al. (eds.), *ICREEM 2022*, Lecture Notes in Mechanical Engineering,
https://doi.org/10.1007/978-981-99-5946-4_17

1 Introduction

The operation of oil and gas sectors provides challenges to keep the assets in good condition and operating at optimal level for sustainable plant performance. High equipment reliability and availability is the pre-requisite for world class plant utilization, in ensuring high volume production and subsequently generating cash and profits. Heat exchanger representing more than 35% of the exposed surface received only 1% of the resources of inspection and produced 50% of the business interruption costs [1]. Therefore, preventing unplanned shutdown due to heat exchanger failure is the main priority of equipment strategy. Previously, statistical analysis has been utilized as a mean to assess leaks and failures of fixed equipment in process industries. Reference [2] has presented a simplified practical approach for the use of statistical tools for inspection planning and integrity assessment, which focused on corrosion-related material degradation of piping on an offshore production facility. The implementation of extreme value analysis (EVA) for assessment of the minimum remaining thickness of tubes in industrial heat exchangers, including air-cooled heat exchangers and shell-and-tube heat exchangers has also been carried out by Wang [3].

Statistical analyses have been utilized as a mean to assess failures of heat exchanger in a process industry. These current typical reliability studies mainly driven by historical reliability and reactive responses for corrective strategies. However, the application of these methods is not common, largely because the use of statistics requires specialist knowledge, and because lack of specific standard reference exists [4]. Apart from that, the existing standard practice of API 581 provides broad guidelines on inspection program using risk-based methods for pressurized fixed equipment including pressure vessel and heat exchanger tube bundles [5]. Nonetheless, the user's typical approaches may vary from run-to-failure and maintenance or repair for turnaround activities. These practices have proven to be ineffective for minimizing the impact on business interruption, such as delay in turnaround as well as unplanned shutdown.

In this study, a tool to assess the inspection data, reliability and remaining life of heat exchanger tube bundles using the statistical approach by probability analysis was proposed. This tool is known as Statistical Tube Exchanger Analytic and Reliability (STELLAR). STELLAR provides the methodology to assess the accuracy of inspection data, reliability and remaining life of heat exchanger tube bundles using the statistical approach by probability analysis. This statistical method, when combined with existing reliability methods, offers the potential for better use of inspection results and for the prediction of the probability of future leaks and the remaining useful life of the equipment.

2 Methodology

STELLAR specifies the historical NDT tube testing data as well as condition of the tube bundles up to the current state. The methodology assessed the historical and current tube condition and subsequently provided the prediction of HEX life. Figure 1 shows the flowchart of STELLAR. The main data input required for the analysis comprises of the historical tube wall loss percentage obtained from non-damage testing (NDT) inspection report. The NDT techniques may vary to include Internal Rotary Inspection System (IRIS), Remote Field Testing (RFET), Acoustic Eye (AE), Eddy Current (ECT) and Magnetic Flux Leakage (MFL).

The fitness of NDT data was statistically evaluated using $h < h_{-crit}$ of Mandel test and Large Extreme Value (LEV) scale and location parameters, as per ASTM G16-13 and ASTM E-691, respectively [6, 7]. The probability analysis of corrosion rate (CR) at the tube bundles was characterized based on 95% confidence interval

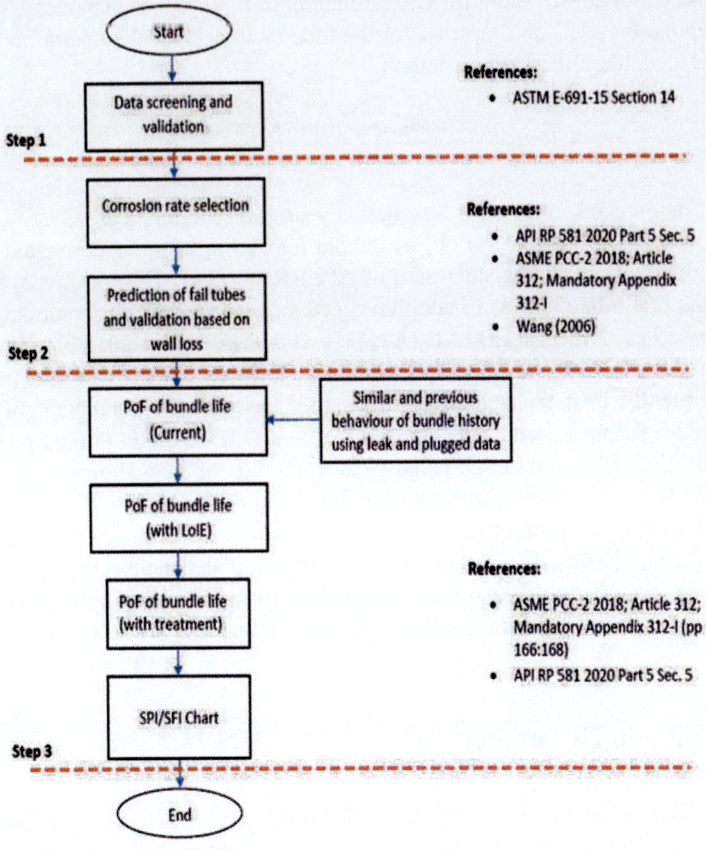

Fig. 1 Methodology in STELLAR analysis

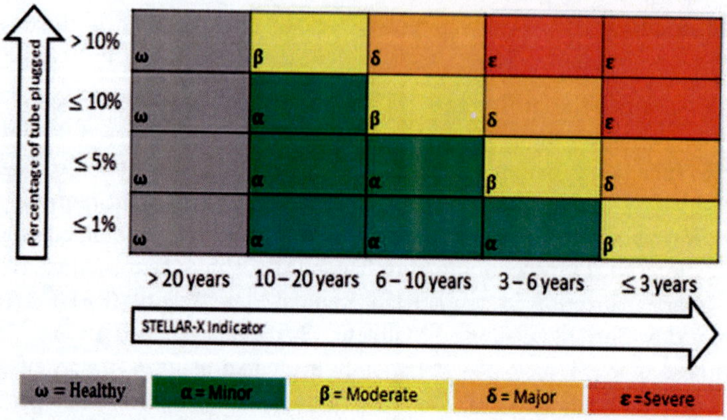

Fig. 2 STELLAR index indicator to classify the severity of heat exchanger

(CI) at 90% probability value for CR estimation to represent the characteristics of tubes population [8]. Long-term CR of the tube bundles over the operation period was used using the following formula:

$$CR = \frac{\text{thickness}_{\text{nominat}} - \text{thickness}_{\text{inspection}}}{\text{Duration}} \tag{1}$$

where duration is taken from the installation year to inspection year.

The probability of failure (PoF) of bundle remaining life was computed using Weibull distribution plot as suggested in API RP 581 3RD ED (A1) (A2) [8], to predict the first tube failures. In addition, Level of Inspection Effectiveness (LoIE) based on API RP 581 3RD ED (A1) (A2) [8] and ASME PCC-2-2018 [9] was taken into consideration to signify the level of inspection uncertainties in the analysis.

Subsequently, heat exchangers were classified based on the percentage of tubes reaching the failure criteria over a period of years. STELLAR Plugging/Failure Index (SPI/SFI) was used for the severity categorization of the equipment, as shown in Fig. 2. SPI is set based on the tube plugging indicator, where the tubes have reached wall loss limit for plugging recommendation in maintenance strategies. On the other hand, SFI provides leak failure indicator, calculated against the Hoop stress at minimum remaining thickness. The criteria are mapped based on the percentage of tube predicted to plug or fail against time when it happens.

3 Result and Discussion

Figures 3 and 4 present the overall classification result of 57 heat exchangers obtained based on the SPI and SFI, respectively. Based on the findings, it was observed that 22 heat exchangers were within high-priority classification of severe, major and moderate categories for SPI. On the other hand, only six heat exchangers were within the severe, major and moderate categories for SFI. The results can be used for definitive worklist for turnaround scopes on the heat exchangers, for the following:

- Plan and resource priority planning (e.g. cleaning tool, spare materials of plugs and tubes).
- Maintenance priority planning (e.g. plugging, re-tubing, tube bundle replacement).

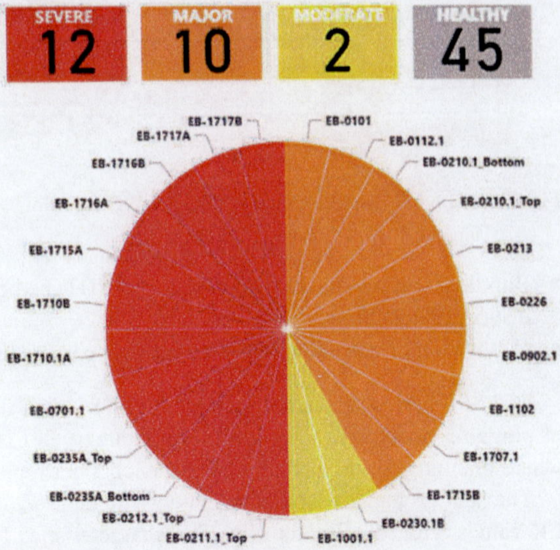

Fig. 3 SPI result for 57 heat exchangers

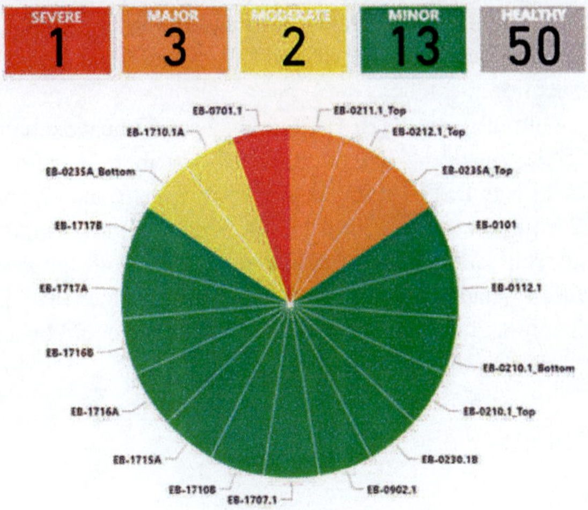

Fig. 4 SFI result for 57 heat exchangers

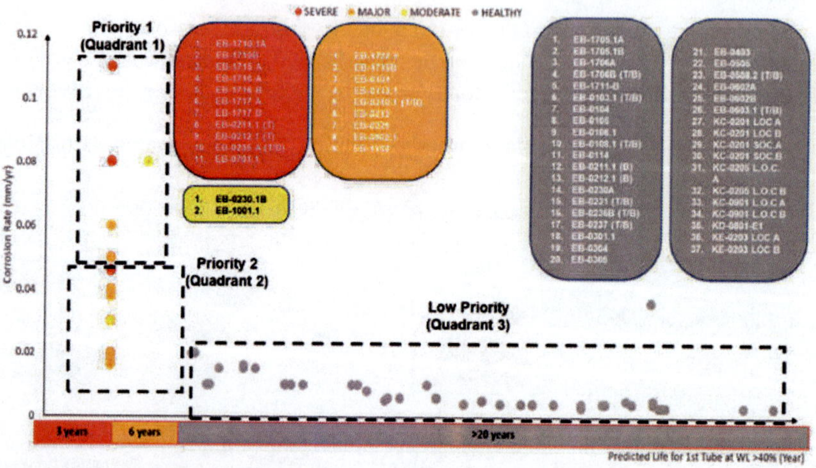

Fig. 5 STELLAR TA and Reliability Strategy (STARS) plot

- Bundle inspection loading planning (e.g. NDT methods, inspection tube coverage, inspection locations.
- Calibration of corrosion rate with process parameters.

The SPI indicator is recommended for short-term strategy for the early planning of plugging or retubing. For long-term strategy, SFI is the most suitable indicator, where planning for tube bundle or heat exchanger replacement could be justified.

The heat exchanger can be further classified based on the correlation of predicted CR values with tube bundle life. The characteristics for each heat exchanger tube bundle define the quadrant grouping for the equipment in STELLAR TA & Reliability Strategy (STARS) Plot, as shown in Figs. 5 and 6. Specific quadrants provide different risk and mitigation strategies to address in improving the operation and equipment life, especially for turnaround (TA) preparation as well as prioritization during TA execution.

Outputs from STELLAR analysis of 57 heat exchangers were mapped within the STARS quadrants, based on the CR distribution against the years of first tube failure for SPI and SFI limits, as shown in Figs. 6 and 7, respectively. It was found that 22 equipment are within the Quadrant 1 and 2, indicative of low average bundle lifecycle (MTBF) with either high failure rate or low failure rate. Those bundles within Quadrant 2 possess potential for high MTBF, if high attention for thorough cleaning, good inspection and maintenance could be addressed.

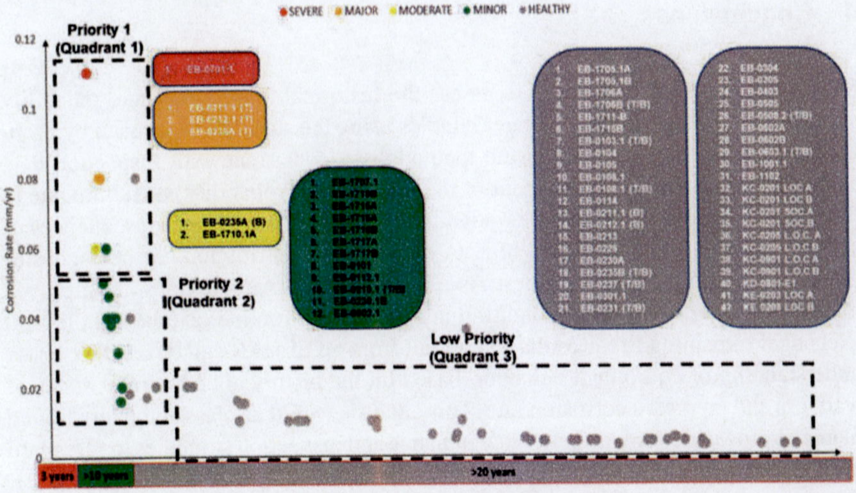

Fig. 6 Classification of corrosion rate with tube bundle life (SPI)

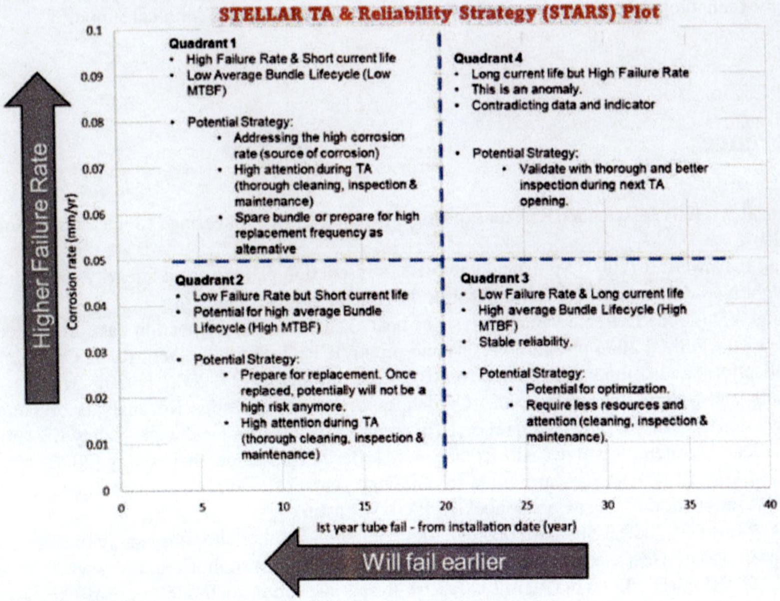

Fig. 7 Classification of corrosion rate with tube bundle life (SFI)

4 Conclusions

STELLAR has been developed to assess the historical inspection data, reliability and remaining life of heat exchanger bundles using the statistical approach by probability analysis. The tool offers the methodology consistent with inspection data accuracy and screening, assessment of the likelihood (probability) of failure due to flaws damage, deterioration or degradation. The assessment of the consequences of such failure and implementation of EVA and Weibull distribution has been considered in predicting the failed tubes. As a result, STELLAR provides an improved methodology for monitoring condition and prediction routine of tube bundle heat exchanger remaining life. Results and output from STELLAR can be used for better understanding of equipment behavior, based on the historical and current states of wall loss and predicted corrosion rate. This analysis would also assist the turnaround planning process to provide better maintenance and inspection strategies to safeguard the equipment as well as the TA window.

Acknowledgements The authors gratefully acknowledge the PETRONAS Chemicals Group Berhad for funding support (cost center: 015MD0-120), as well as Digital Analytics Structural Integrity Technology (DASIT), Universiti Teknologi PETRONAS, for technical support.

References

1. Gonzalez R (2019) Extended RBI for exchanger bundles. Inspectioneering J (September/October 2019)
2. Khan F, Howard R (2007) Statistical approach to inspection planning and integrity assessment. Insight Non-destructive Test Cond Monitor 49:26–36
3. Wang WD (2006) Extreme value analysis of heat exchanger tube inspection data. In Proceedings of the ASME 2006 pressure vessels and piping/ICPVT-11 conference, vol 7: Operations, applications, and components, Vancouver, BC, Canada, July 23–27, 2006, pp 305–313
4. Health and Safety Executive (2002) Guidelines for use of statistics for analysis of sample inspection of corrosion. Research report 016, prepared by TWI for Health and Safety Executive
5. American Petroleum Institute, API RP 581—Risk Based Inspection Technology (2016)
6. ASTM G16-13 (2019) Standard guide for applying statistics to analysis of corrosion data. In: ASTM international, West Conshohocken, PA. www.astm.org
7. ASTM E691-20 (2020) Standard practice for conducting an interlaboratory study to determine the precision of a test method. In: ASTM international, West Conshohocken, PA. www.astm.org
8. API RP 581 3RD ED (A1) (A2) (2020) Risk-based inspection methodology, 3rd edn., April 2016; Addendum 1, April 2019; Addendum 2, October 2020
9. ASME PCC-2–2018 (2018) Repair of pressure equipment and piping, December 2018

Materials and Automotive

Corrosion of Copper in Water-In-Biodiesel Diesel Emulsion Fuel

Davannendran Chandran and Revathi Raviadaran

Abstract This study aimed to investigate the corrosion effect of water-in-biodiesel diesel emulsion fuel (W5B5) prepared using three different emulsifiers towards copper. Copper was immersed in W5B5 for 500 h at 25 °C. For comparison, additional copper coupons were immersed in diesel fuel and biodiesel fuel for 500 h at 25 °C. The W5B5 consists of 2 vol% of surfactant, 5 vol% of water and 93 vol% of biodiesel diesel fuel blend. The utilized biodiesel diesel fuel blend was B5, which consists of 5 vol% of biodiesel fuel with 95 vol% of diesel fuel. The evaluated emulsifiers were polyglycerol polyricinoleate (PGPR), Xanthan gum, as well as combined Span 80/Tween 80 of equal concentration. Among the performed analysis were mass loss, sedimentation index and total acid number (TAN). Highest corrosion rate was measured for copper exposed to W5B5 prepared with combined Span80/Tween80 followed with W5B5 prepared with PGPR, W5B5 prepared with Xanthan, biodiesel fuel and finally diesel fuel at 0.0011, 0.0009, 0.0007, 0.0006 and 0.0003 mm/yr, respectively. The trend of corrosion rate corroborated with the TAN value. To benefit from the possible reduction in exhaust emission of using water-in-diesel emulsion fuel or water-in-biodiesel diesel emulsion fuel, more studies focusing on formulation of stable emulsion fuel using less acidic emulsifiers and its effect towards compatibility with fuel system metals are required.

Keywords Corrosion · Copper · Diesel fuel · Biodiesel fuel · Water · Emulsifier

D. Chandran (✉)
Department of Mechanical Engineering, Universiti Teknologi PETRONAS, 32610 Seri Iskandar, Perak, Malaysia
e-mail: davannendran@utp.edu.my

R. Raviadaran
Department of Chemical Engineering, Universiti Teknologi PETRONAS, 32610 Seri Iskandar, Perak, Malaysia
e-mail: revathi.raviadaran@utp.edu.my

F. Ahmad et al. (eds.), *ICREEM 2022*, Lecture Notes in Mechanical Engineering,
https://doi.org/10.1007/978-981-99-5946-4_18

1 Introduction

Global temperature is anticipated to increase up to 4.8 °C by the end of the twenty-first century due to rising greenhouse gas (GHG) emissions. This situation can cause the sea level to rise by 0.82 m resulting in increased heat waves, floods and lack of food supply causing crisis to human. Among the contributor to GHG emissions is due to the combustion of fossil fuel in the transportation industry. In this context, efforts have been taken to reduce the consumption of fossil fuel by adopting renewable fuel such as biodiesel fuel and ethanol, to produce cleaner combustion by adding water in diesel fuel known as water-in-diesel emulsion fuel, as well as development of electric vehicle. To date, biodiesel fuel has been approved for use in blended form with diesel fuel up to 30 vol% of biodiesel fuel with 70 vol% of diesel fuel. For water-in-diesel emulsion fuel, it is yet to reach adoption stage despite several studies have reported reduced exhaust emission by using water-in-diesel emulsion fuel. Ithnin et al. [1] reported reduced nitrogen oxide by 32% and particulate matter by 16% when used water-in-diesel emulsion fuel with 5 vol% of water than diesel fuel. Khalife et al. [2] reported reduced carbon dioxide, nitrogen oxide and unburned hydrocarbon emissions by 8.5%, 24% and 28%, respectively using water-in-biodiesel diesel emulsion fuel, which consists of B5 (5 vol% biodiesel fuel with 95 vol% diesel fuel) with 4 vol% of water as compared to diesel fuel. Few other existing studies have also reported reduction in exhaust emission by using water-in-diesel emulsion fuel than diesel fuel [3–5]. However, despite the potential of water-in-diesel emulsion fuel in reducing exhaust emission, compatibility of water-in-diesel emulsion fuel with fuel system materials needs to be established before adopting it. This is important since the use of incompatible fuel can lead to significantly early fuel delivery components failure [6–9]. Till date, only two works have typically reported concerning compatibility of water-in-diesel emulsion fuel with fuel system materials [10, 11]. In these studies, the authors have reported that copper and aluminum exposed to water-in-diesel emulsion fuel consisting of 5 vol% water had higher corrosion rate than diesel fuel by 12 and 2 times, respectively. The authors associated the obtained corrosion rate with the higher acidity value of water-in-diesel emulsion fuel than diesel fuel, due to the addition of emulsifiers in forming stable emulsion. This study aims to investigate corrosion effect of water-in-biodiesel diesel emulsion fuel (W5B5) prepared using different emulsifiers towards copper under laboratory immersion.

2 Materials and Methods

0.02 m in diameter and 0.002 m in thickness of coupon with ~ 99 wt% of copper were utilized. W5B5 consisted of 5 vol% water, 2 vol% emulsifier and 93 vol% B5 (5 vol% biodiesel fuel with 95 vol% diesel fuel). The utilized emulsifiers were polyglycerol polyricinoleate (PGPR), Xanthan gum and combined Span 80/Tween 80 (equal concentration). XiangYi Instruments UH-IID 950 W Ultrasonic Homogenizer

Fig. 1 Immersion of copper coupon in **a** diesel fuel, **b** biodiesel fuel and **c** water-in-biodiesel diesel emulsion fuel

was used to prepare the emulsion fuel set at 60 W power, 4 min total run time (15 s run and 15 s stop) and temperature set at 30 °C. Immersion study was performed in accordance with ASTM G31 method by immersing coupon in fuel as shown in Fig. 1, stored in dark and closed environment for 500 h at 25 °C. Copper coupons were also immersed in diesel fuel and biodiesel fuel at similar conditions for comparison. For each condition, an additional coupon was immersed as duplicate. Corrosion rate was calculated based on the measured mass loss of the coupon. Sedimentation index (SI) was used to express the stability of formed emulsion, where the emulsion fuel was stored for 500 h. Equation (1) was used for calculating the SI, where H is the upper oil phase and H0 is the initial height of emulsion fuel. The total acid number (TAN) analysis was performed for the fuel samples in accordance with ASTM D664 using Mettler Toledo T70 titrator.

$$SI = H/H0 \times 100\% \tag{1}$$

3 Results and Discussion

The stability of the prepared W5B5 was evaluated using SI. The SI value after 500 h of storage was 3.75, 6.25 and 10.00% for W5B5 prepared using combined Span 80/Tween 80, PGPR and Xanthan, respectively. All three formed emulsion fuels were considered stable as the obtained SI values were small and no free water formation was observed. From Fig. 2, copper immersed in W5B5 prepared using combined Span 80/Tween 80 had the highest corrosion rate at 0.0011 mm/yr, followed within W5B5 prepared using PGPR, W5B5 prepared using Xanthan, biodiesel fuel and finally diesel fuel at 0.0009, 0.0007, 0.0006 and 0.0003, respectively. Higher corrosion rate by 3.4, 2.9 and 2.3 times was measured for W5B5 prepared using combined Span 80/Tween 80, PGPR and Xanthan, respectively, as compared to diesel fuel. Corrosion rate of copper immersed in alternative fuel such W5B5 was compared with the corrosion rate obtained for copper immersed in diesel fuel because metals in the fuel delivery system of a diesel engine were selected based on its compatibility with diesel fuel. Therefore, the difference between the corrosion rate of metal in alternative fuel such as W5B5 and in diesel fuel reveals its compatibility for use in diesel engine. Higher corrosion of copper in W5B5 than diesel fuel corroborates with existing studies. Chandran et al. [10] reported 11.5 times higher corrosion rate of copper in water-in-diesel emulsion fuel than diesel fuel after immersion at 25 °C for 1200 h of duration. In another study, the authors [11] reported 2.2 times higher corrosion rate of aluminum immersed in water-in-diesel emulsion fuel than diesel fuel after 1200 h of immersion at 25 °C. In both the existing studies, they used combined Span80/Tween 80 to prepare the water-in-diesel emulsion fuel. The observed higher corrosion for copper in biodiesel fuel than diesel fuel by 1.9 times also agrees with existing studies. Corrosion of copper in biodiesel fuel was 2.9 times higher than in diesel fuel was reported by Fazal et al. [12] due to immersion investigation at 26 °C for 2880 h of duration. In another study. Haseeb et al. [13] reported that copper had 1.6 times higher corrosion in biodiesel fuel than diesel fuel at 28 °C for 2640 h.

Figure 3 shows the TAN value for diesel fuel, biodiesel fuel and W5B5 before and after exposed to copper for 500 h at 25 °C. Before exposure to copper, diesel fuel had the lowest initial TAN value than biodiesel fuel and W5B5. Initial TAN value for W5B5 prepared using Span80/Tween 80, PGPR and Xanthan was higher than diesel fuel by 8.7, 6.6 and 4.7 times, respectively. Among the W5B5, the emulsion fuel prepared using Span80/Tween 80 had up to 1.8 times higher initial TAN value than prepared using PGPR and Xanthan. While, biodiesel fuel had 4.1 times higher initial TAN value than diesel fuel. Allowable TAN value for diesel fuel is not specified in ASTM D975-08a and EN 590:2004 while the allowable maximum TAN value for

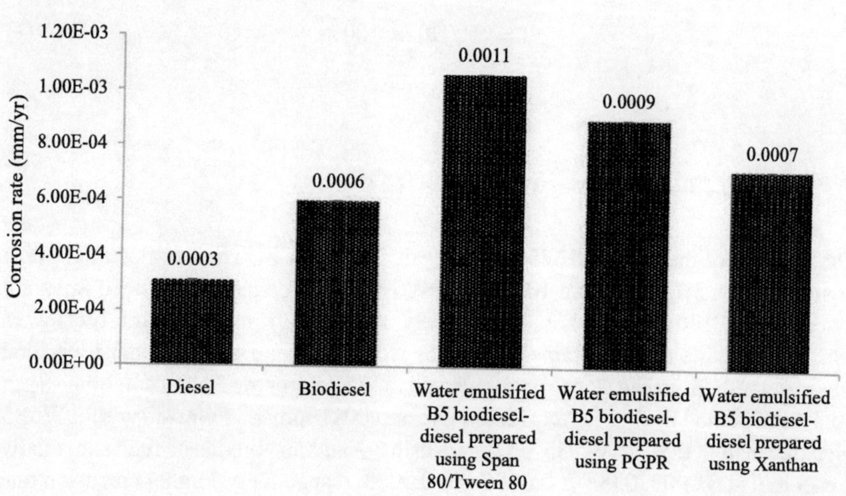

Fig. 2 Corrosion rate of copper in water-in-biodiesel diesel emulsion fuel, diesel fuel and biodiesel fuel for 500 h immersion at 25 °C

biodiesel fuel is specified to 0.5 mgKOH/g in ASTM D6751-12 and EN 14,214:2012. For water-in-diesel emulsion fuel and water-in-biodiesel diesel emulsion fuel, limited information is available concerning the allowable maximum TAN value since these fuels are still in research stage. After exposed to copper, the TAN value for diesel fuel was still lower than for biodiesel fuel and W5B5. TAN value for W5B5 prepared using Span80/Tween 80, PGPR and Xanthan after exposed to copper was higher than diesel fuel by 25.7, 10.2 and 6.4 times, respectively. Among the W5B5, the emulsion fuel prepared using Span80/Tween 80 underwent the highest change in TAN value by 5.3 times followed with prepared using PGPR and finally Xanthan by 2.8 and 2.4 times, respectively. While biodiesel fuel had 3.4 times higher TAN value than diesel fuel, the increased TAN value of fuel after exposed to metal agrees with existing studies [12, 14, 15]. The increase in TAN value suggests that the fuel has degraded during exposure to metal [16]. The higher increase in TAN value for biodiesel fuel than diesel fuel could be attributed to biodiesel fuel's higher vulnerability to deterioration related to its lower stability than diesel fuel.

Since stable emulsion was formed based on the obtained SI values and free water was not observed, the acidity of the emulsion fuel is suggested to have influenced the higher corrosion rate of copper in W5B5 than biodiesel fuel and diesel fuel. Considering that the higher TAN value of W5B5 than diesel fuel is suggested to be influenced by the addition of high acidity emulsifiers to formulate stable emulsion fuel, utilization of less acidic emulsifier in the preparation of emulsion fuel might result in lower corrosion rate of metal as compared to the metal exposed to emulsion fuel prepared using more acidic emulsifier. For this study, the measured TAN values for Span 80/Tween 80, PGPR and Xanthan are 7.9/1.9, 2.1 and 1.4 mgKOH/

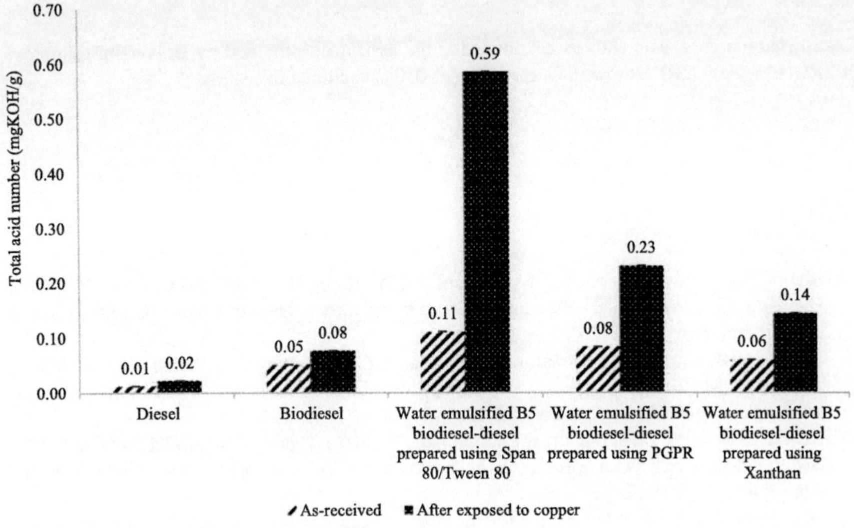

Fig. 3 Changes in acidity value for water-in-biodiesel diesel emulsion fuel, diesel fuel and biodiesel fuel exposed to copper for 500 h immersion at 25 °C

g, respectively. The highest TAN value for Span 80/Tween 80 indeed resulted in the W5B5 with the highest initial TAN value followed with prepared using PGPR and Xanthan. The obtained corrosion rate indeed correlated with the TAN value of W5B5, biodiesel fuel and diesel fuel. Relationship between acidity of fuel and metal corrosion has been reported by Tsuchiya et al. [17]. The authors highlighted higher corrosion of metal in fuel with higher TAN value. Several other existing studies have also highlighted the relationship between acidity of fuel and metal corrosion [10, 11]. Therefore, it is indeed critical to select emulsifier with low TAN value during formulation of emulsion fuel while ensuring stable emulsion at the same time to prevent high corrosion of metal.

4 Conclusion

Corrosion behavior of diesel fuel, biodiesel fuel and W5B5 prepared using combined Span 80/Tween 80, PGPR and Xanthan towards copper was investigated. From the obtained SI values of less than 10%, stable emulsion was formed and no free water formation was observed. Copper immersed in W5B5 experienced higher corrosion than in diesel fuel and biodiesel fuel by up to 4 times. The obtained corrosion rate correlated with the TAN value of the fuels. Formulation of emulsion fuel using less acidic emulsifier resulted in emulsion fuel with low initial TAN value and culminated in lower corrosion rate than the emulsion fuel with higher initial TAN value. More studies exploring emulsifiers with low acidity are required to better understand the stability of emulsion fuel and the compatibility of fuel system metals with emulsion fuel.

Acknowledgements The authors acknowledge the assistance provided by Universiti Teknologi PETRONAS via STIRF Research Grant 015LA0-030 to facilitate this study.

References

1. Ithnin AM, Yahya WJ, Ahmad MA, Ramlan NA, Kadir HA, Sidik NAC, Koga T (2018) Emulsifier-free Water-in-Diesel emulsion fuel: Its stability behaviour, engine performance and exhaust emission. Fuel 215:454–462
2. Khalife E, Kazerooni H, Mirsalim M, Shojaei TR, Mohammadi P, Salleh AM, Najafi B, Tabatabaei M (2017) Experimental investigation of low-level water in waste-oil produced biodiesel-diesel fuel blend. Energy 121:331–340
3. Xu Y, Hellier P, Purton S, Baganz F, Ladommatos N (2016) Algal biomass and diesel emulsions: an alternative approach for utilizing the energy content of microalgal biomass in diesel engines. Appl Energy 172:80–95
4. Ogunkoya D, Li S, Rojas OJ, Fang T (2015) Performance, combustion, and emissions in a diesel engine operated with fuel-in-water emulsions based on lignin. Appl Energy 154:851–861

5. Mazlan NA, Yahya WJ, Ithnin AM, Hasannuddin A, Ramlan NA, Sugeng DA, Adib AM, Koga T, Mamat R, Azwadi N (2018) Effects of different water percentages in non-surfactant emulsion fuel on performance and exhaust emissions of a light-duty truck. J Clean Prod 179:559–566
6. Chandran D (2020) Compatibility of diesel engine materials with biodiesel fuel. Renew Energy 147:89–99
7. Chandran D, Kiat N, Nang H, Gan S, Choo Y, Jahis S (2016) Compatibility of biodiesel fuel with metals and elastomers in fuel delivery system of a diesel engine. J Oil Palm Res 28:64–73
8. Chandran D, Ng H, Lau H, Gan S, Choo Y (2017) Deterioration of palm biodiesel fuel under common rail diesel engine operation. Energy 120:854–863
9. Chandran D, Gan S, Lau H, Raviadaran R, Salim M, Khalid M (2018) Critical relationship between biodiesel fuel properties and degradation of fuel delivery materials of a diesel engine. Therm Sci Eng Prog 7:20–26
10. Chandran D, Khalid M, Raviadaran R, Lau HLN, Yung CL, Kanesan D, Salim M (2019) Sustainability of water in diesel emulsion fuel: an assessment of its corrosion behaviour towards copper. J Clean Prod 220:1005–1013
11. Chandran D, Khalid M, Raviadaran R, Jagadish P (2019) Corrosion of aluminium in water-in-diesel-emulsion fuel. IOP Conf Ser Mat Sci Eng 542:012004
12. Fazal MA, Haseeb ASMA, Masjuki HH (2012) Degradation of automotive materials in palm biodiesel. Energy 40:76–83
13. Haseeb ASMA, Masjuki HH, Ann LJ, Fazal MA (2010) Corrosion characteristics of copper and leaded bronze in palm biodiesel. Fuel Process Technol 91:329–334
14. Fazal MA, Haseeb ASMA, Masjuki HH (2013) Corrosion mechanism of copper in palm biodiesel. Corros Sci 67:50–59
15. Fazal MA, Haseeb ASMA, Masjuki HH (2010) Comparative corrosive characteristics of petroleum diesel and palm biodiesel for automotive materials. Fuel Process Technol 91:1308–1315
16. Chandran D, Ng HK, Lau HLN, Gan S, Choo YM (2016) Investigation of the effects of palm biodiesel dissolved oxygen and conductivity on metal corrosion and elastomer degradation under novel immersion method. Appl Therm Eng 104:294–308
17. Tsuchiya T, Shiotani H, Goto S, Sugiyama G, Maeda A (2006) Japanese standards for diesel fuel containing 5% FAME: investigation of acid generation in FAME blended diesel fuels and its impact on corrosion. SAE Technical Paper, pp 2006-01-3303

Development and Validation of a Free Piston Engine Linear Generator Simulation Model Including Cycle-To-Cycle Variation and Ignition Timing Sub-Models

Ahmed T. Raheem, A. Rashid A. Aziz, Saiful A. Zulkifli, Wasiu B. Ayandotun, and Masri B. Baharom

Abstract Free piston engine linear generator (FPELG) is a novel and promising engine for the next generation of hybrid engines. It has several benefits over conventional engines, including minimal friction, a multi-fuel engine, fewer parts, minimal emissions, and excellent thermal efficiency. FPELG is an engine that lacks a crankshaft, and cycle-to-cycle variation (CCV) is viewed as the critical challenge affecting its stability and performance. Therefore, in the present research, the simulation model of FPELG including two sub-models was developed. The CCV and ignition timing (IG) were proposed as sub-models. The FPELG model was demonstrated and run via MATLAB/Simulink. The experimental data were used to validate the model. The validation results indicate that the error was about 5% for the most of cycles, which is within an acceptable range, indicating that the model is valid for accurately predicting future outcomes. Analysis of the simulation model's result showed that the CCV occurs when certain parameters vary during engine running,

A. T. Raheem (✉) · A. R. A. Aziz · S. A. Zulkifli · W. B. Ayandotun · M. B. Baharom
Center for Automotive Research and Electric Mobility (CAREM), University Teknologi PETRONAS, 32610 Seri Iskandar, Perak, Malaysia
e-mail: Ahmed_19000341@utp.edu.my; a.t.utp3@gmail.com

A. R. A. Aziz
e-mail: rashid@utp.edu.my

A. T. Raheem · A. R. A. Aziz · W. B. Ayandotun · M. B. Baharom
Department of Mechatronics Engineering, Al-Khwarizmi College of Engineering, University of Baghdad, 10071 Baghdad, Iraq

S. A. Zulkifli
Department of Mechanical Engineering, University Teknologi PETRONAS, 32610 Seri Iskandar, Perak, Malaysia

A. T. Raheem
Department of Electrical and Electronica Engineering, University Teknologi PETRONAS, 32610 Seri Iskandar, Perak, Malaysia

F. Ahmad et al. (eds.), *ICREEM 2022*, Lecture Notes in Mechanical Engineering,
https://doi.org/10.1007/978-981-99-5946-4_19

one of which is the IG variation. Two scenarios were executed for the IG variation in order that CCV variation could be generated using IG variation. In the first scenario where IG occurs after top dead centre (TDC), the modeling results indicated the formation of two peaks and low in-cylinder pressure. While in the second scenario which involves IG occurring early before TDC, the single peak shape and high in-cylinder pressure were produced. Therefore, the IG significantly impacts on combustion characteristics, which is reflecting on the stability and performance of the FPELG. Therefore, further investigation into the effect of the ignition timing on the combustion characteristics is required. In addition, the optimal IG and control of IG are required for achieving high engine performance with stability, this could be achieved in our next work.

Keywords Free piston engine · Numerical model · Engine performance · Ignition timing

1 Introduction

The free piston engine linear generator (FPELG) is a modern engine design that addresses the limitations of conventional engines, such as the fuel consumption used, friction, emissions, heat release, and pressure, which are unproductive relative to the engine's power output, i.e., performance, stability, and emissions. Piston movement in FPELG is the unrestricted movement within the cylinder because there is no crankshaft [1–5]. Numerous technologies have been utilized to reduce engine size, particularly automobile engines. In addition to improving the engine efficiency of automobiles, therefore, as mentioned above related to the advantages of FPELG, when a few engine components are manufactured, friction loss, thermal loss, and emissions are minimized. All of these benefits are provided in the FPELG [1–6]. In 1928, Pescara built the first FPE prototype. This prototype is called a single-piston air compressor [7, 8]. Other researchers made the engine for advanced applications, including hydraulic free piston engines and free-piston linear generators [9–14]. In addition, they addressed some of the constraints that were described in [2, 13].

The architecture and working principles of this kind of free piston engine are illustrated clearly in previous studies [2, 6]. Briefing, while the engine is operating, two forces are released, these forces (each force for each side of the cylinders) working alternately to push the shaft (translator) right and left, giving resonance movement, and finally generating the current [2, 6, 12, 15, 16]. However, the FPELG has some issues, including piston motion control (PMC), generator performance, and combustion characteristics [2, 5, 9, 10, 13].

In the first phase of FPELG operation, increasing the piston velocity from zero to combustion velocity and overtaking the compression force that permits combustion to happen in a stable and continuous manner is considered the most significant technical challenge [2, 17]. All the problems stated are a result of the engine operating without a crankshaft, which is producing the free movement of pistons inside cylinders. As an outcome, the characteristics of piston motion can be regulated by controlling parameters such as load, injection position, amount of injection fuel, and ignition timing [2, 6, 18]. Moreover, the energy released by the cylinder is determined by the amount of fuel injected [19, 20].

The expected maximum piston acceleration in FPELG is greater than that of conventional engines. The reason is that the high in-cylinder pressure and temperature in FPELG cause increased expansion. In addition, low losses in the heat transfer of the FPELG's cylinder compared to that of a conventional engine [21]. In fact, there are further benefits, like excellent thermal efficiency, less friction, and low emissions. However, as a result of the free piston movement within the cylinder, FPELG engines show a greater change in combustion pressure at each cycle compared to conventional engines [2, 12, 18, 22]. Recent studies have analysed FPELGs from different viewpoints, including combustion characteristics, optimization, injection, ignition, emission analysis, heat transfer, modelling, and control. However, there are comparatively fewer studies compared to the vast number of issues in FPELG [6, 12, 18, 23–25]. Some authors found that because unrestrained piston movement, the implementation of an electronic control system to avoid pistons from striking the cylinder head is required [6, 12, 26]. Despite the fact that adding a damping system in the cylinder could avoid such an action [12, 27]. However, a control system is considered the main part that is utilized to reduce the cycle-to-cycle fluctuation of the FPELG [2, 18]. Many studies [2, 5, 26, 28–31] have designed a control method to overcome control complexity in the FPELG. However, they discovered that adjusting the movement of the piston (stability of piston movement) between the BDC and TDC is a major problem for linear FPEs because unconstrained piston motions have no mechanical limitations.

Other researchers discovered that due to the pistons moving freely, it is impossible to guarantee of the piston motion and dead centre positions (TDC and BDC positions), resulting in cycle-to-cycle variations [32, 33]. Thus, several authors have conducted extensive research on the control approaches of FPELG to achieve stable engine running. Jia and Roskilly et al. [15, 26, 34] developed a predictive piston motion control approach based on the piston's predicted displacement and velocity. Authors [35–37] presented an accurate piston trajectory control to control the piston motion of a hydraulic free-piston engine by following predetermined trajectories with such an active motion controller that utilizes a strong repeated monitoring system. In addition, feedforward controllers were studied to enhance tracking accuracy [35]. In conclusion, according to the literature, one method of control approach is trajectory tracking control. While the engine is running, the piston movement is controlled to follow a predefined trajectory [35–37]. Another kind of control approach involves balancing the load and adjusting combustion parameters [19, 26, 38]. Moreover, controlling the output current with steady engine operation using a predefined reference current profile can be considered another kind of control approach [39–41]. However, the development of models and simulations that include the cycle-to-cycle variation (CCV) for the FPELG is one of the most crucial aspects of understanding and addressing PMC issues. On the other hand, to design and develop effective control strategies for adjusting the parameters that have a direct impact on the CCV issue.

Therefore, a simulation model for FPELG including CCV is proposed in this study. In addition, the ignition timing (IG) is also proposed as a sub-model. MATLAB/Simulink software was used to build and run the FPELG model. This model was validated using experimental data. The finding demonstrates that the model and test are in good agreement. The effect of different ignition timing on the in-cylinder pressure is studied. The simulation results showed that the IG has a significant effect on the in-cylinder pressure.

2 FPELG Operation Principle and Methodology

2.1 Experiment Setup

The FPELG is depicted in Fig. 1. This schematic clearly shows the engine components and connections. Using locally available materials, the Centre for Automotive Research and Electrical Mobility (CAREM) at Universiti Teknologi PETRONAS (UTP) developed and built the direct injection two-stroke FPELG (GX-5). Briefing, the linear generator (LG) is linked between two cylinders, one of these cylinders for the combustion and the second one working as a rebound device. In addition, high-pressure pipelines are linked between these cylinders and the air tank, which is filled by the compressor. There are also sensors to monitor such as airflow, pressure, injection, temperature, and piston position. Data are obtained using the Data Acquisition System (DAS). Eventually, the engine is operating via the LabVIEW interface.

Table 1 contains the FPELG prototype parameters and working conditions, involving input parameters such as the mass of connecting rod and pistons, ambient conditions, engine settings, geometrical specifications, and LEM settings. The Engine Control Unit (ECU) was utilized to regulate the engine's input and output signals, as described in detail in the methodology. In addition, the instruments utilized, and the procedures implemented to test this prototype are described.

2.2 Operation Principle

Figure 2 shows the test rig for the FPELG (GX-5) experiment. The GX-5 runs in a manner similar to that of a two-stroke FPELG. Basically, when the engine is operating, two forces are released. The first force generated in the first cylinder works alternately with the second force generated in the second cylinder to push the translator into the left and right sides, resulting in resonance motion. Thus, electricity (current) is generated at the final stage of this process. By assuming the left cylinder represents the bounce side and the right cylinder for the combustion side. Therefore, when the engine starts to run, the left chamber will be supplied by sufficient inlet

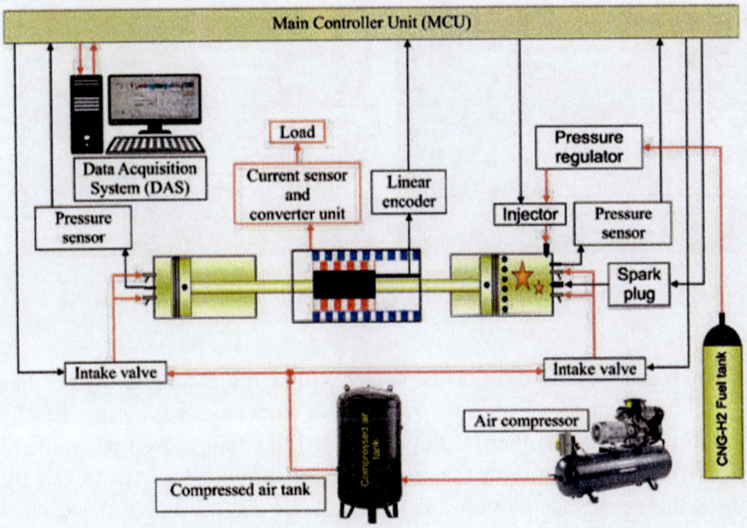

Fig. 1 Schematic diagram of the FPELG experimental setup

Table 1 Engine parameters and working conditions

Parameter	Value	Unit
Combustion chamber bore	56	(mm)
Bounce chamber bore	56	(mm)
Maximum stroke	96	(mm)
Effective stroke	84	(mm)
Piston and connecting rod mass	7	(kg)
External load resistance	4.7	(ohm)
Number of cylinders	2	–
Intake pressure	8	(bar)
Intake valve opening time	5–100	(ms)
Cylinder displacement volume	221	(cc)
Engine frequency	30	(Hz)
Fuel injection duration	16.79	(ms)
Injection position BTDC	59	(mm)
Ignition position BTDC	7	(mm)

pressure to push the piston and rod component into the right side. At this time, the air valve and fuel valve will be released, allowing the air–fuel mixture to enter the combustion chamber. When the piston reaches TDC, the Engine Control Unit (ECU) must send an ignition signal to the ignition system, the spark will be provided, and combustion will be achieved. The massive pressure resulting in the release of heat during combustion will drive the piston and rod components back to BDC. As soon

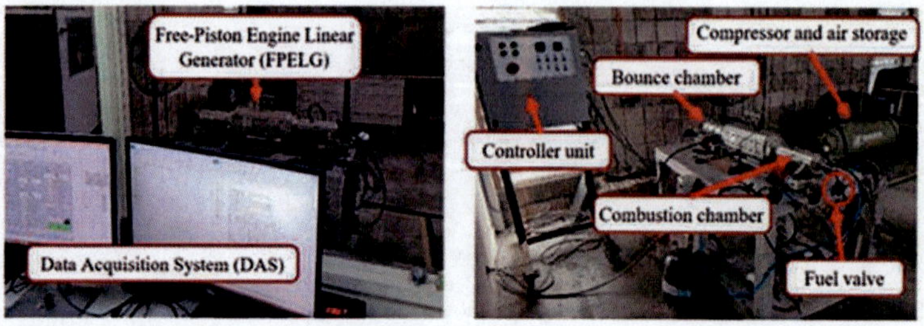

Fig. 2 FPELG prototype developed by CAREM in Universiti Teknologi PETRONAS

as the piston reaches BDC, the exhaust valves will be released, and any residual gas mixture will be removed. As a result, the left side works as a gas spring. Briefly, the gas within the bounce chamber (left cylinder) will be compressed when the piston reaches at BDC in order to force the piston and rod component back to TDC again. Throughout this resonance motion, LG will generate electricity (current). The DAS will be utilized for experimental data collection.

2.3 Methodology

2.3.1 Mathematical Model and Simulation

The forces exerted on pistons are calculated using the major factors, e.g., LEM, friction, in-cylinder pressure, and moving mass [12]. In particular, based on Newton's second law, Eq. (1) [17] derives the movement dynamic formula and piston movement for the motoring mode. While Eq. (2) [42] derives the movement dynamic formula and piston movement for the generating mode.

$$F_m + F_l - F_r - F_f = m\frac{d^2x}{dt^2} \tag{1}$$

$$F_l - F_r - F_f - F_g = m\frac{d^2x}{dt^2} \tag{2}$$

Here, F_m symbolizes the electrical force generated by the LEM when functioning as a motor (unit: N); F_l and F_r symbolize the pressure force of the left and right cylinders (N), respectively; F_f symbolizes the frictional resistance between the piston parts and cylinders (N); m symbolizes the rod and piston components mass (kg); x symbolizes the displacement of piston movement (m); and F_g symbolizes the LEM resistance force when functioning as a generator (N).

Figure 3a is a free-body diagram illustrating the dynamic model of the FPELG while the LEM is functioning as a motor in motoring mode. Where the driving force helps in pulling the piston into the right side, assuming the motion begins on the left

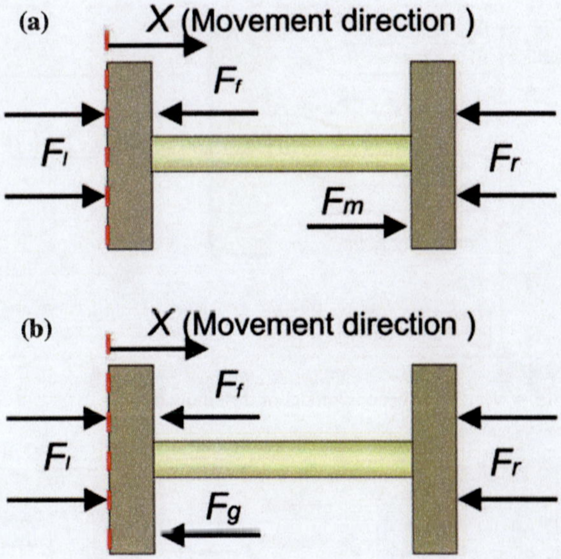

Fig. 3 Free body diagram of FPELG **a** LEM is working as a motor in the motoring mode, and **b** LEM is working as a generator in the generating mode

side (LS) and travels to the right side (RS). In addition, the force generated by gas pressure from the left cylinder will force the piston to move in the same direction as the motoring force, exactly from the LS to the RS. The force of friction between the piston components such as piston rings and cylinders wall, in addition to the force generated by gas pressure from the right cylinder, will force the piston moves in the opposite direction. While Fig. 3b illustrates the FPELG dynamic model when the LEM is turning into generating mode and functioning as a generator. Using the same assumptions as previously, the motion begins from the LS moving toward the RS. Whereas the force produced by the combustion (in-cylinder gas pressure) in the left cylinder will force the piston to move toward RS. While the force produced by the generator, force produced by friction, and force coming from gas pressure in the right cylinder will force the piston to move in the opposite direction.

The dynamic equation of FPELG was therefore expressed in this study based on Newton's second law (Eq. 2) according to generator mode or stable engine operating. However, the sub-equations and extra information, such as geometry equations and the complexities of the derivation of some main equations used in this study, were published previously. For further details, please see the published studies by the scientist groups [12, 42]. Figure 4 illustrates the MATLAB/Simulink simulation modelling for the dynamic equation.

However, there are sub-models, one for each force that is used in the dynamic model as input factors namely left cylinder force, right cylinder force, friction force, and generator force. For the bounce side, LEM, and friction, the equations used in the current study are similar to those detailed and developed by Ahmed T. et al. [12]. While, the combustion side equations used in this study are similar to those utilized by Boru Jia et al. [42]. Therefore Fig. 5 a, b, and c illustrates the MATLAB/

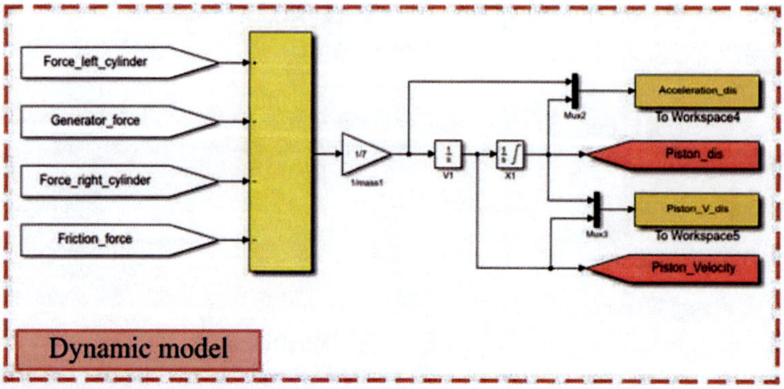

Fig. 4 Detailed dynamic model of the simulation

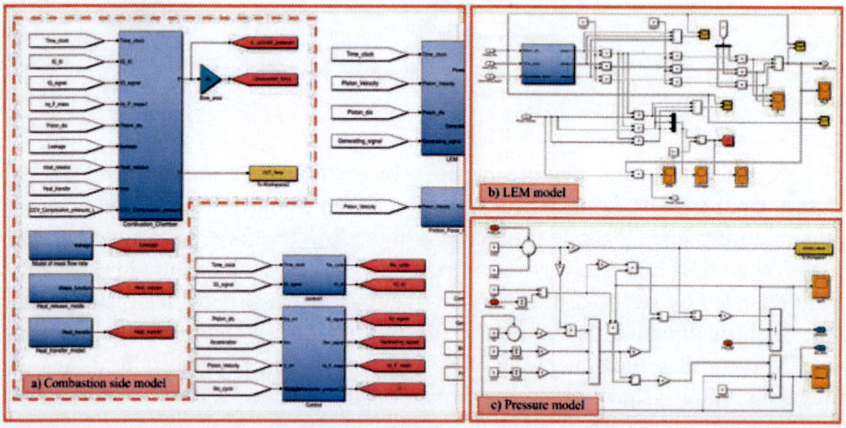

Fig. 5 FPELG simulation model using MATLAB/Simulink **a** Combustion side model, **b** LEM model, and **c** pressure model

Simulink simulation modelling for the sub-model as an example Fig. 5a represents the combustion side simulation model, Fig. 5b shows the LEM simulation model, and Fig. 5c shows the pressure inside cylinder model.

2.3.2 Cycle to Cycle Variation Model

This study aims to investigate the cyclic behaviour of the FPELG, including cycle-to-cycle variation (CCV). Where the combustion pressure (in-cylinder pressure) value changes from cycle to cycle as a natural phenomenon. CCV in combustion pressures can be measured using the coefficient of variation (COV). Therefore, it is crucial to demonstrate CCV in the final simulation model of the FPELG. This is accomplished in the current study via adding a randomness sub-model into the compression pressure

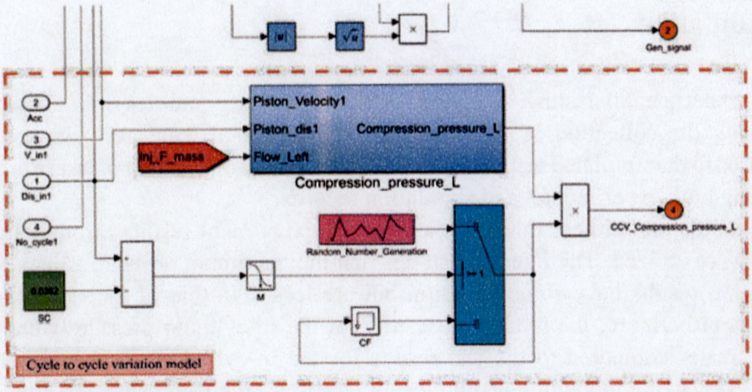

Fig. 6 Cycle to cycle variation model

values, which is represented by the randomness factor (RF). By applying RF within a specific range on the compression pressure values, different peak pressure values will be generated for each cycle, resulting in a CCV shape. Figure 6 illustrates the Simulink model built to achieve this goal.

Thereafter, the outputs of the CCV model (deferent peak values of compression pressure) and the output of the thermodynamic model (pressure after ignition occurs) will produce the final values of in-cylinder pressure with CCV shape. Thus, this model can be easily used for predicting the in-cylinder pressure with different ignition timing (IG). The second objective of this work is achieved by predicting the in-cylinder pressure under various ignition timing. Figure 7 illustrates the in-cylinder pressure against time with different scenarios of ignition timing.

Fig. 7 Diagram of the in-cylinder pressure against time using different ignition timing scenarios

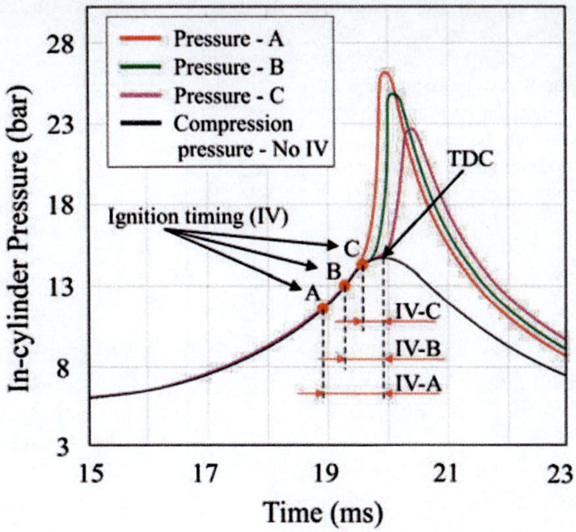

3 Validation

In the experimental testing, a particular set of unique qualities was confirmed. Following the collection of data, then the experimental data are examined and compared to the simulated results. Figure 8 illustrates the in-cylinder pressure profiles including both experimental and simulation records.

For validation, either simulation as well as experiment results throughout five cycles were utilized. The figure illustrates that the maximum pressure values of the simulation results have some variation but are less than that of the experimental results. Furthermore, the modelling results for the in-cylinder pressure were relatively greater compared to the test results for the in-cylinder pressure. Although the difference, the modelling result closely resembles the test results, and the error percentage was within an acceptable range. Figure 8 demonstrates that the highest error percentage is lower than 10%. However, the error that observed between the model results and test results for most cycles values for the in-cylinder pressure was about 5% and below, confirming that the model is valid. While Fig. 9 illustrates the linear position of the piston inside cylinder as a function of time for both modelling and test results.

The test results indicate a number of variances for the piston position profile. Thus, the percentage of error was further evaluated after analysing and comparing modelling results with test results. As illustrated in the figure, the error percentage of most piston position data that obtained from both modelling and test was also about 5%. However, the highest error was under 10%, which is satisfactory. This percentage of error can be linked to the precision of the experiment's sensors and instruments. In addition, the leakage between the piston ring and the cylinder wall, as well as specific ambient factors, could have impacted the results of the test. In the modelling phase of this study, however, specific assumptions were taken into account for the simulation model. Nonetheless, the comparison demonstrates that

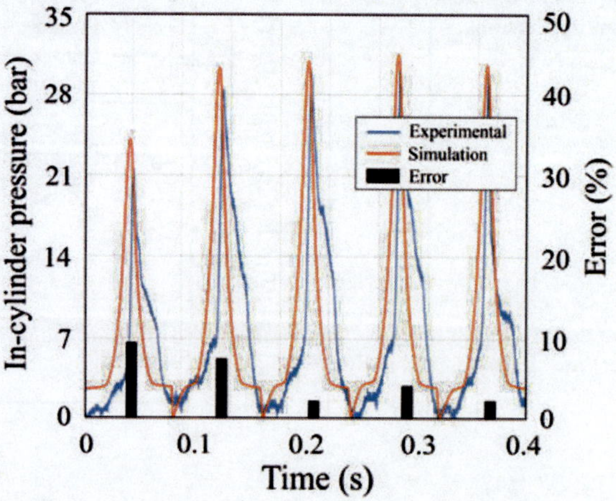

Fig. 8 Comparisons of in-cylinder pressures between the test and modelling results with error percentage

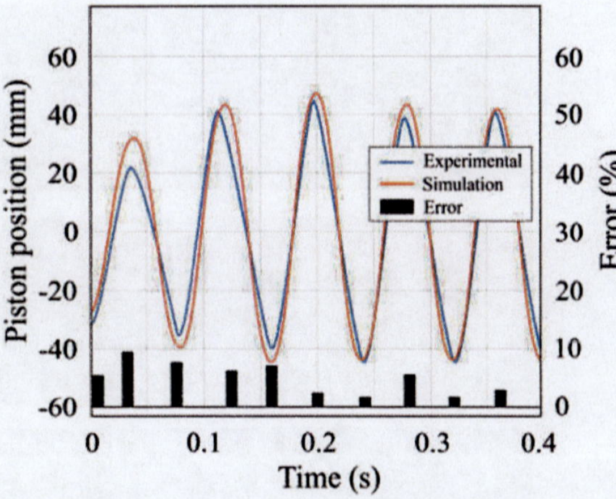

Fig. 9 Comparisons of piston displacement profiles between the test and modelling results with error percentage

the simulation model is accurate (valid) and can be utilized to predict the outcomes (accurate results).

4 Results and Discussion

The goal of the ignition system is to provide a high voltage for the spark plug to ignite the compressed mixture of air–fuel inside the cylinder within a specific time. However, the cycle-to-cycle variation is produced, not only because of ignition time variation but there are many parameters difficult to control in the FPELG because there is no crankshaft. Therefore, in this study, three objectives are achieved, firstly, the direct injection two stroke FPELG model is developed. The CCV model is also designed and proposed. Thereafter, using the CCV model and connected with ignition timing, the ignition timing model is proposed. The simulation model including CCV and ignition timing sub-models is validated with experimental data that indicated the model is valid and can predict the results with good accuracy. The major benefit of this model is that it can be used to investigate the effect of ignition time on combustion characteristics in detail. That will be the next of our work. However, in this study, the influence of ignition timing on the in-cylinder pressure is examined and analysed. Figure 10 illustrates the simulation results for some parameters namely from top to down; piston position (piston displacement profile), piston velocity, pressure of bounce chamber, and pressure of combustion chamber (in-cylinder pressure). Must be taken into consideration all these parameters' curves are illustrated against time (the time is used as an x-axis).

Firstly piston displacement results, Fig. 10 shows that the piston moves smoothly between TDC and BDC. Moreover, the TDC position is + 42 mm and the BDC position is -42 mm, according to the operation condition listed in Table 1. However,

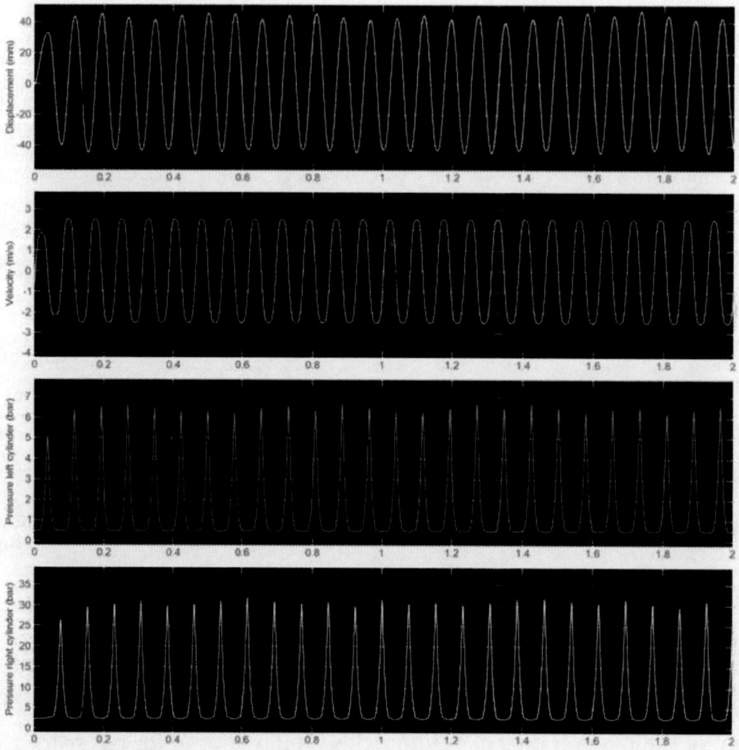

Fig. 10 Simulation results of the FPELG including the profiles of the piston displacement, piston velocity, bounce pressure, and combustion pressure

the results show that there are variations in the peaks of piston positions. This variation is represented by COV. In the current study, the simulation is run with a COV of 9%. Several reasons for the TDC and BDC variations, but the main reason is that the FPELG works without a crankshaft, and because of the different combustion characteristics that occur for each cycle resulting in different in-cylinder pressure, which in turn causes the piston to be pushed by different forces for each cycle. On the other hand, Fig. 10 illustrates the pressure against time for both bounce and combustion cylinders. The results show that the pressure of both the bounce and combustion sides fluctuates. The reasons for that can be summarized as follows the amount of fuel injected into the cylinder, the fuel amount that actually burnt in the cylinder, leakages, heat transfer, the heat released, and ignition timing. These parameters are considered as the main parameters that can influence combustion behaviours, leading to the production of different pressure values. For further details, Fig. 11 demonstrates the peaks of pressure for both cylinders (bounce and combustion). In addition, the variations of TDC and BDC positions for the 26 cycles can be seen clearly in Fig. 11.

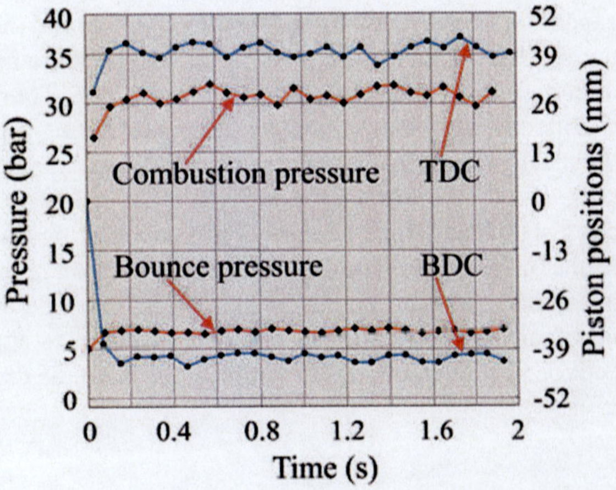

Fig. 11 Simulation results of the FPELG including peaks of pressure for both cylinders (bounce and combustion), and the TDC and BDC positions for the 26 cycles

Further investigations were carried out in this study. In order to investigate the influence of ignition timing on the in-cylinder pressure, Fig. 12 illustrates the results of the in-cylinder pressure against time for different ignition timing scenarios.

The results indicate that the ignition timing has a significant influence on the in-cylinder pressure, hence affecting the stability and performance of FPELG. It can be seen clearly from Fig. 12 that the in-cylinder pressure is high when the early or advanced ignition timing (i.e., IG before TDC) occurs. While low and double peaks of in-cylinder pressure are formed when the IG is retarded. This behaviour is similar to the experimental results published in a previous study by Ahmed T et al. [18]. However, if the ignition timing occurs too early, the most of mixture (air–fuel) is burned before the piston reaches TDC. Thus, low pressure is produced during the

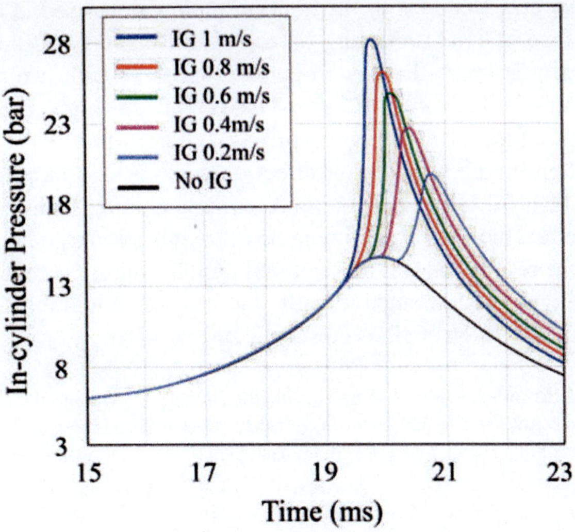

Fig. 12 Simulation results of the in-cylinder pressure against time using different ignition timing

expansion stroke, resulting in low output power and influencing the performance of the FPELG. While, if the ignition timing occurs retard or after TDC, the most of mixture must burn during the expansion stroke. Thus, because the volume of the cylinder expands and the pressure decreases during expansion stroke in accord with the ideal gas law, the low pressure is produced when the combustion occurs in the same direction as piston movement to BDC. Can be concluded that the ignition timing has a significant effect on the stability and performance of the FPELG. Therefore, further investigation into the effect of ignition timing on combustion characteristics is required. In addition, the ignition timing must be optimized in order to achieve maximum engine performance. Finally, in order to attain high engine stability, the control of ignition timing is required, which will be the topic of our next study.

5 Conclusion

FPELG is an engine that lacks a crankshaft and the cycle to cycle variation is considered as the most significant factor influencing on its stability and performance. Therefore, in the present work, the FPELG simulation model was designed. The cycle to cycle variation sub-model was proposed. In addition, the ignition timing sub-model was also proposed. MATLAB/Simulink software was used to build and run the FPELG model. The simulation model was validated using experimental data, and the findings of validation indicate that the model is valid for accurately predicting results. The piston displacement, the piston velocity, and the pressure for both bounce and combustion cylinders were measured using model results in order to investigate the CCV and its effect on the stability and performance of FPELG. Analysis of the simulation model's result showed that the CCV occurs when certain parameters change during engine operation, one of which is the IG variation. Thus, several scenarios of the ignition timing were used in order to investigate the influence of ignition timing on the stability and performance of FPELG. The results demonstrated that low in-cylinder pressure with two peaks was obtained when the retard IG occurred (IG after TDC). While the high in-cylinder pressure was provided when the advanced IG occurred (IG before TDC). The results indicate that the ignition timing has a major effect on combustion characteristics; thus, this reflects on the FPELG's performance. According to these findings, additional research into the effect of the ignition timing on the combustion characteristics is needed. In addition, the results showed that optimal ignition timing is required in order to achieve high engine performance. Finally, the control of ignition timing is required in order to achieve high engine stability.

Acknowledgements The authors gratefully acknowledge the support of this work by the Centre for Automobile and Electric Mobility (CAREM), Universiti Teknologi PETRONAS, 32610 Seri Iskandar, Perak Darul Ridzuan, Malaysia Perak, Malaysia.

References

1. Keller M, Jochim B, Abel D, Beeckmann J, Pitsch H, Albin T (2018) Nonlinear model predictive control for the starting process of a free-piston linear generator. IFAC-PapersOnLine 51(31):632–639
2. Raheem AT, Aziz ARA, Zulkifli SA, Rahem AT, Ayandotun WB (2022) A review of free piston engine control literature—taxonomy and techniques. Alex Eng J
3. Gusev S, Ziviani D, De Paepe M (2019) Model predictive control of a free piston compressor/expander with an integrated linear motor/alternator. In IOP Conference Series: Materials Science and Engineering, 2019, 604(1): IOP Publishing, p. 012076
4. Wang Y, Chen L, Jia B, Roskilly AP (2017) Experimental study of the operation characteristics of an air-driven free-piston linear expander. Appl Energy 195:93–99
5. Hung NB, Lim O (2016) A review of free-piston linear engines. Appl Energy 178:78–97
6. Raheem AT, Aziz ARA, Zulkifli SA, Baharom MB, Rahem AT, Ayandotun WB (2022) Optimisation of operating parameters on the performance characteristics of a free piston engine linear generator fuelled by CNG–H2 blends using the response surface methodology (RSM). Int J Hydrogen Energy 47(3):1996–2016
7. Pateras PR (1928) Motor-compressor apparatus. ed: Google Patents
8. Pateras PR (1941) Motor compressor of the free piston type. ed: Google Patents
9. Hung NB, Lim OT (2014) A study of a two-stroke free piston linear engine using numerical analysis. J Mech Sci Technol 28(4):1545–1557
10. Zhao Z, Wu D, Zhang Z, Zhang F, Zhao C (2014) Experimental investigation of the cycle-to-cycle variations in combustion process of a hydraulic free-piston engine. Energy 78:257–265
11. Hanipah MR, Mikalsen R, Roskilly A (2015) Recent commercial free-piston engine developments for automotive applications. Appl Therm Eng 75:493–503
12. Raheem AT, Aziz ARA, Zulkifli SA, Rahem AT, Ayandotun WB (2022) Development, validation, and performance evaluation of An Air-Driven Free-Piston linear expander numerical model
13. Mikalsen R, Roskilly AP (2007) A review of free-piston engine history and applications. Appl Therm Eng 27(14–15):2339–2352
14. Farmer H (1947) Free-piston compressor-engines. Proceedings of the Institution of Mechanical Engineers 156(1):253–271
15. Jia B, Zuo Z, Feng H, Tian G, Smallbone A, Roskilly A (2016) Effect of closed-loop controlled resonance based mechanism to start free piston engine generator: Simulation and test results. Appl Energy 164:532–539
16. Feng H et al (2016) Research on combustion process of a free piston diesel linear generator. Appl Energy 161:395–403
17. Jia B, Zuo Z, Feng H, Tian G, Roskilly A (2014) Investigation of the starting process of free-piston engine generator by mechanical resonance. Energy Procedia 61:572–577
18. Raheem AT, Aziz ARA, Zulkifli SA, Ayandotun WB, Zainal EZ, Elfakki SM (2021) Experimental analysis for the influence of ignition time on combustion characteristics of a free piston engine linear generator. In Journal of Physics: Conference Series, 1793(1): IOP Publishing, p 012051
19. Jia B, Mikalsen R, Smallbone A, Zuo Z, Feng H, Roskilly AP (2016) Piston motion control of a free-piston engine generator: A new approach using cascade control. Appl Energy 179:1166–1175
20. Yu H, Xu Z, Zhang Q, Liu L, Hua R (2019) Two-stroke thermodynamic cycle optimization of a single-cylinder free-piston engine generator. Adv Mater Sci Eng, 2019
21. Mikalsen R, Roskilly AP (2008) The design and simulation of a two-stroke free-piston compression ignition engine for electrical power generation. Appl Therm Eng 28(5–6):589–600
22. Ba HN, Lim O, Iida N (2014) Simulation study of si-hcci transition in a two-stroke free piston engine fuelled with propane. SAE Tech Pap 1

23. Ayandotun WB, Aziz ARA, Karim ZAA, Mohammed SE, Raheem AT, Ismael MA (2021) Investigation on the combustion and performance characteristics of a DI free piston linear generator engine fuelled with CNG-CO2 blend. Appl Therm Eng 198:117441
24. Ismael MA et al (2021) Experimental study on combustion stability and performance of hydrogen-enriched compressed natural gas of a free-piston linear generator. Int J Hydrogen Energy 46(79):39536–39547
25. Ayandotun WB, Aziz ARA, Ismael MA, Mohammed SE, Raheem AT (2022) Comparative assessment of the performance and combustion attributes of a free piston linear generator fuelled with CNG and CNG-CO2. Case Stud Therm Eng, p 102345
26. Mikalsen R, Roskilly A (2010) The control of a free-piston engine generator. Part 2: Engine dynamics and piston motion control. Appl Energy 87(4):1281–1287
27. Robinson MC, Clark N (2014) Fundamental analysis of spring-varied, free piston, Otto engine device. SAE Int J Engines 7(1):195–220
28. Goto S et al. (2014) Development of free piston engine linear generator system part 2-investigation of control system for generator. SAE Tech Pap 0148–7191
29. Ibrahim AA, Aziz ARA, Zainal Abidin EZ, Zulkifli SA (2011) Operation of free-piston linear-generator engine using mosfet and igbt drivers. J Appl Sci, pp 1–6
30. Zhou Y, Sofianopoulos A, Lawler B, Mamalis S (2020) Advanced combustion free-piston engines: A comprehensive review. Int J Engine Res 21(7):1205–1230
31. Wang X, Chen F, Zhu R, Yang G, Zhang C (2018) A review of the design and control of free-piston linear generator. Energies 11(8):2179
32. Zhang C et al (2018) A free-piston linear generator control strategy for improving output power. Energies 11(1):135
33. Chen F et al. (2017) A novel stable control strategy of single cylinder free-piston linear generator. In 2017 IEEE International Conference on Cybernetics and Intelligent Systems (CIS) and IEEE Conference on Robotics, Automation and Mechatronics (RAM), IEEE, pp 587–592
34. Mikalsen R, Jones E, Roskilly A (2010) Predictive piston motion control in a free-piston internal combustion engine. Appl Energy 87(5):1722–1728
35. Li K, Zhang C, Sun Z (2015) Precise piston trajectory control for a free piston engine. Control Eng Pract 34:30–38
36. Zhang C, Li K, Sun Z (2015) Modeling of piston trajectory-based HCCI combustion enabled by a free piston engine. Appl Energy 139:313–326
37. Zhang C, Sun Z (2016) Using variable piston trajectory to reduce engine-out emissions. Appl Energy 170:403–414
38. Feng H, Guo Y, Song Y, Guo C, Zuo Z (2016) Study of the injection control strategies of a compression ignition free piston engine linear generator in a one-stroke starting process. Energies 9(6):453
39. Xu Z, Chang S (2009) Hierarchical hybrid control of a four-stroke free-piston engine for electrical power generation. In 2009 International Conference on Mechatronics and Automation, 2009: IEEE, pp 4045–4049
40. Feng H, Song Y, Zuo Z, Shang J, Wang Y, Roskilly AP (2015) Stable operation and electricity generating characteristics of a single-cylinder free piston engine linear generator: Simulation and experiments. Energies 8(2):765–785
41. Xu Z, Chang S (2010) Improved moving coil electric machine for internal combustion linear generator. IEEE Trans Energy Convers 25(2):281–286
42. Jia B, Zuo Z, Tian G, Feng H, Roskilly AP (2015) Development and validation of a free-piston engine generator numerical model. Energy Convers Manage 91:333–341

Corrosion Behavior of SS304 in Hydrogen Chloride-Containing Environments

Ahmad Zaki Abas, Azmi Mohammed Nor, Muhammad Firdaus Suhor, and Nik Mohd Radi Nik Mohamed Daud

Abstract The corrosion behavior of stainless steel SS304 was investigated in three major aqueous environments, namely, dry, water-saturated, and free-water conditions, and in two different methanol environments, namely, dry and water-saturated environments with 99% methanol + 1% water. In all the experiments, hydrogen chloride gas was generated through the reaction between sulfuric acid and sodium chloride. An inert carrier gas such as argon then pushed the hydrogen chloride gas to flow over desiccants to make the latter dry. The dry hydrogen chloride gas would then flow into the second flask to which a pre-weighted stainless-steel specimen was exposed at ambient conditions for 48 h. The specimen was retrieved and subjected to surface and chemical analysis using scanning electron microscopy (SEM)/EDAX. The presence of chloride would affirm the adsorption of chloride ions on the metal surface and vice versa. The specimen was then weighted to determine corrosion rate. The results suggested that stainless steel would not corrode in dry hydrogen chloride gas. However, when saturated with moisture, its corrosion rate would increase by about one order of magnitude. In free-water environment, corrosion rate would basically triple. This was because hydrogen chloride gas would react with moisture or water, turning into hydrochloric acid. Being a reducing acid, hydrochloric acid would attack the Cr_2O_3 layer, leading to general and/or localized corrosion. The presence of methanol seems to have increased the corrosion rate of stainless steels by a factor of three to four.

Keywords Methanol · Stainless steel · Localized corrosion · Methyl chloride

---

A. Z. Abas · A. Mohammed Nor (✉) · M. F. Suhor · N. M. R. N. M. Daud
PETRONAS Research Sdn Bhd, 43000 Kajang Selangor, Malaysia
e-mail: azminr@petronas.com

A. Z. Abas
e-mail: ahmadzakiabas@petronas.com

M. F. Suhor
e-mail: mfirdaus_suhor@petronas.com

N. M. R. N. M. Daud
e-mail: radi.daud@petronas.com

F. Ahmad et al. (eds.), *ICREEM 2022*, Lecture Notes in Mechanical Engineering,
https://doi.org/10.1007/978-981-99-5946-4_20

1 Introduction

Stainless steel is a versatile material with widespread applications in numerous corrosive conditions. Austenitic stainless steels are an extraordinary family of stainless-steel alloys that have exceptional corrosion resistance and equally impressive mechanical properties which allow them to be used as a construction material in various aggressive environments such as heat exchanger systems, drilling platforms, wastewater treatment, and desalination plants. The most popular austenitic grades have been typed 304 [1].

Stainless steel austenitic 304 are used in gas processing plants for their corrosion resistance. Although stainless steels have extremely good general resistance, stainless steels are nevertheless susceptible to pitting corrosion. This localized dissolution of an oxide-covered metal in specific aggressive environments is one of the most common and catastrophic causes of failure of metallic structures.

Pitting is a localized form of corrosion in which metallic dissolutions occur at vulnerable sites and/or regions on the metal surface resulting in the formation of cavities in passivated metals or alloys that are exposed to aqueous, nearly neutral solutions containing aggressive anions. It generally occurs in chloride, halide, or bromide solutions. If a fault in the passive layer or a surface defect results in the local destruction of the former, dissolution of the steel underneath leads to a build-up of positively charged metallic ions, which in turn causes negative charges (e.g. chloride ions) to migrate near the defect. A significant number of researches have been done in order to study the initiation and the growth of pitting on traditional metallic materials [2].

This paper studies the corrosion behavior of stainless steel SS304 in various conditions of hydrogen chloride-containing environments to understand the mechanism by which hydrogen chloride gas led to general corrosion and pitting of gas processing plant. Understanding the mechanism is vital before a suitable corrosion control method could be proposed.

2 Methodology

2.1 Materials

The material used in this work was stainless steel SS 304, which was obtained from affected gas processing plant, as shown in Fig. 1. The dimensions were 70 mm in length, 7 mm in width, and 3 mm in thickness. The chemical composition is given in Table 1.

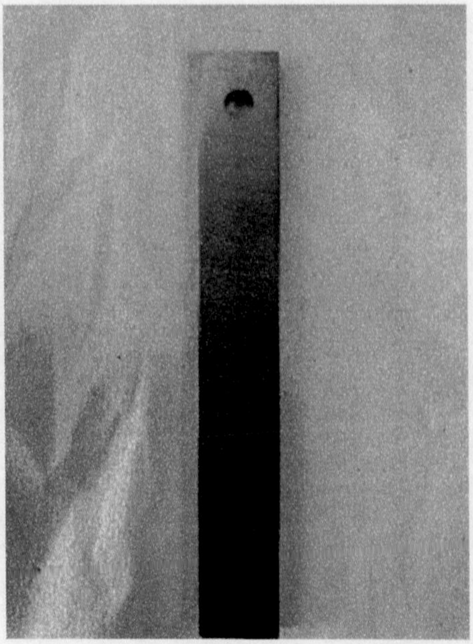

Fig. 1 Sample used in experiment

Table 1 Chemical compositions of stainless steel SS304 used in the present study (balance Fe)

	C	Cr	Ni	Mn	S	Si
Weight percentage, wt %	0.055	18.2	8.10	8.10	0.001	0.35

3 Equipment and Test Conditions

Figure 2 shows the general experimental setup for the study. In the experiments, hydrogen chloride gas (HCl) was generated through the reaction of sulfuric acid and sodium chloride. The concentration of HCl gas generated was 675 ppm. Relatively dry argon gas was used to push the HCl gas over the pre-weighted stainless-steel specimen hung in a second flask and finally into a beaker containing water for bubbling. The experimental setup was adjusted accordingly for the different test conditions. In all the experiments, the specimens were exposed to hydrogen chloride for 48 h.

As shown in Fig. 2, dry HCl was achieved by pushing the generated HCl gas to flow through silica gel before entering the second flask that contained the specimen. Water-saturated HCl was obtained by flowing argon gas into a flask that contained water at its bottom before the argon gas flowed into the flask in which HCl was generated and subsequently into the flask containing the specimen as illustrated in Fig. 3. As for free-water conditions, a separation funnel filled up with deionized water was mounted on the flask that contained the specimen; water droplets continuously dripped onto the specimen inside the flask as depicted in Fig. 4. On the other hand,

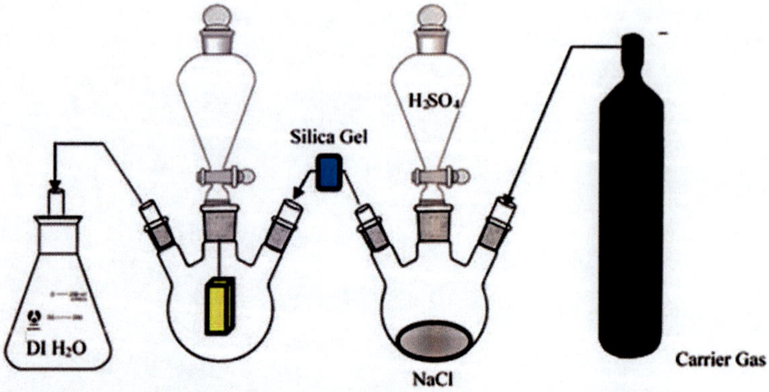

Fig. 2 Experimental setup for the effect of dry hydrogen chloride gas on stainless steel

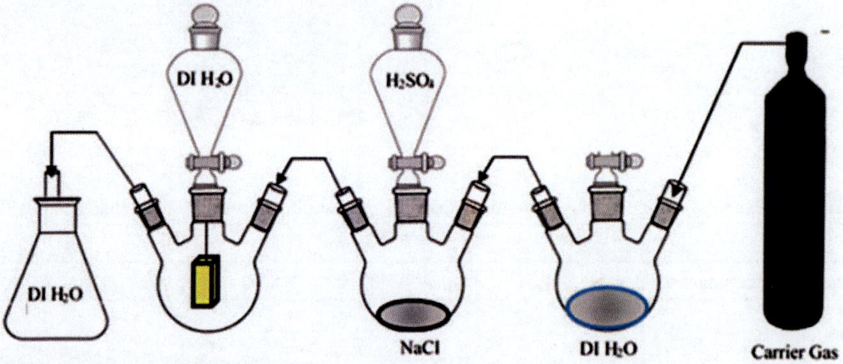

Fig. 3 Experimental setup for the effect of water-saturated hydrogen chloride gas on stainless steel

methanolic conditions were obtained by immersing the specimen-containing flask with methanol as schematically shown in Figs. 5 and 6.

Upon completion of the tests, the specimens were photographed and analyzed using Scanning Electron Microscopy (SEM) and Energy Dispersive X-Ray Analysis (EDAX). Since the corrosion products seemed fragile, the specimens were dried using tissues for wet test conditions. The specimens were also cleaned before they were weighted again.

4 Results and Discussion

Figure 7 shows the stainless-steel specimen after exposure to dry hydrogen chloride condition. The specimen did not show any sign of corrosion, and the calculated corrosion rate was 0.01 mm/yr. As shown in Fig. 8, the SEM image also reveals that

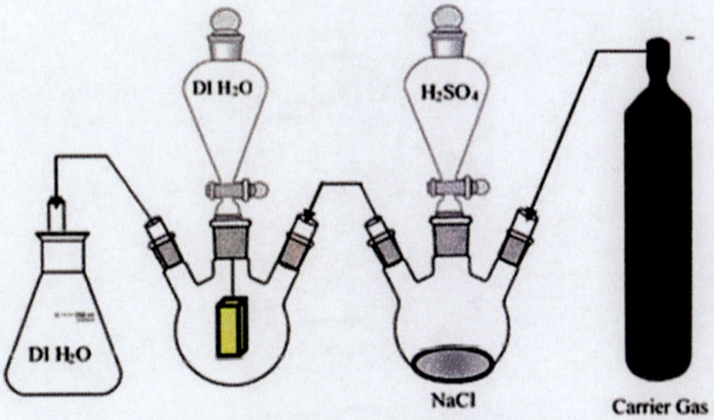

Fig. 4 Experimental setup for the effect of free water hydrogen chloride gas on stainless steel

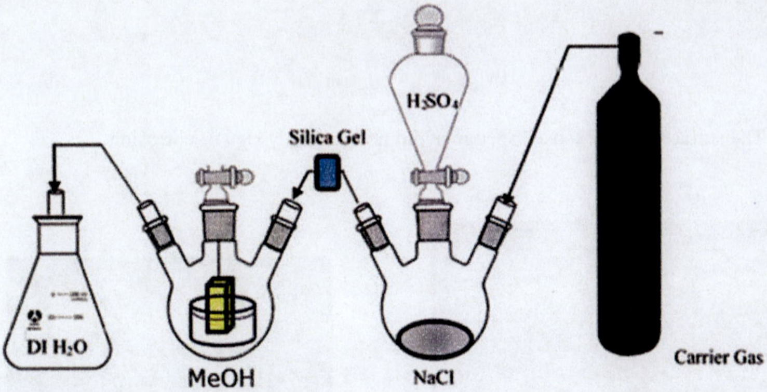

Fig. 5 Experimental setup for the effect of dry hydrogen chloride gas and methanol on stainless steel

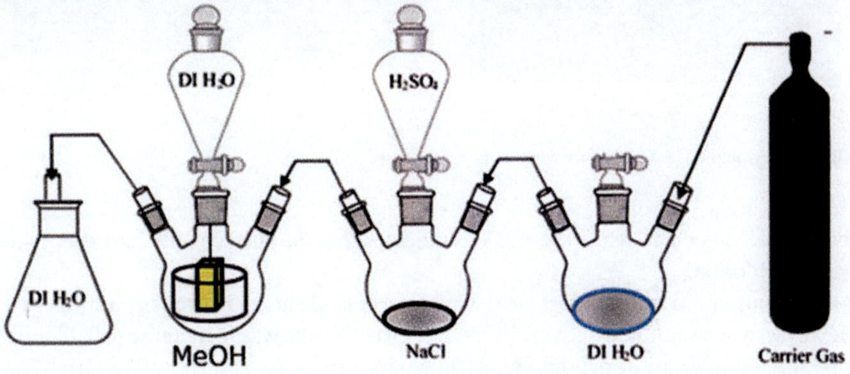

Fig. 6 Experimental setup for the effect of water-saturated hydrogen chloride gas and methanol on stainless steel

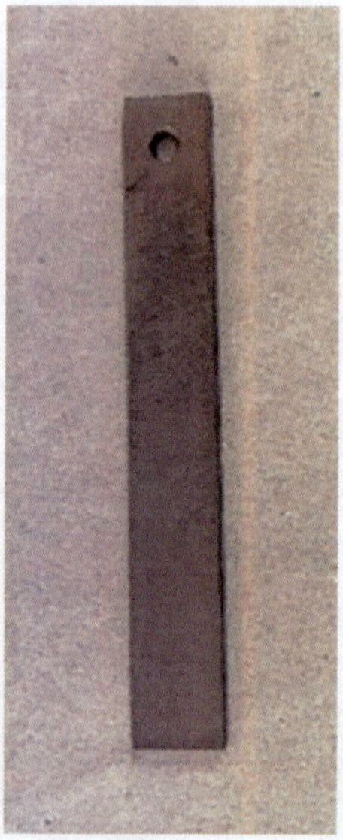

Fig. 7 The surface of the exposed specimen did not show any sign of corrosion

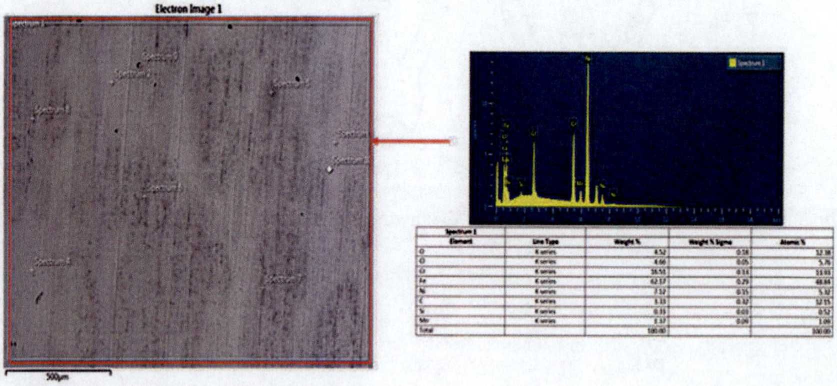

Fig. 8 SEM and EDAX show 4.6wt % concentration of chlorides

the surface was clean while the EDAX indicates that the surface only had about 4.6 wt% of chloride.

It was observed that stainless steel would not corrode in dry hydrogen chloride gas. However, when saturated with moisture, its corrosion rate would increase to 0.14 mm/y or about one order of magnitude as shown by Fig. 9. As shown by Fig. 10, SEM also revealed that the surface was white layer while EDAX indicated that the surface only had about 33.5 wt% of chloride.

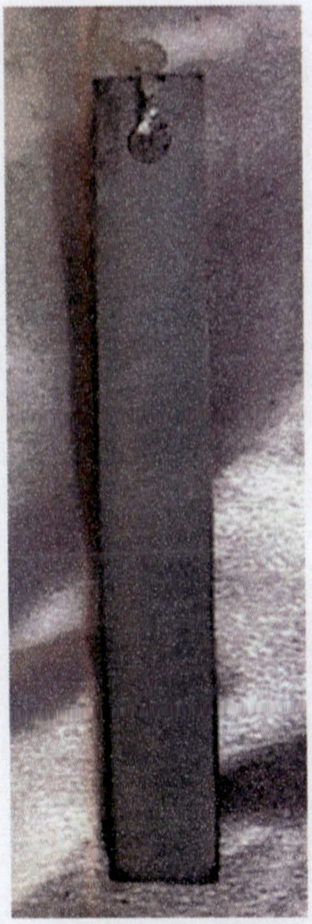

Fig. 9 White layer on the surface of stainless-steel specimen

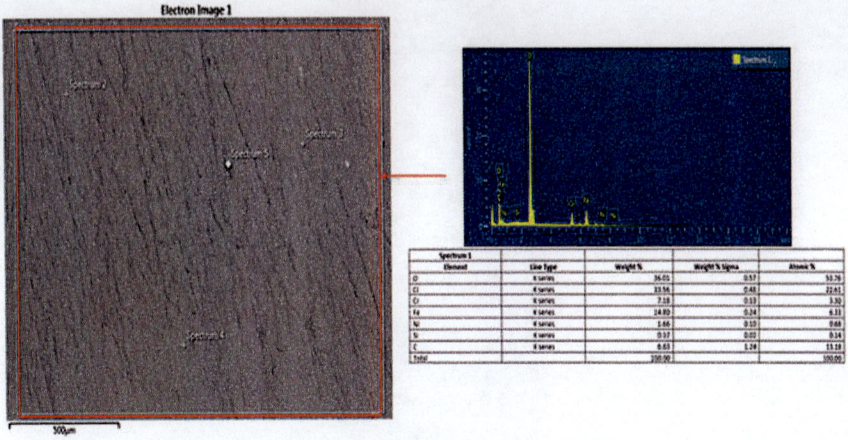

Spectrum 1				
Element	Line Type	Weight %	Weight % Sigma	Atomic %
O	K series	36.01	0.57	53.76
Cl	K series	33.56	0.48	22.61
Cr	K series	7.18	0.13	3.30
Fe	K series	14.80	0.24	6.31
Ni	K series	1.66	0.10	0.68
Si	K series	0.17	0.02	0.14
C	K series	6.63	1.24	13.19
Total		100.00		100.00

Fig. 10 SEM and EDEX show 33.5 wt % concentration of chlorides

Figure 11 shows the stainless-steel specimen after exposure to free-water condition. The specimen shows general corrosion surface, and the calculated corrosion rate was 0.33 mm/yr. This was because HCl gas would react with moisture or water, turning into hydrochloric acid. Being a reducing acid, hydrochloric acid would attack the Cr_2O_3 layer, leading to general and/or localized corrosion. As shown in Fig. 12, the SEM image also reveals that the surface was general corrosion while the EDAX indicates that the surface only had about 6.5 wt% of chloride.

Figure 13 shows the stainless-steel specimen after exposure to dry hydrogen chloride with methanol. The specimen shows pitting on the surface and the calculated corrosion rate was 0.56 mm/yr. As shown in Figs. 14, the SEM also reveals that the surface was attacked by pitting while the EDAX indicates that the surface only had about 9.6 wt% of chloride.

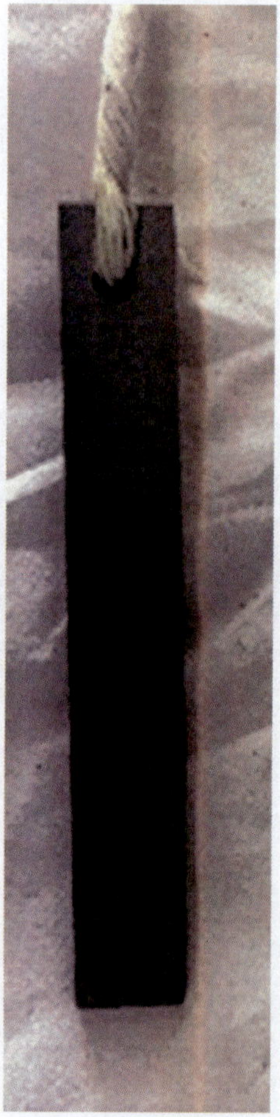

Fig. 11 General corrosion on the surface of stainless-steel specimen

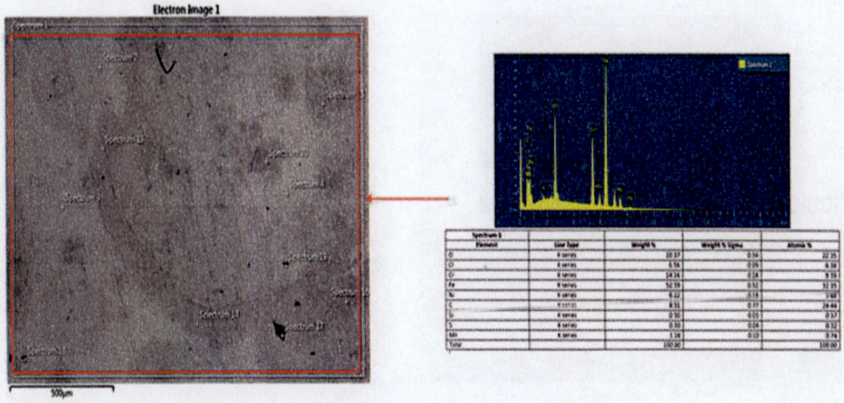

Fig. 12 SEM and EDEX show 6.5 wt % concentration of chlorides

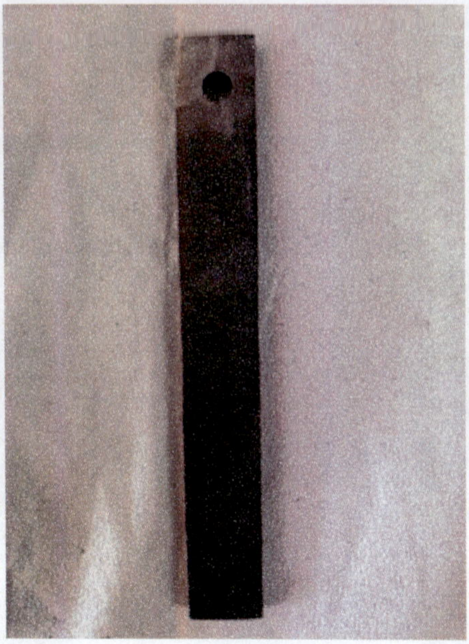

Fig. 13 Pitting corrosion on the surface of stainless-steel specimen

The presence of methanol seems to have increased the corrosion rate of stainless steels by a factor of four. The possible reason for the increase was that HCl gas also dissolved in methanol (CH_3OH), producing corrosive methyl chloride by the following chemical reaction:

$$CH_3OH + HCl = H_2O + CH_3Cl$$

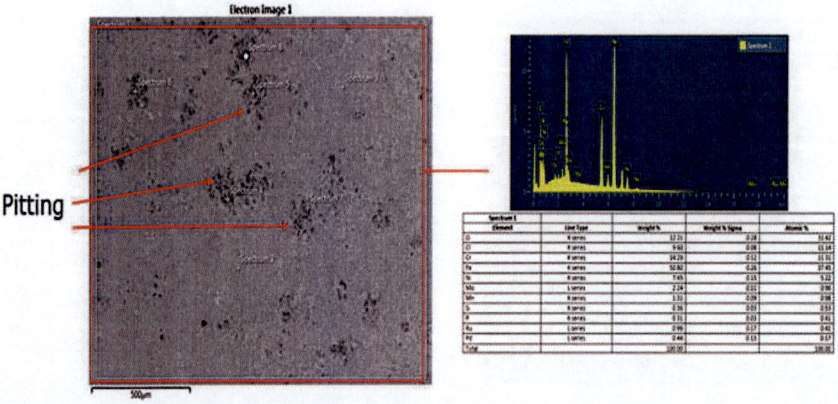

Fig. 14 SEM image shows pitting on surface of the specimen and EDEX shows 9.6 wt % concentration of chlorides

This was evidenced from the fact that the pH of methanol decreased from pH5 to pH1 in Fig. 15 as HCl gas flowed into the bottom flask where the specimens were hung. Interestingly, the color of methanol also changed from dark yellow to colorless; *methyl chloride is colorless*. That was why stainless steel still corroded in methanol dissolved with HCl gas even in the virtual absence of water.

Figure 16 shows the stainless-steel specimen after exposure to water-saturated hydrogen chloride with methanol. The specimen shows general corrosion on the surface and the calculated corrosion rate was 0.35 mm/yr. As shown by Fig. 17, the

Before the test:pH 5 After the test: pH 1

Fig. 15 The color of methanol turned from dark yellow to colorless solution when hydrogen chloride gas dissolved in it, suggesting the formation of methyl chloride

Fig. 16 General corrosion on the surface of stainless-steel specimen

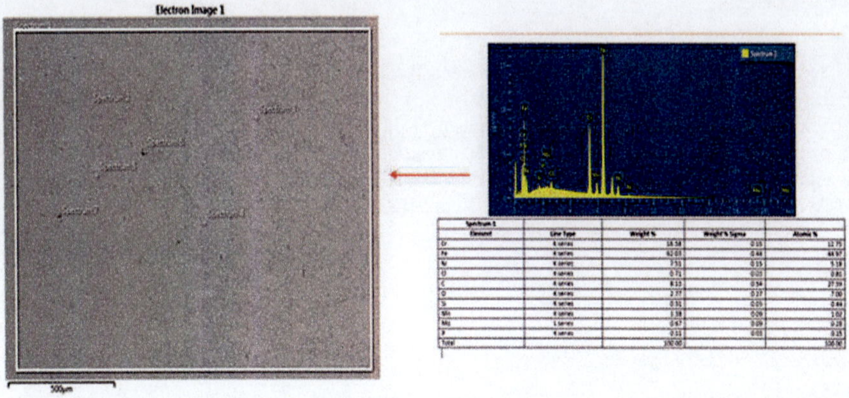

Fig. 17 SEM image shows general corrosion on surface of the specimen and EDEX shows 0.7 wt % concentration of chlorides

SEM image also reveals that the surface was covered by general corrosion while the EDAX indicates that the surface only had about 0.7 wt% of chloride. The pH of methanol decreased from pH5 to pH1.

It is also interesting to note that dry HCl gas dissolved in methanol caused a slightly higher corrosion rate than water-saturated HCl dissolved in methanol. This could have probably been due to the lower concentration of HCl gas that dissolved in methanol as much was already absorbed in moisture.

5 Conclusions

Thus, the results of the above experiments indicate the following:

1. The corrosion rates of stainless steels in dry hydrogen chloride (HCl) gas were relatively very low and virtually negligible.
2. The corrosion rates of stainless steels in dissolved HCl in MeOH + 1 wt% water were higher than those without MeOH.
3. The corrosion rates of stainless steels in HCl-dissolved methanol were generally of the same order of magnitude.

Acknowledgements The authors wish to acknowledge the contributions made by Mr. Amiruddin Hassan, Dr M Syamzareen Rafeen, and Mr. Sharizal M Azam Shah Wong in the execution of the project.

References

1. Sedriks J (1996) Corrosion of stainless steels." John Wiley & Sons, 2nd Edition, New York, USA
2. Szklarska-Smialowska Z (2002) Mechanism of pit nucleation by electrical breakdown of the passive film. Corros Sci 44(5):1143-1149. https://doi.org/10.1016/S0010-938X(01)00113-5
3. Amin MA, Rehim SSAE, Sherbini EEFE (2006) AC and DC studies of the pitting corrosion of Al in perchlorate solutions. Electrochim Acta 51(22):4754–4764

Analysis of Tensile and Hardness of Empty Fruit Bunch/Chitosan Reinforced with Graphene Oxide Composite Film

Aein Afina Mohd Redzuan, Muhammad Imthiaz Daud Mohamad Zamani, Adel Mohammed Al-Dhahebi, and Mohamed Shuaib Mohamed Saheed

Abstract Natural fibre-reinforced polymer is gaining attraction in automotive industry recently due to environmental concerns. It is also driven by the need to manufacture lightweight and fuel-efficient vehicles. Herein, empty fruit bunch/chitosan reinforced with graphene oxide (EFB/C/GO) composite's properties were investigated. The developed EFB/C/GO through hot compression moulding shows good dispersion of fillers with the presence of voids on the surface. Functional groups were also observed in the FTIR analysis. The tensile strength analysis shows the EFB/C/GO registered lower value (11.42 MPa) as compared to EFB/C composite (31.19 MPa). However, the elastic modulus is more than 340 times larger as compared to EFB/C. The Rockwell hardness of EFB/C/GO is also four times higher than the EFB/C composite. This indicates EFB/C/GO has a higher resistance to change in shape but brittle while EFB/C is stiffer and ductile with lower resistance to shape changes.

Keywords Mechanical properties · Nanostructured materials · Polymer

A. A. M. Redzuan · M. I. D. M. Zamani · A. M. Al-Dhahebi · M. S. M. Saheed (✉)
Department of Mechanical Engineering, Universiti Teknologi PETRONAS, 32610 Seri Iskandar, Perak, Malaysia
e-mail: shuaib.saheed@utp.edu.my

A. A. M. Redzuan · A. M. Al-Dhahebi · M. S. M. Saheed
Center of Innovative Nanostructures and Nanodevices (COINN), Universiti Teknologi PETRONAS, 32610 Seri Iskandar, Perak, Malaysia

M. I. D. M. Zamani
Infineon Technologies, Free Trade Zone Batu Berendam, 75350 Batu Berendam, Melaka, Malaysia

F. Ahmad et al. (eds.), *ICREEM 2022*, Lecture Notes in Mechanical Engineering,
https://doi.org/10.1007/978-981-99-5946-4_21

1 Introduction

Natural fibres such as cotton, flax, jute, and empty fruit bunch (EFB) have attracted the interest in recent years due to its renewable nature and eco-friendly. It has various advantages such as low cost, better thermal and insulating properties, low density, biodegradability, and acceptable specific properties. Naturally available fibres are proficient ingredients that are potent to replace the synthetic materials such as glass and carbon. It is proved by involving its yields for the lesser weightiness and energy preservation claims. The outstanding characteristics of natural fibres, such as high specific modulus, lightweight, and excellent resistance to wear and tear, have given rise to their utilisation as reinforcing agents in polymer composites, including thermoset and thermoplastics [1–3]. Graphene and its derivatives such as graphene oxide (GO) has been shown to have multiple desired properties, including strong molecular barrier abilities, electrical conductivity, and mechanical strength. The oxygenated groups in GO possess several advantages over pristine graphene, including improved solubility and surface functionalisation, of which supports the dispersibility in polymer matrices. As such, GO is an effective fillers in polymer nanocomposites [4]. The reinforcement of GO in the composite materials for sector such as automotive helps in weight reduction up to 50% compared to the traditional steel components. It also enhances durability, flexibility, and strength of the car components [5]. In this work, mechanical properties such as tensile and hardness of polymer nanocomposite based on EFB/chitosan reinforced with GO is studied. Chitosan is considered as a biopolymer that is synthesised naturally, cheap, and biodegradable. Its drawbacks such as poor mechanical strength, low thermal stability and barrier properties could be overcome with the inclusion of fillers such GO. The composite films were prepared by using hot compression moulding technique.

2 Methodology

2.1 Materials

The EFB fibres and GO were purchased from Szetech Engineering Sdn. Bhd., Selangor Malaysia and Ugent Tech Sdn. Bhd., Selangor, Malaysia, respectively. For the other materials such as sodium hydroxide (NaOH), acetic acid and chitosan were all bought from the Sigma Avantis Sdn. Bhd., Malaysia. Long EFB fibres were granulated to short fibre using a granulator machine.

2.2 Sodium Hydroxide Treatment of EFB Fibres

Sodium hydroxide (NaOH) pellets are weighted and mixed in deionised water to achieve a solution and stirred using a magnetic stirrer at 200 rpm to dissolve the pellets. Once the NaOH dissolved completely, the granulated EFB fibres were weighed and soaked in 0.1M NaOH solution before combining with the chitosan and GO. Pre-treatment of EFB fibres was done by using 5% NaOH with the ratio of 50 g/L of EFB fibres and 50 g/L of 0.1M NaOH with 1 L of deionised water in a 1L beaker. The beaker was put on a thermocouple hot plate mechanical stirrer at 200 rpm for 1 h at 80 °C. After an hour, the soaked EFB fibres were later washed with deionised water 3 times to fully remove the excess NaOH solution and put on a glass tray. The fibres were dried in an oven for 24 h at 80 °C with 30% fan speed [6, 7].

2.3 Ball Milling and Sieving of EFB Fibres

After the pre-treatment and drying, the short EFB fibres were further subjected to ball mill machining to achieve micron size fibres. The planetary ball mill machine specifications were set to 390 rpm, four repetitions with a 5-min interval to mill the EFB fibres to micron size. This process was done multiple times until all the EFB fibres were milled to the desired size. The EFB fibres were to be put in corundum ball mill jar with 8 ceramic balls of different diameters. The ball mill was filled with ball to material (BMR) ratio of 2:1. The milled EFB fibres are later then sieved using a sieving machine with a 63 μm sieve. The process time of the sieving machine was set to 10 min to fully sieve the fibres through the sieve and is collected at the bottom plate [6, 8, 9].

2.4 Preparation of EFB/chitosan Reinforced with GO

Two types of samples were prepared, namely, EFB/Chitosan (EFB/C) and EFB/Chitosan reinforced with GO (EFB/C/GO). 33.33 g of chitosan was added into 100 ml of DI water for 5 min on a hot plate mechanical stirrer with rotation speed of 200 rpm at 80 °C. 25 g of EFB fibres are then added into the solution and stirred for 10 min. For EFB/C/GO sample, 4 g of GO was added as well and stirred for another 10 min. Then, 10 ml of acetic acid was added into both beakers with EFB/C and EFB/C/GO and stirred until homogenous state was achieved. Both samples were compounded using hot compression moulding machine. The samples were hot-pressed in three steps. Samples were hot-pressed at 120 °C, 25 tonnes for 15 min and cooled at room temperature for 3 min, then second hot-pressed at 110 °C, 25 tonnes for 10 min and cooled down at room temperature for 3 min. Lastly, samples were then hot-pressed for the third time at 110 °C, 25 tonnes for 10 min to fully remove the moisture from the composites [10].

2.5 *Tensile Strength and Hardness Measurements*

The hot-pressed composites were moulded with dimensions of 135 × 10 × 2.8 mm^3 and subsequently tested using GOTECH Universal Testing Machine 01/2530 in accordance with ASTM D638 for tensile strength analysis. Each type of composite was tested for three batch [11]. Composite hardness was measured by using Indentec Hardness Testing Machine which is also a Rockwell hardness machine and gives the reading of HRR. The measurement protocol was done in accordance with ASTM D785. The machine was set up with a steel ball indenter with a precise diameter of 0.5 inches and forced onto the surface of the sample with a preload of 10 kg. The test measures the depth of penetration by the steel ball indenter by a huge load, which is 60 kg compared to the preload's depth of penetration [12].

3 FESEM and FTIR Analyses

The characterisations of the composite film of the surface morphology were observed by the FESEM Supra V55. For Fourier Transform Infrared analysis, the FTIR Perkin Elmer Model machine was used to obtain the functional groups located in both types of the composites and the method used is Potassium Bromide (KBr) that acts as a carrier with 100% transmittance [13].

4 Result and Discussion

Figure 1 shows the FESEM images of the prepared composites with both types of samples having a clean and smooth surface. This is attributed to the alkali treatment of the EFB fibres. Hassan et al. [14] stated that a cleaner and rougher surface in theory would improve the adhesion between the EFB fibres and the polymer matrix. Agglomeration of fibres was not observed as short fibres were used instead of long fibres. This improves the homogenous mixing of fibres within chitosan. For the EFB/C/GO composite, voids were observed on the surface as shown in Fig. 1d. It should be noted that the presence of voids makes the fibre and matrix bonding to decrease in terms of the mechanical strength of the composite [15].

It can be observed that the band at 3268 and 2920 cm^{-1} corresponds to the –OH (Hydroxyl) and –CH (Alkanes) groups, respectively for the EFB/C FTIR spectrum. At the peak of 1558 and 1410 cm^{-1} corresponds to the N–H (Amide II) and $-CH_2$ (Methylene) groups, respectively. Moreover, the 1410 cm^{-1} characteristics on the deformation of the cellulose or –CH (Alkenes) deformation from lignin of the EFB fibres. The peak of 1016 cm^{-1} that is between 1000 to 1150 cm^{-1} indicates the C–O (Carbonyl) and C–O–C groups [16].

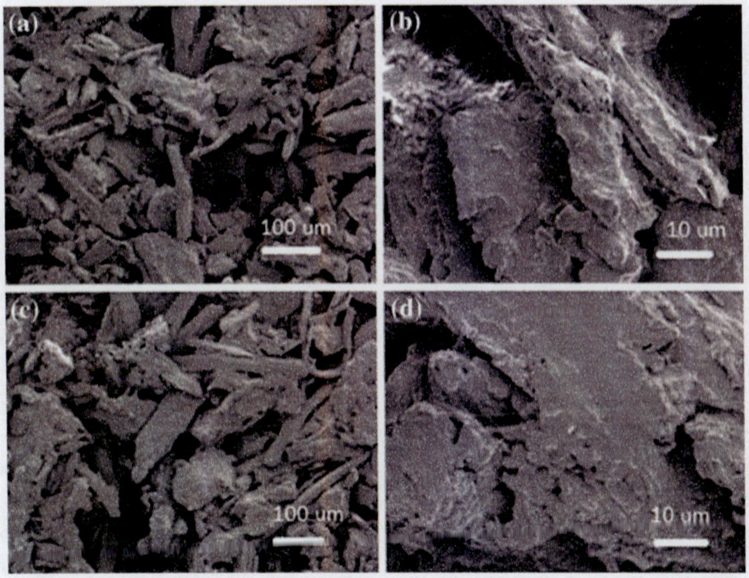

Fig. 1 FESEM images **a** EFB/C at 200x magnification, **b** EFB/C at 2 kx magnification, **c** EFB/C/GO at 200x magnification, **d** EFB/C/GO at 2 kx magnification

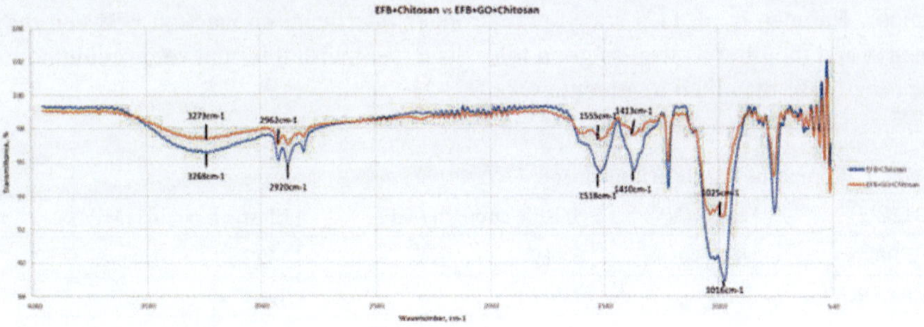

Fig. 2 FTIR spectrum for EFB/C and EFB/C/GO

FTIR spectrum for EFB/GO/CS composite shows the presence of new peak at 3273 cm^{-1}, which describes –OH (Hydroxyl) but had a lower intensity of transmittance. At the peak of 2962 cm^{-1}, corresponds to the –CH (Alkanes) group. The peak of transmittance at 1555 and 1413 cm^{-1} indicates the existence of N–H (Amide II) and $-CH_2$ (Methylene), respectively. At 1025 cm^{-1} peak, exhibits the C–O–H (Aldehyde) group which is in the graphene oxide [17]. The higher intensity transmittance of EFB/GO/CS composite compared to EFB/CS composite is due to the cross-linking of the graphene oxide with the polymer matrix chitosan and empty fruit bunch (Fig. 2).

Table 1 shows the load, tensile strength, and elastic modulus for the composite tensile measurement. The tensile strength and load of EFB/C composite is higher compared to the EFB/C/GO composite but is opposite for the elastic modulus. The tensile strength and the elastic modulus of the EFB/C/GO composite were lower than expected and the possible explanation is due to the method of manufacturing. The elastic modulus indicates the elasticity of the composite, which also relates to the deformation of the composite and the load needed to deform it whereas the tensile strength indicates the amount of maximum stress that the composite can handle. It is also implying that the tensile strength is the limit between the rupture zone and plasticity zone [18]. EFB/C composite which has a low elastic modulus means that the composite properties are the less resistance in stretching, but with a high tensile means the composite has ductile properties. As for EFB/C/GO composite, high elastic modulus means the composite has a higher resistance to stretching and the low tensile strength means the composite has brittle properties. Costa et al. [11] described that from the method of fabrication, by the type of metal mould and specifications could affect the properties of composite.

The average hardness of the sample that contains GO is higher compared to the sample without GO as indicated in Table 2. The average hardness for EFB/C/GO increased up to 395% from the EFB/C. The enhanced mechanical properties of EFB/C/GO could be from the uniform dispersion of GO particles within the matrix, which results in increase in resistance to indentation. Both samples have undergone through NaOH treatment, which can also contribute to the difference of hardness values. Rahman et al. [12] stated that stronger interfacial connection between the matrix and the filler as well as good filler dispersion with minimal voids could both be because of the alkali treatment.

Table 1 Composite tensile measurement

Sample	Load (N)	Tensile strength (MPa)	Elastic modulus (MPa)
EFB/C	873	31.19	427
EFB/C/GO	320	11.42	3904

Table 2 Rockwell hardness measurements

Sample	Hardness (HRR)	
	EFB/C	EFB/C/GO
1	5.1	16.5
2	4.9	21
3	5.1	21
4	6.2	25.7
Average	5.33	21.05

5 Conclusions

In summary, polymeric nanocomposites have been successfully prepared and its mechanical properties were investigated through the analyses of tensile and hardness. The prepared nanocomposites, EFB/C and EFB/C/GO, show good dispersion of fillers with the polymer matrix, but voids are also observed to be present within the surface. Functional groups such as –OH (Hydroxyl), –CH (Alkanes), N–H (Amide II), $-CH_2$ (Methylene), and C–O (Carboxyl) are shown in the FTIR analysis. Tensile strength and elastic modulus for both composites show contrasting outcome. EFB/CS has higher tensile strength at 31.19 MPa compared to EFB/GO/CS, which is at 11.42 MPa. This indicates the ductility of EFB/C composite. For elastic modulus, EFB/GO/CS obtained a value of 3904 MPa compared to EFB/CS, which only obtained a value of 427 MPa, indicating the higher resistance to stretching of EFB/GO/CS compared to the EFB/CS. Furthermore, EFB/GO/CS has a higher average hardness value of 21.05 HRR compared to EFB/CS at 5.33 HRR. These differing observations could be due to high variability in the natural fibre properties, moulding technique employed in this study, and the use of biopolymer. In the future research direction, improvement will be carried out in terms of processes and techniques in order to achieve the requirement of industries.

Acknowledgements The authors gratefully appreciate the technical assistance by Mr Musa M Yusuf and financial support for the research activities through YUTP-FRG (015LC0-275).

References

1. Jawaid M, Abdul Khalil HPS, Abu Bakar A (2010) Mechanical performance of oil palm empty fruit bunches/jute fibres reinforced epoxy hybrid composites. Mater Sci Eng A 527:7944–7949
2. Gayathri N, Shanmuganathan VK, Joyson A et al Mechanical properties investigation on natural fiber reinforced epoxy polymer composite. Mater Today Proceed. https://doi.org/10.1016/j.matpr.2022.10.121
3. Haris NIN et al (2022) Dynamic mechanical properties of natural fiber reinforced hybrid polymer composites: a review. J Market Res 19:167–182
4. Smith AT, LaChance AM, Zeng S, Liu B, Sun L (2019) Synthesis, properties, and applications of graphene oxide/reduced graphene oxide and their nanocomposites. Nano Mater Sci 1:31–47
5. Elmarakbi A, Azoti W (2018) State of the art on graphene lightweighting nanocomposites for automotive applications. In: Micro and nano technologies. Elsevier, pp 1–23
6. Supian MAF, Amin KNM, Jamari SS, Mohamad S (2020) Production of cellulose nanofiber (CNF) from empty fruit bunch (EFB) via mechanical method. J Environ Chem Eng 8:103024(1–5)
7. Foo ML et al (2019) Surface-modified nanocrystalline cellulose from oil palm empty fruit bunch for effective binding of curcumin. Int J Biol Macromol 138:1064–1071
8. Setyaningsih D et al (2018) IOP conference series: earth and environmental science. Institute of Physics Publishing, vol 141
9. Zeng J, Liu L, Li J, Dong J, Cheng Z (2020) Properties of cellulose nanofibril produced from wet ball milling after enzymatic treatment vs. mechanical grinding of bleached softwood kraft fibers. BioResources 15:3809–3820

10. Guerrero P, Muxika A, Zarandona I, de la Caba K (2019) Crosslinking of chitosan films processed by compression molding. Carbohyd Polym 206:820–826
11. Costa UO et al (2020) Mechanical properties of composites with graphene oxide functionalization of either epoxy matrix or curaua fiber reinforcement. J Market Res 9:13390–13401
12. Rahman MR, Islam MN, Huque MM (2010) Influence of fiber treatment on the mechanical and morphological properties of sawdust reinforced polypropylene composites. J Polym Environ 18:443–450
13. Ravindran L, Sreekala MS, Thomas S (2019) Novel processing parameters for the extraction of cellulose nanofibres (CNF) from environmentally benign pineapple leaf fibres (PALF): structure property relationships. Int J Biol Macromol 131:858–870. https://doi.org/10.1016/j.carbpol.2018.11.064
14. Hassan A, Isa MM, Ishak ZM, Ishak N, Rahman NA, Salleh FM (2018) Characterization of sodium hydroxide-treated kenaf fibres for biodegradable composite application. High Perform Polym 30:890–899
15. Ayu RS et al (2020) Characterization study of Empty Fruit Bunch (EFB) Fibers reinforcement in Poly(Butylene) Succinate (PBS)/starch/glycerol composite sheet. Polymers 12:1571
16. Chan MY, Husseinsyah S, Ahmad Amirudin MAA (2016) Ensile and thermal properties of crosslinked chitosan/empty fruit bunch biofilms by phthalic anhydride. J Phys Sci 27 67–81
17. Gul W, Alrobei J (2021) Effect of graphene oxide nanoparticles on the physical and mechanical properties of medium density fiberboard. Polymers 13:1818
18. Callister WD (2015) Materials science and engineering: an introduction, 9th edn. John Wiley & Sons (Asia) Pte Ltd, pp 118–139

Development of Corrosion Management Framework with Equipment Susceptible to Creep

Mokhtar Che Ismail, Syamim Akhtar, and Faiq Sulieman

Abstract Petroleum refineries are complicated systems that undertake many phases of operation at a high-risk level. Over the years, major accidents in petroleum refineries led to the loss of billions of dollars, impacting asset damage, production loss and harm to the environment. Hence, it is crucial to be taken care of because of the significant impact on operational and maintenance expenditure spending, such as Capital Expenditure (CAPEX) and Operating Expenditure (OPEX). Therefore, a thorough approach is required for the safe operation of the refinery. The objective of this study is to evaluate, identify the failure and develop a corrosion management framework for the high pressure superheated steam section (HPSS2, convection section) of Ethylene Crack Gas Furnace 3 known as F-1303 in Olefins Plant, PCG Kerteh Integrated Petrochemical Complex, Terengganu. This study analyzes the failure of furnace convection coil through the material's microstructure by using several mechanical testing methods, such as metallurgical examination, chemical analysis, hardness testing and tensile testing to evaluate the damage mechanism. It is followed by a development of a framework with the detection of all possible threats, consequences and recovery plan. This will reduce the high risk of equipment failure and sustain the asset integrity of furnace convection coil.

1 Introduction

API 571, *Damage Mechanisms Affecting Fixed Equipment in the Refining Industry, Third Edition*, stated that over 60 different damage mechanisms tend to occur to process equipment in refineries. The aftermath of this damage mechanism can cause injuries, deaths, and a huge loss to a company due to the explosion caused by the failure. For Olefins Plant, the revenue loss in January 2017 due to unplanned plant shutdown has contributed to RM 4.8 million of production loss. Creep falls under

M. C. Ismail (✉) · S. Akhtar · F. Sulieman
Center for Corrosion Research, Universiti Teknologi Petronas, 32610 Perak, Seri Iskandar, Malaysia
e-mail: mokhtis@utp.edu.my

F. Ahmad et al. (eds.), *ICREEM 2022*, Lecture Notes in Mechanical Engineering,
https://doi.org/10.1007/978-981-99-5946-4_22

one of the damage mechanisms. Creep occurs due to the presence of stresses below the yield strength of metal at elevated temperatures. The temperature at which creep begins depends on the alloy composition. This metal deformation is detected through the material's microstructure with the growth of cavities and creep voids. Apart from that, the relaxation of residual stress caused by service operation at high temperatures also leads to this deformation [1]. Creep deformation is recognized in the ASME Boiler and Pressure Vessel Code, where the operating life correlates with the presence of this deformation. Creep failures are characterized by (1) the bulging or blisters in the tube, (2) thick-edged fractures often with very little obvious ductility, (3) longitudinal "stress cracks" in either or both ID and OD oxide scales, (4) external or internal oxide-scale thicknesses that suggest higher-than-expected temperatures, and (5) intergranular voids and cracks in the microstructure. Thus, it is crucial for this damage mechanism to be assessed through a corrosion management framework to ensure the safety of the plant and integrity of static equipment is maintained.

A corrosion management system (CMS) is the documentation of processes and procedures in sequence to cater the threat of the presence of corrosion on existing and future assets and assets systems [2]. This system (CMS) is developed through a framework, which intends to be applicable to a variety of asset types and industries. Corrosion management takes into consideration the risks and consequences of corrosion incidents. Corrosion threats should be diminished to a point where the benefits gained are balanced against the expenditure of resources.

Bowtie analysis is a risk evaluation method. The method takes its name from the shape of the diagram that you create, which looks like a man's bowtie. According to de Ruijter and Guldenmund [3], there are two types of bowties known as quantitative and qualitative bowties. In high-hazard industries such as oil and gas, bowtie analysis is widely used. Bowtie method is widely known for risk identification on safety. Figure 1 shows the bowtie method that covers all aspects needed in designing a framework, thus suitable for designing a corrosion management framework.

2 Methodology

2.1 Data Gathering

The Piping & Instrumentation Diagram (P&ID), GA Drawing of furnace convection coil, Manufacturing Datasheet and Mill Certificate are received from the inspection engineer. Other data are collected verbally through online meetings by communicating with inspection engineer such as the description of the incident, maintenance plan of the team, management plan on the asset integrity and other preventive measures taken after the incident of coil rupture. The most crucial data are the experimental data based on the mechanical testing.

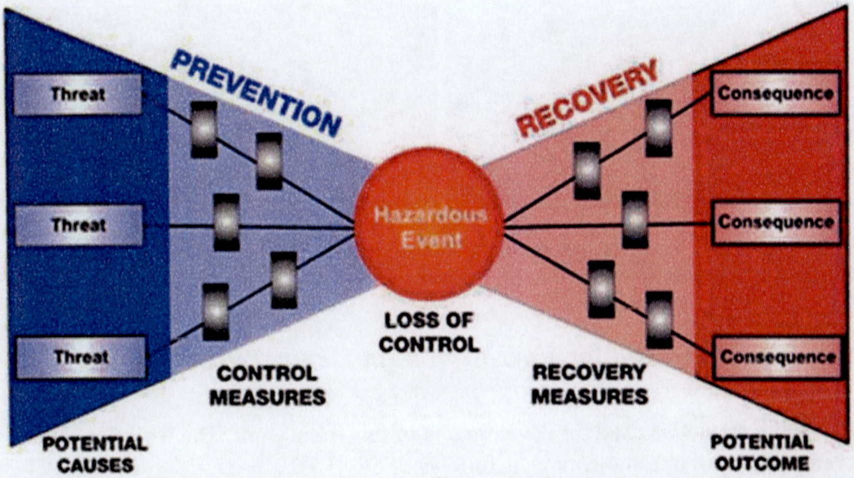

Fig. 1 Bowtie template for the development of framework

2.2 *Scanning Electron Microscopy (SEM)*

A section of the bottom lip of the fracture was examined using a FEI Quanta Scanning Electron Microscope (SEM).

2.3 *Scanning Analysis Through Optical Emission Spectroscopy (OES)*

The sample was chemically analyzed by Optical Emission Spectroscopy (OES). This is to ensure the elemental composition of diversity metals and alloys.

2.4 *Tensile Testing*

A total of three test specimens were machines from the coil and the specimens are orientated in the longitudinal direction. The tensile test was carried out with a range of temperature between 20 and 650 °C in accordance with ASTM E8/E8M.

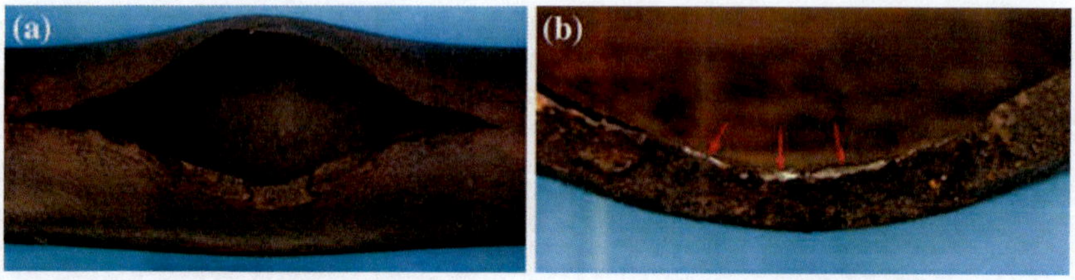

Fig. 2 **a** 1 General view of fracture on coil due to bulging, F-1303, **b** Detail of mechanical damage observed on the bottom lip of fracture

2.5 *Bowtie Method for Corrosion Management Framework*

The bowtie method is used for development of this framework. The workflow of the development of the framework is as follows:

1. Data gathering and detailed study on creep failure
2. Tabulate the Excel sheet prior to bowtie method
3. Place information into framework model.

3 Result and Discussion

3.1 *Mechanical Testing Methods*

Figure 2 shows the view of fracture on coil, F-1303. The "fish-mouth" rupture was observed on the surface, which is also called as bulging. There are a few creep failures that have been characterized and one of it is "bulging or blisters in tube". A similar failure of a coil in a petrochemical plant located in Mexico is reported with bulging due to overheating damage [4], indicating that this type of failure commonly occurs in petrochemical industries.

3.2 *Scanning Electron Microscopy*

Scanning Electron Microscope (SEM) showed that the fracture surface is covered with a surface oxide. The material's microstructure can be seen from Fig. 3. Possible micro-voids were observed on the fracture surface while Fig. 3b shows the micro-cracking neat the fracture surface. Roy et al. [5] discovered that the initiation of formation of voids is through suspension of carbides, accelerates the creep. The presence of micro-voids is one of the early signs of creep.

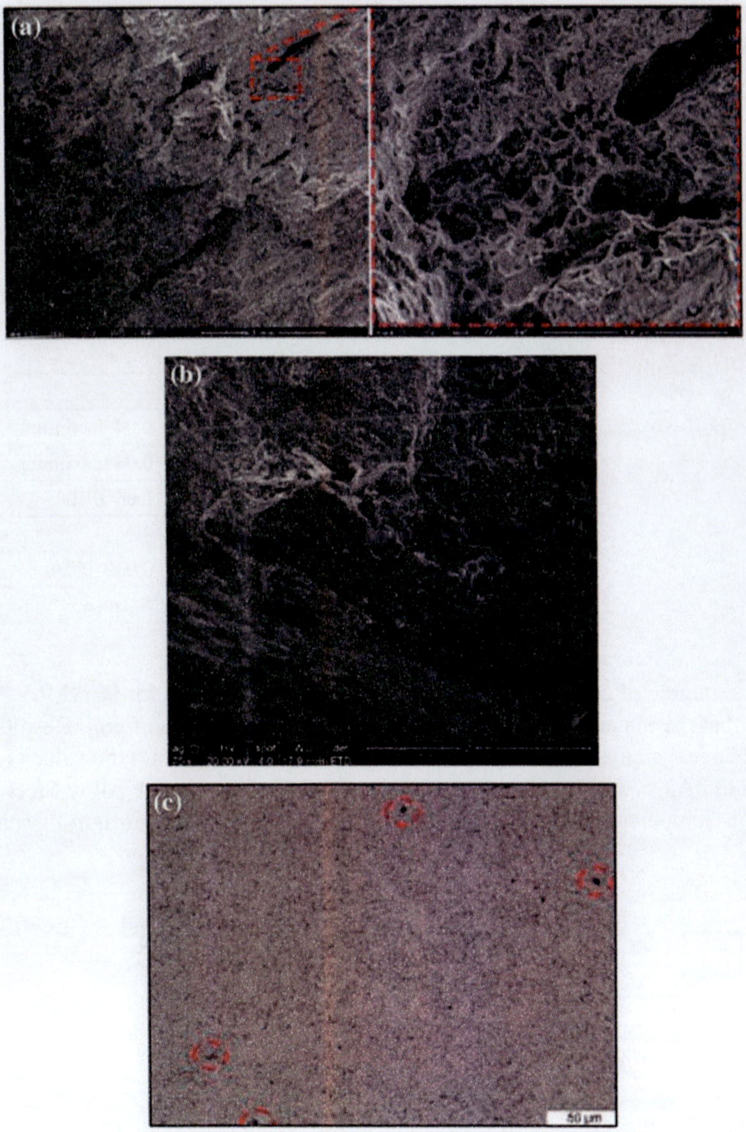

Fig. 3 **a** Possible micro-voids observed on the fracture surface, **b** micro-cracking observed near the fracture surface, **c** formation of creep voids at 3 o'clock sample of F-1303

3.3 Scanning Analysis Through Optical Emission Spectroscopy (OES)

The weight for every element is compared with the actual sample of 9Cr1MoV with accordance standard of SA-335/SA-335M P91. The weight (%) for each element

Table 1 OEM chemical analysis results

Element weight (%)	Coil 3	SA-335/P91
C	0.12	0.08–0.12
Si	0.27	0.20–0.50
Mn	0.46	0.30–0.60
S	<0.005	0.010 maximum
P	0.010	0.020 maximum
Ni	0.17	0.40 maximum
Cr	8.49	8.00–9.50
Mo	0.92	0.85–9.50
Cu	0.10	–
Co	0.02	–
Al	0.02	0.04 maximum
V	0.02	0.04 maximum
Nb	0.064	0.06–0.10
Ti	<0.005	–
N	0.04	0.030–0.070
Fe	Balance	Balance

from the sample of coil F-1303 is within the actual weight (%) for 9Cr1MoV. This proves that the material composition does not affect the failure of coil. Results for OES chemical analysis are summarized in Table 1. Figure 4 shows the values taken from A335/A335M—Standard Specification for Seamless Ferritic Alloy-Steel Pire for High-Temperature Service for comparison with the element weight of sample coil 3.

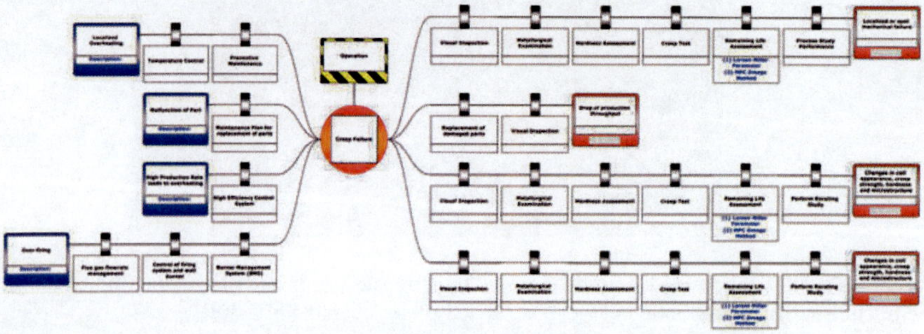

Fig. 4 Corrosion management framework using BowtieXP Software for operation scope

Table 2 Tensile testing results

Sample	Temperature °C	0.2% yield strength	Ultimate tensile strength	% Reduction of area	% Elongation
Coil 3—TI	20	322	555	71.5	23.4
ASME/API 530:2008 minimum	20	415	585	–	20
Coil 3—T2	608	176	190	81.7	27.5
ASME/API 530:2008 minimum	608	207	270	–	–
Coil 3—T3	650	158	180	84.7	21.2
ASME/API 530:2008 minimum	650	165	205	–	–

3.4 *Tensile Testing*

The tensile tests were carried out over a range between 20 and 650 °C to evaluate the maximum strength in tensile stresses. This test shows how the material will behave during operation based on operating parameters. If the ultimate tensile strength is below the minimum specification for that material, it might lead to material degradation and cause a huge failure. The tensile test has been done on three temperatures 20, 608 and 650 °C. Table 2 shows the tensile properties obtained for Coil 3 at 608 °C that were below the minimum specification for SA 335/335M P91 with 190 MPa, which is below 270 MPa.

From the mechanical testing methods, the scanning electron microscopy and metallurgical examination show the presence of creep voids on the microstructure. The Scanning Analysis through Optical Emission Spectroscopy (OES) proves that the weight for each material element follows the standard. Lastly, for tensile testing, the ultimate tensile strength (MPa) for F-1303 at 3 o'clock is lower than the minimum with 190 MPa. The material is recommended to undergo heat treatment to increase the ultimate tensile strength for material 9Cr1MoV. Based on the appearance of creep voids, the damage mechanism causing failure of Coil 3 is creep as a result of medium to long-term overheating.

3.5 *Development of Bowtie Method for Corrosion Management Framework*

Corrosion management framework was created using Bowtie analysis method (Fig. 4).

During the operation stage at the furnace, the coil is exposed to various hazards. After taking all possibilities into account, there are four threats identified, which are localized overheating, high production rate leading to overheating, over-firing and malfunctioning part. Localized overheating can be prevented with temperature control by implementing furnace temperature control and programming. Fuzzy Logic Controller (FLC) can be used for temperature control system. FLC can heat a system up to a targeted temperature and hold it constantly [6]. Preventive maintenance is also important in preventing localized overheating. Preventive maintenance includes:

- Scheduled maintenance for distributed control system (DCS) calibration
- Decoking cycles and inspection of furnace tubes
- Dilution steam injection

High production rate can lead to overheating, changes in coil appearance, creep strength, hardness and microstructure across the coil. The microstructure failure will lead to other impacts. The creep strength, ductility, hardness and coil appearance are affected due to microstructure. This can be prevented with high efficiency control system. The use of higher sensitivity transmitter can detect fast temperature change.

Right side of the diagram (consequences) are localized or spot microstructural failure and changes in coil appearance, creep strength, hardness and microstructure across the coil. Visual inspection (VI) for examination of cracking, bulging, bowing or blistering can be proposed. Then, further non-destructive testing (NDT) will be implemented. Metallurgical assessment according to ASTM E 1351-01 is one of the most important methods in nondestructive testing (NDT) for microstructural analysis. It discloses creep at an earlier stage compared to other non-destructive testing (NDT). This assessment is to determine the remaining service lifetime. Vickers Hardness Test is a type of test to measure hardness of a material in microhardness. If the hardness properties do not meet the requirements of the material and operating parameters, it will lead to fracture of equipment. This may lead to creep fracture. Creep test to determine deformation in metallic materials at elevated temperature. Larson Miller Parameter is known as an extrapolation method. This parameter predicts the lifetime of a steel. MPC Omega Method is also able to predict the remaining life of the material. Process study is performed to identify fouling factor or coke growth across coil length to prevent hot spot. The acceptable fouling factor or coke growth should be reflected in the operation to maintain the coil integrity. Creep factor is the ratio between creep life at reference condition and actual creep life [7].

Part malfunction is a threat that can lead to creep failure, which can lead to drop of production output. Therefore, it is vital to create maintenance plan for part replacement. For example, broken attemperator nozzle tip of F-1303 can cause flow restriction leading to overheating and creep damage. Previous repair record must be evaluated, and the frequency of part failure can be determined. Then new maintenance schedule for part replacement can be created to replace the part before the part failure even occur.

4 Conclusion

Based on the appearance of creep voids, the damage mechanism causing failure of Coil 3 is creep as a result of medium to long-term overheating. Bowtie method can be used as model to build corrosion management framework. The flow of a bowtie diagram provides a very effective means of illustrating the overall factors at work in relation to a major failure. The diagram can be used to analyze and communicate corrosion failure scenarios, in this case, equipment failure caused by creep.

Acknowledgements The authors wish to thank Yayasan Universiti Teknologi Petronas Fundamental Research Grant (YUTP-FRG) for funding the research.

References

1. Bouchard PJ (2001) Residual stresses in lifetime and structural integrity assessment. In: Encyclopedia of materials: science and technology. Elsevier, pp 8134–8142. https://doi.org/10.1016/B0-08-043152-6/01459-5
2. Abuzreda A (2020) Assets integrity management through corrosion mitigation, monitoring and inspection to enhance safety and reliability in refineries (a case study of AGOCO refineries)
3. de Ruijter A, Guldenmund F (2016) The bowtie method: a review. Saf Sci 88:211–218. https://doi.org/10.1016/j.ssci.2016.03.001
4. Salgado J (2011) In situ metallography as non-destructive test to analyze the microstructural damage in the petrochemical industry
5. Roy AK, Hasan MH, Pal J (2009) Creep deformation of Alloys 617 and 276 at 750–950 °C. Mater Sci Eng A 520(1–2):184–188. https://doi.org/10.1016/j.msea.2009.05.029
6. Singhala P, Shah D, Patel B (2014) Temperature control using fuzzy logic. Int J Instrum Control Syst 4(1):1–10. https://doi.org/10.5121/ijics.2014.4101
7. Ogiriki EA, Li YG, Nikolaidis T, Isaiah TE, Sule G (2015) Effect of fouling, thermal barrier coating degradation and film cooling holes blockage on gas turbine engine creep life. Procedia CIRP 38:228–233. https://doi.org/10.1016/j.procir.2015.07.017

Electronic and Electrochemical Properties of Novel Cathode Material $NaFeSO_4OH$ by First-Principle Calculations

Check for updates

Aqeel Idrus, Fadhlul Wafi Badrudin, Siti Nur Amira Shaffee, Oskar Hasdinor Hassan, Fatin Nabilah Sazman, Nur Hamizah Mohd Zaki, Mohd Zaid Zolkiffly, Ab Malik Marwan Ali, Shahrul Izwan Ahmad, Rahimi Baharom, Mohamad Fariz Mohamad Taib, and Muhd Zu Azhan Yahya

Abstract We reported first-principle calculations on the structural and electronic properties of layered sodium iron(II) hydroxysulfate, $NaFeSO_4OH$ as a potential cathode for sodium-ion battery. Layered $NaFeSO_4OH$ was virtually derived from the experimentally reported layered $LiFeSO_4OH$ by replacing Li with Na. A redox voltage of 3.00 V was computed along with the theoretical capacity of 140 mAh/g, yielding an energy density of 420 Wh/kg. The electronic band gap is predicted to be similar to the Li counterpart of the material. This material has some valuable positive attributes, including high capacity and energy density, sturdy host structure as well as being made up of abundant elements.

A. Idrus · M. Zu Azhan Yahya
Faculty of Defence Science and Technology, Universiti Pertahanan Nasional Malaysia, 57000 Kuala Lumpur, Malaysia

F. W. Badrudin · S. I. Ahmad
Centre for Defence Foundation Studies, Universiti Pertahanan Nasional Malaysia, 57000 Kuala Lumpur, Malaysia

S. N. A. Shaffee · M. Z. Zolkiffly
Renewables Technology Program, PETRONAS Group Research and Technology, Jln Ayer Hitam, Kawasan Institusi Bangi, 43000 Bandar Baru Bangi, Selangor, Malaysia

O. H. Hassan
Faculty of Arts and Design, Universiti Teknologi MARA, 40450 Shah Alam, Selangor, Malaysia

F. N. Sazman · N. H. M. Zaki · A. M. M. Ali · M. F. M. Taib (✉)
Ionic Materials and Devices (iMADE), Institute of Science, Universiti Teknologi MARA, 40450 Shah Alam, Selangor, Malaysia
e-mail: mfariz@uitm.edu.my

R. Baharom
Faculty of Electrical Engineering, Universiti Teknologi MARA, 40450 Shah Alam, Selangor, Malaysia

F. Ahmad et al. (eds.), *ICREEM 2022*, Lecture Notes in Mechanical Engineering,
https://doi.org/10.1007/978-981-99-5946-4_23

Keywords Layered $NaFeSO_4OH$ · Layered $LiFeSO_4OH$ · First-principle calculation · Electronic properties · Electrochemical properties

1 Introduction

Harnessing intermittent, renewable energy resources such as sunlight and wind as alternatives to fossil fuels in generating electricity and building green transportations that emit very little to no pollution are among the main strategies we commonly fall back to in the fight against the global warming crisis the whole world is experiencing today. The key to those two strategies is undoubtedly energy storage technologies [1]. Among the options available, rechargeable lithium (Li) ion batteries (LIBs) are touted as one of the best candidates due to their high energy density and reusability [2]. The chemistries that make up such technology have a long history filled with discoveries and breakthroughs [3] and are continuously being explored extensively to this day [4]. Additionally, with theoretical and technological advances, first-principle calculations have now become a ubiquitous tool to aid the experimental side in materials science research [5].

In 1997, Padhi et al. introduced the iron-based polyanionic cathode, $LiFePO_4$, which operated at a voltage of 3.5 V versus Li with reversible capacity of 100–110 mAh/g [6], and it had been shown to be very stable thermodynamically [7], an issue which $LiCoO_2$—the cathode of the first ever commercialized LIB—struggled a lot with [8], alongside the issue of toxicity of Co [9]. As such, $LiFePO_4$ is rightly regarded as among the best contenders to replace $LiCoO_2$. Yet $LiFePO_4$ suffers from poor ionic and electronic conductivities [10] and thus requires modifications [11, 12], which could increase production cost. From $LiFePO_4$, through various motivations such as wanting to increase the open circuit voltage while maintaining the cost advantage and others, the proposed cathodes had seen a shift to, among other candidates, $LiFeSO_4F$ [13–15] and then to $LiFeSO_4OH$ [16–18]. In this work, we have based our material on the layered $LiFeSO_4OH$ due to its higher redox voltage (3.6 V) [15] compared to the tavorite polymorph (3.2 V) [17, 18], in addition to the framework of its host structure which allows for better ionic migration [19] along with its better thermodynamic stability [20].

In parallel, cathode materials for sodium (Na) ion batteries (SIB) have also been explored. The larger and heavier Na poses its own challenges to researchers designing new cathode material. However, Na has its appeal from abundance point of view, with 23,000 parts per million (ppm) abundance in the Earth's crust compared to just 17 ppm for Li. Furthermore, it is structurally similar with Li in having a single valence electron, which in principle should allow it to operate within known LIB chemistries. $NaCoO_2$, $NaFePO_4$ and $NaFeSO_4F$ have all been explored as the Na counterparts of the cathodes mentioned above, via both experimental and computational routes [20–26]. However, to date, the $NaMSO_4OH$ family of cathodes is yet to be discovered [27]. The focus has instead shifted to, among other ventures, increasing the content of the polyanion SO_4^{2-} group [28, 29]; investigation of Prussian blue cathode materials [30]; and even exploring potassium ion battery (KIB) chemistries [31].

The reason of $NaMSO_4OH$ family of cathodes has not yet been able to be synthesized is certainly an interesting research problem. However, that will not be addressed here. Instead, we seek to answer what properties a cathode would have if we replaced Li ions with Na ions in the layered $LiFeSO_4OH$ framework which is made up of highly abundant and environmentally friendly components. We recently reported the predicted structural properties of layered $NaFeSO_4OH$ from first principles calculations [32]. In this paper, we present our calculation results on the electrochemical and the electronic properties of this material. They include the redox voltage, gravimetric capacity, energy density, bond order and bond length, Mulliken charge population analysis, electronic band structure and the electronic density of states for Na.

2 Computational Details

All calculations were performed using the Cambridge Serial Total Energy Package (CASTEP) code [33] that is based on the plane-wave basis set of the density functional theory (DFT). Generalized gradient approximation (GGA) [34] to the exchange–correlation functional in DFT, treated with the formalism of Perdew-Burke-Ernzerhof (PBE) [35] was used, along with ultrasoft pseudopotential [36] representation of the reciprocal space. A minimum energy cut-off of 380 eV for $LiFeSO_4OH$ and 410 eV for $NaFeSO_4OH$ was suggested by the computational code, which was confirmed via convergence test to be sufficient to converge the single point energy of the unit cell of each material to the lowest consistent level. However, the energy cut-off for $LiFeSO_4OH$ was tested up to 500 eV until the accuracy of the results reaches a satisfactory level, and the corresponding energy cut-off for $NaFeSO_4OH$ was adjusted accordingly. For the $FeSO_4OH$ phase, a minimum cut-off energy of 500 eV was required to reproduce the experimentally measured lattice parameters. The Monkhorst–Pack [37] k-point grid for the Brillouin Zone sampling was set at 2 $\times$ 3 $\times$ 2 for all three materials.

We have applied the spin-polarized option due to the presence of iron (Fe), and an antiferromagnetic (AFM) ordering of the spin on the four Fe atoms ($\uparrow\downarrow\downarrow\uparrow$) was found to be lower in energy than a ferromagnetic (FM) ordering. This has been observed in many other reports [18, 38–40]. Additionally, Hubbard U correction was needed for electronic property calculations (electronic band structure, density of states) to increase the separation between the conduction and the valence bands to give a realistic electronic band gap value. In this work, $U_{eff} = 4$ eV was used for the Fe 3*d*-orbitals, in line with other research on Fe-based cathodes [18, 41–43].

For a layered phase like the materials investigated in this study, the van der Waals (vdW) interlayer dispersion interactions are essential to be considered, as attested by other studies [18, 44–46]. Thus, the semi-empirical dispersion-corrected DFT (DFT-D) [47] was applied to the GGA-PBE functional to deal with the vdW interaction effectively. In this work, we have chosen the DFT-D2 scheme developed by Grimme (in some papers, denoted as G06) [48], just as we did in our group's previous work on layered $LiFeSO_4OH$ [49].

Using Broyden-Fletcher-Goldfarb-Shanno (BFGS) algorithm, we optimized each structure until the self-consistent field (SCF) converged to the ultra-fine thresholds: energy change per atom is 5×10^{-6} eV, maximum force at 0.01 eV/Å, maximum stress at 0.02 GPA and maximum displacement of atoms at 5×10^{-4} Å. Layered $NaFeSO_4OH$ was built using experimental crystallographic data of layered $LiFeSO_4OH$ [15] and placing Na at each original Li position since there is no other basis to place the Na atoms anywhere else inside the structure. The process of geometry optimization then adjusted the inputted atomic positions as well as the lattice parameters accordingly until the energy of the material converged to the minimum. For the dealkylated $FeSO_4OH$ phase, we simply removed all the alkali metal atoms in the unit cell.

3 Results and Discussion

We will first present our results on the electrochemical aspect of the material comprising of redox voltage, gravimetric capacity and energy density. The redox voltage, also known as the open-circuit voltage (V_"OC") or intercalation voltage, can be calculated using the following (1):

$$V(x) = \frac{E_{tot}(A_{n-x}FeSO_4OH) - E_{tot}(A_{n-x}FeSO_4OH) + xE_{tot}(A)}{x} \quad (1)$$

Equation (1), $E_{tot}(Y)$ is the total energy of composition Y, $E_{tot}(A)$ is the energy of one alkali metal atom, A, and x is the number of A atoms per formula unit removed. In principle, $x = 1$ to get from $AFeSO_4OH$ to the end member $FeSO_4OH$; in practice $x = 4$ in our work following the symmetry of the unit cell whereby the multiplicity due to Wyckoff position of each atom is 4. To calculate the cell voltage, the optimized final energy of each structure was inputted into Eq. (1). Our group had previously calculated the redox voltage of layered $LiFeSO_4OH$ using several approaches. The closest value obtained was 3.85 V from the GGA-PBE + DFT-D2 approach [49], which was +0.25 V more than the experimental value [15], and similar to the value obtained by Eames et al. (3.87 V) [18]. Using the same approach while assuming the following (2) and (3) redox reaction during discharge:

$$\text{Anode}: Na \rightarrow Na^+ + e^- \quad (2)$$

$$\text{Cathode}: FeSO_4OH + Na^+ \rightarrow NaFeSO_4OH \quad (3)$$

the redox voltage for layered $NaFeSO_4OH$ was calculated to be at 3.00 V versus Na. Figure 1 shows the redox voltage of layered $NaFeSO_4OH$ in comparison to other closely related cathodes. A lower voltage than the lithiated counterpart is expected due to the higher standard reduction potential of Na ($E^{\circ}_{(Na/Na^{+})} = -2.71$ V vs. standard hydrogen electrode, SHE) compared to that of Li ($E^{\circ}_{(Li/Li^{+})} = -3.04$ V vs. SHE) [50]. From computational perspective, the lower voltage of layered $NaFeSO_4OH$ can be explained from its lower energies compared to the Li counterparts as shown in Fig. 2. The larger difference in energy between $NaFeSO_4OH$ and Na compared to between $LiFeSO_4OH$ and Li may also be a factor. Interestingly, one sees that the higher standard reduction potential of Na somehow or other is manifested in the lower energetics of Na and its compounds compared to the Li counterparts, resulting in the lower voltage.

The theoretical gravimetric/specific capacity, Q_{Th} of layered $NaFeSO_4OH$ for complete Na extraction is 140 mAh/g, which is rather decent compared to other sodiated cathodes as shown in Fig. 3. Of course, the practical capacity could well be lower. Based on the literature data of closely related cathode systems that we have collected to plot Fig. 3, the experimental capacity of the other cathodes is less than their theoretical values by about 31 mAh/g, on average. If we arbitrarily apply this difference to layered $NaFeSO_4OH$, this puts its hypothetical practical/experimental capacity, Q_{Ex} at a still-decent value of 109 mAh/g which corresponds to an extraction

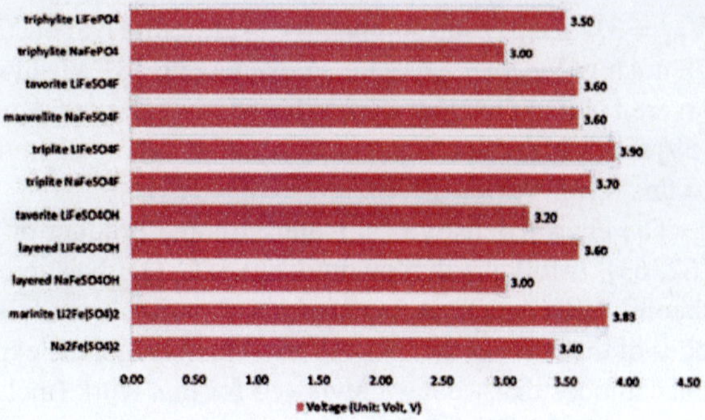

Fig. 1 Redox voltage of layered $NaFeSO_4OH$ and various other lithiated and sodiated cathodes. References: [6, 25, 51–53]

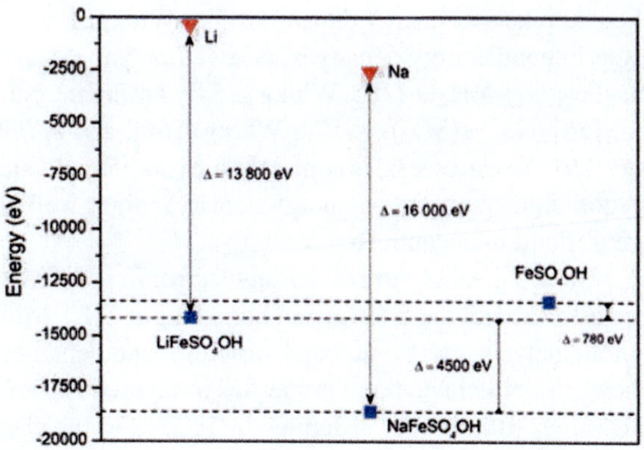

Fig. 2 The final, relaxed energies of the compounds, used in predicting the redox voltage of the cathodes investigated in this study

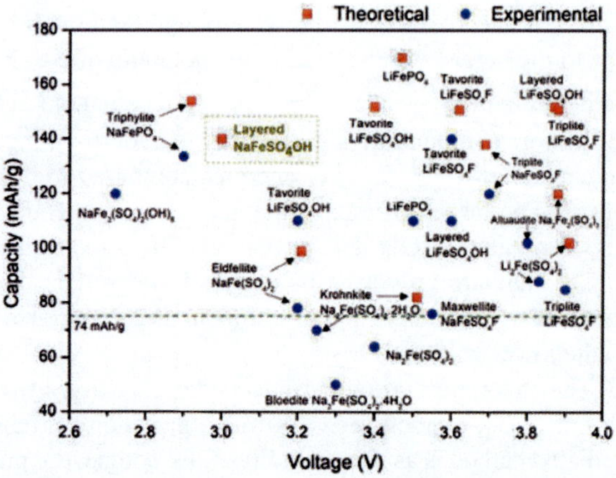

Fig. 3 Theoretical and experimental redox voltage and gravimetric capacity of various lithiated and sodiated cathodes. The dashed line indicates the capacity for layered $NaFeSO_4OH$ corresponding to the extraction of 0.5 Na ions. References: [6, 28, 29, 51–53, 59–61]

of about 0.76 Na ions. If at least 0.5 Na ions can be extracted and inserted from/to layered $NaFeSO_4OH$, the capacity is about 74 mAh/g. This is still better than *bloedite* $Na_2Fe(SO_4)_2{\cdot}4H_2O$ (Q_{Th} = 73 mAh/g, Q_{Ex} = 50 mAh/g, V_{Ex} = 3.3 V) [54], *kröhnkite* $Na_2Fe(SO_4)_2{\cdot}2H_2O$ (Q_{Th} = 82 mAh/g, Q_{Ex} = 70 mAh/g, V_{Th} = 3.51 V, V_{Ex} = 3.25 V) [55, 56], $Na_2Fe(SO_4)_2$ (Q_{Th} = 91 mAh/g, Q_{Ex} = 64 mAh/g, V_{Ex} = 3.4 V) [57] and comparable to eldfellite $NaFe(SO_4)_2$ (Q_{Th} = 99 mAh/g, Q_{Ex} = 78 mAh/g, V_{Th} = 3.21 V, V_{Ex} = 3.2 V) [29, 58]. Meanwhile, the predicted voltage of layered $NaFeSO_4OH$ is 0.28 V higher than the redox voltage of $NaFe_3(SO_4)_2(OH)_6$ [59] the only sodium-containing hydroxysulfate cathode reported thus far according to this recent review [27].

The energy density of a material is the product of the capacity and the voltage [62, 63]. It follows that layered $NaFeSO_4OH$ has an energy density of 420 Wh/kg theoretically. Again, as depicted in Fig. 4, it is one of the highest values among cathodes of the Na–Fe–S(P)–O system. On average, the experimental energy density for the cathodes that we have reviewed for this work (inclusive of the LIB chemistries) is around 114 Wh/kg less than the theoretical values (with standard deviation of 75 Wh/kg). Applying such a large difference to layered $NaFeSO_4OH$ yields a hypothetical practical energy density of 306 Wh/kg. Even so, it is still better than actual experimental energy density of *maxwellite* $NaFeSO_4F$ (269.8 Wh/kg) [60], *bloedite* $Na_2Fe(SO_4)_2{\cdot}4H_2O$ (165 Wh/kg) [54]; *kröhnkite* $Na_2Fe(SO_4)_2{\cdot}2H_2O$ (227.5 Wh/kg) [55], $Na_2Fe(SO_4)_2$ (217.6 Wh/kg) [64], and *eldfellite* $NaFe(SO_4)_2$ (249.6 Wh/kg) [29]. Nevertheless, it is merely a hypothetical value; the actual practical values (redox voltage, capacity, energy density) might well be different due to factors that are difficult to account for.

Now we discuss our results on electronic properties calculations where we have performed bond order (BO) and bond length (BL), Mulliken charge (electron) population analysis, electronic band structure, and density of states calculations. Table 1 shows the calculated bond properties of layered $NaFeSO_4OH$ comprising of BO and BL values. Bond order is defined by IUPAC as the electron population in the region

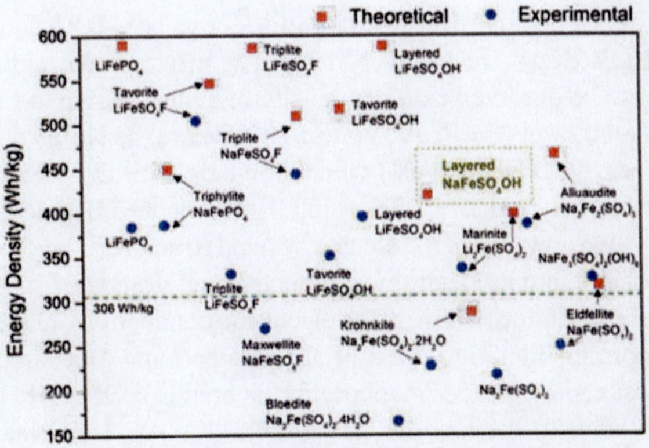

Fig. 4 Energy density of various cathodes computed using the data from references listed in Fig. 3. The dashed line indicates a hypothetical practical energy density for layered $NaFeSO_4OH$

between atoms A and B—a large positive value signifies strong bonding between the atoms [65], while a purely ionic bond like that of Li–O and Na–O should have BO values close to zero. As one can see from Table 1, however, some Na ions have high BO values similar to Fe–O bonds, which are covalent in character. But as we argued in our previous report [32], the Na–O bond is still ionic for two reasons: (1) the average bond length of Na–O is longer than that of Li–O (2.3152 Å vs. 1.9735 Å), which is known to be fully ionic [49]; (2) charge density map calculations proved there is no overlap in the region between Na and O [32]. The high Na–O BO values, therefore, should mean that some of the Na ions are strongly held in place by the O atoms—a conclusion we adapt from Tripathi et al., who found the experimental bond sums of Na–O in a closely related compound $NaFeSO_4F$ (1.16) to be greater than that for Li–O in $LiFeSO_4F$ (0.99) [60].

The Mulliken charge (electron) population analysis shown in Table 2 helps to explain the high Na–O bond values. It was observed that in layered $NaFeSO_4OH$, the oxygens are more positive by 0.24e in total. These charges are not missing, though; rather, they are redistributed mainly to sodium (+0.12e) and sulfur (+0.10e)—that is, the total charge is still conserved (the total charge difference of NFSOH − LFSOH = 0). In comparison with other work, even though different material was investigated and different technique was used, it was noted that Banerjee et al. had calculated

Table 1 Bond order, BO and bond length, BL (Å) analysis of layered $NaFeSO_4OH$

	Na–O		Fe–O		S–O		H–O	
	BO	BL	BO	BL	BO	BL	BO	BL
Sodiated $NaFeSO_4OH$	0.09	2.19	0.15	2.34	0.57	1.51	0.05	1.88
	0.09	2.55	0.18	2.17	0.61	1.49	0.65	0.99
	0.13	2.32	0.19	2.22	0.62	1.49		
	0.21	2.23	0.20	2.19	0.64	1.47		
	0.24	2.30	0.25	2.15				
			0.27	2.09				

the average net Bader charge for Na to be +0.85 and +0.87 in $Na_2Fe(SO_4)_2$ and $NaFe(SO_4)_2$, respectively [58]. The effect of the redistributed charges on Na can also be observed from its density of states, shown in Fig. 5 for the interval after the Fermi level (E = 0 eV) up to 10 eV, where the Na-2*p* orbital forms a significant peak (nearly 2 electrons/eV) which coincide with O-2*p* within the same energy interval, while the peak of Li-2*s* orbital barely reached 0.4 electrons/eV [49].

We now turn to the electronic band structure of layered $NaFeSO_4OH$ to gain a little insight into its electronic conductivity. Before we discuss our results, let us take note of some information on the electronic conductivity of the base material, $LiFeSO_4OH$, from the literature. First of all, no electronic (nor ionic) conductivity measurement was reported experimentally for layered $LiFeSO_4OH$ [15]. Moreover, this point also extends to the *tavorite* polymorph [17, 19]. However, there were indications that

Table 2 Mulliken charge transfer analysis of layered $AFeSO_4OH$ versus A^+/A

Species	Charge (*e*)			NFSOH–LFSOH
	A = *Li*			
	$LiFeSO_4OH$	*Li_0FeSO_4OH*	Δ_1	
A	0.97	–	–	−0.12
Fe	1.07	1.34	+0.27	−0.01
S	2.35	2.37	+0.02	−0.10
O1	−0.89	−0.78	+0.11	+0.03
O2	−0.89	−0.86	+0.03	+0.01
O3	−0.94	−0.81	+0.13	+0.03
O4	−0.99	−0.83	+0.16	+0.07
O5	−1.04	−0.81	+0.23	+0.10
H	0.37	0.38	+0.01	−0.01
Sum	0.01	0.00	0.96	0.00
	A = *Na*			
	$NaFeSO_4OH$	*Na_0FeSO_4OH*	Δ_2	
A	0.85	–	–	−0.12
Fe	1.06	1.32	+0.26	−0.01
S	2.25	2.37	+0.12	−0.10
O1	−0.86	−0.80	+0.06	+0.03
O2	−0.88	−0.86	+0.02	+0.01
O3	−0.91	−0.80	+0.11	+0.03
O4	−0.92	−0.81	+0.11	+0.07
O5	−0.94	−0.79	+0.15	+0.10
H	0.36	0.38	+0.02	−0.01
Sum	0.01	0.01	0.85	0.00

NFSOH = $NaFeSO_4OH$, LFSOH = $LiFeSO_4OH$

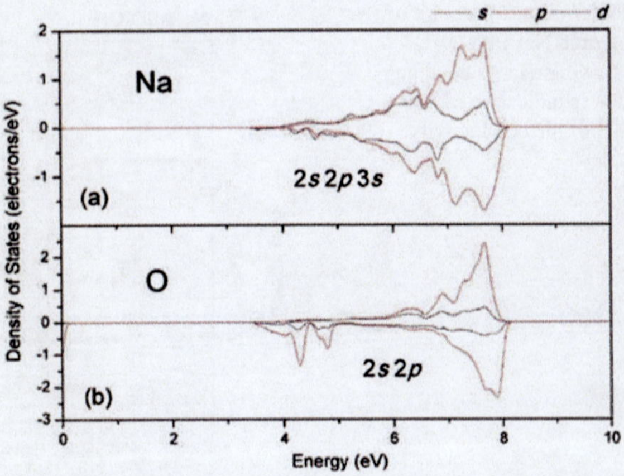

Fig. 5 Density of state of Na and O within the 0–10 eV energy interval in layered $NaFeSO_4OH$

the electronic conductivity of the $FeSO_4OH$ host structure, in general, is low. For instance, for layered $LiFeSO_4OH$, the specific capacity quickly reduced to well below 60 mAh/g before reaching 10C rate [15], while for *tavorite* $LiFeSO_4OH$, Reddy et al. observed that the number of Li ions inserted and extracted decreased with increasing C rate (Fig. 6), which they said is "characteristic of compounds with poor electronic conductivity" [19].

Based on these indirect experimental observations, one could perhaps safely conclude that the electronic conductivity of the $FeSO_4OH$ host structure seems to be intrinsically low regardless of morphology. Computationally, this could be reflected by a large electronic band gap. Indeed, this was what we found for layered $LiFeSO_4OH$ (band gap = 3.483 eV) [49]. It is therefore not surprising that the same is true for layered $NaFeSO_4OH$. Figure 7 displays the band structure of layered

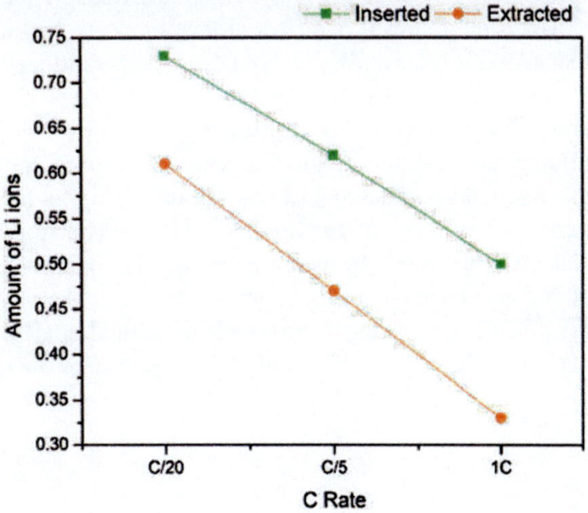

Fig. 6 Experimental observation by [19] on the decreasing amount of Li ions reversibly inserted into the $FeOHSO_4$ host structure with increasing C rate, suggesting low electronic conductivity of the host structure

Fig. 7 Band structure of layered $NaFeSO_4OH$, showing a large band gap that is indicative of low electronic conductivity

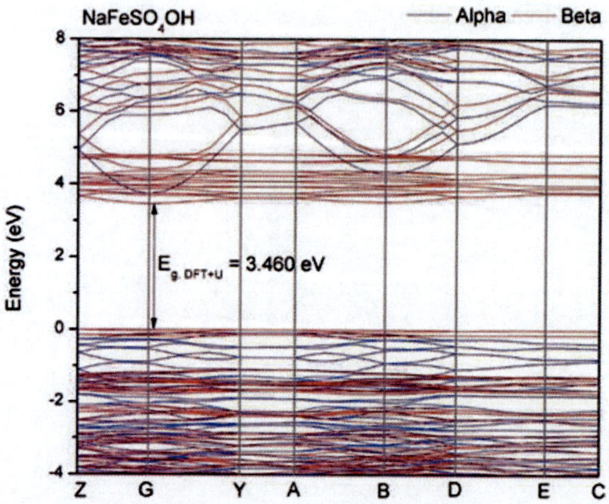

$NaFeSO_4OH$ as obtained from GGA-PBE + DFT-D2 + U combination, which shows that the valence band maximum (VBM) and the conduction band minimum (CBM) are located almost exactly along the G line within the Z–Y points of the irreducible Brillouin Zone (IBZ), and the separation between them, i.e., the band gap, e.g., is 3.460 eV. Without the Hubbard U correction, the band gap is very much lower (<1 eV), which cannot be true because of what we know from experimental findings on $LiFeSO_4OH$ that we mentioned earlier.

Based on this finding, we find that there is not much difference between the band gap of layered $NaFeSO_4OH$ and that of layered $LiFeSO_4OH$ (3.460 eV vs. 3.483 eV). Thus, the electronic conductivity of the two may be similar and is not affected by the change in alkali metal ions (Li^+, Na^+). The electrons released by them at the anode during discharge are received by the transition metal (Fe in this case) at the cathode, which is mostly bonded to oxygen, giving rise to the Fe^{3+}/Fe^{2+} reduction couple.

Upon the removal of Li/Na atoms, our calculations found that the band gap is reduced to 1.226 eV. We see that $\overline{BL}_{Fe\text{–}O}$ has shortened to 2.0442 Å, coinciding with the reduced band gap. Similar findings can also be observed for *triphylite* $LiFePO_4$ and its delithiated phase, *heterosite* $FePO_4$ as shown in Table 3. Removing the sodium/lithium atom causes the metallic characteristic to enhance as the Fe–O interaction increased, thus reducing the band gap [69]. This could lead this material to exhibit as a charge-transfer insulator. Based on this characteristic, the electronic conductivity of $NaFeSO_4OH$ should be increased during the sodiation process. Perhaps this material possesses a large band gap. There is a need to find a way to reduce the large band gap. For this, we are looking into a couple of methodologies, which are partial substitution of the transition metal [70] and applying strain to the lattice [71].

Table 3 Bond length and band gap of $LiFePO_4$ and $FePO_4$

Cathode materials	$\overline{BL}_{Fe-O}$ (Å)		Band gap (eV)		References
	Exp	Calculated	Exp	Calculated	
Triphylite $LiFePO_4$	2.157	2.15–2.18	3.7	3.84	[66–68]
Heterosite $FePO_4$	2.015	2.06–2.07	1.9	1.88	

4 Conclusions

We have presented the electronic and electrochemical properties of a novel, low-cost cathode, namely, layered $NaFeSO_4OH$ as obtained from theoretical calculations. The cathode is predicted to give a redox voltage of 3.00 V versus Na metal. The difference in voltage between layered $LiFeSO_4OH$ and layered $NaFeSO_4OH$ was discussed. Charge population analysis helps explain the bond properties of the material. The theoretical gravimetric capacity of layered $NaFeSO_4OH$ is calculated to be 140 mAh/g, leading to a theoretical energy density of 420 Wh/kg, and we showed that these values are quite promising compared to other experimentally reported sodiated cathodes. However, this cathode is predicted to have a large electronic band gap very similar to its lithiated counterpart, a feature seemingly inherent to the host structure. If that can be reduced, it will add to its positive attributes, which include robust host structure, decent voltage, high capacity, good energy density, and is made up of abundant atomic components.

Acknowledgements This work is supported by internal UPNM grant (Grant code: UPNM/2021/GPJP/STG/2). The authors would like to express their gratitude to Core Group for Academic Development and Research Experts (CADRE), in Centre for Defence Foundation Studies, Universiti Pertahanan Nasional Malaysia and Institute of Science, Universiti Teknologi MARA (UiTM) for their support in providing research.

References

1. Kebede AA, Kalogiannis T, Van Mierlo J, Berecibar M (2022) A comprehensive review of stationary energy storage devices for large scale renewable energy sources grid integration. Renew Sustain Energy Rev 159:112213
2. Jeevanantham B, Shobana MK (2022) Enhanced cathode materials for advanced lithium-ion batteries using nickel-rich and lithium/manganese-rich $LiNi_xMn_yCo_zO_2$. J Energy Storage 54:105353
3. Li M, Lu J, Chen Z, Amine K (2018) 30 Years of lithium-ion batteries. Adv Mater 30:1800561
4. Manthiram A (2020) A reflection on lithium-ion battery cathode chemistry. Nat Commun 11:1550
5. Su LYJ, Sun C (2014) Review of the first principles calculations and the design of cathode materials for Li-ion batteries. 358–368
6. Padhi AK, Nanjundaswamy KS, Goodenough JB (1997) Phospho-olivines as positive-electrode materials for rechargeable lithium batteries. J Electrochem Soc 144:2–8

7. Ji Y, Jafvert CT, Zyaykina NN, Zhao F (2022) Decomposition of PVDF to delaminate cathode materials from end-of-life lithium-ion battery cathodes. J Clean Prod 367:133112
8. Kasnatscheew J, Röser S, Börner M, Winter M (2019) Do increased Ni contents in $LiNi_xMn_yCo_zO_2$ (NMC) electrodes decrease structural and thermal stability of Li ion batteries? A thorough look by consideration of the Li^+ extraction ratio. ACS Appl Energy Mater 2:7733–7737
9. Sironval V, Reylandt L, Chaurand P, Ibouraadaten S, Palmai-Pallag M, Yakoub Y, Ucakar B, Rose J, Poleunis C, Vanbever R, Marbaix E, Lison D, van den Brule S (2018) Respiratory hazard of Li-ion battery components: elective toxicity of lithium cobalt oxide ($LiCoO_2$) particles in a mouse bioassay. Arch Toxicol 92:1673–1684
10. Nie Z, McCormack P, Bilheux HZ, Bilheux JC, Robinson JP, Nanda J, Koenig GM (2019) Probing lithiation and delithiation of thick sintered lithium-ion battery electrodes with neutron imaging. J Power Sources 419:127–136
11. Xiao H, Pender JP, Meece-Rayle MA, de Souza JP, Klavetter KC, Ha H, Lin J, Heller A, Ellison CJ, Mullins CB (2017) Reduced-graphene oxide/poly(acrylic acid) aerogels as a three-dimensional replacement for metal-foil current collectors in lithium-ion batteries. ACS Appl Mater Interfaces 9:22641–22651
12. Seung-Hyun Lee D, Bin Im W, Liang X (2019) High density conductive $LiFePO_4$ cathode with enhanced high-rate and high temperature performance. Mater Chem Phys 232:367–373
13. Islam MS, Fisher CAJ (2014) Lithium and sodium battery cathode materials: computational insights into voltage, diffusion and nanostructural properties. Chem Soc Rev 43:185–204
14. Liu L, Zhang B, Huang X (2011) A 3.9 V polyanion-type cathode material for Li-ion batteries. Prog Nat Sci Mater Int 21:211–215
15. Subban CV, Ati M, Rousse G, Abakumov AM, Van Tendeloo G, Janot R, Tarascon J-M (2013) Preparation, structure, and electrochemistry of layered polyanionic hydroxysulfates: $LiMSO_4OH$ (M = Fe Co, Mn) electrodes for Li-ion batteries. J Am Chem Soc 135:3653–3661
16. Kitajou A, Ishado Y, Inoishi A, Okada S (2018) Amorphous xLiF-$FeSO_4$ ($1 \leq x \leq 2$) composites as a cathode material for lithium ion batteries. Solid State Ion 326:48–51
17. Ati M, Sougrati MT, Rousse G, Recham N, Doublet ML, Jumas JC, Tarascon JM (2012) Single-step synthesis of $FeSO_4F_{1-y}OH_y$ ($0 \leq y \leq 1$) positive electrodes for Li-based batteries. Chem Mater 24:1472–1485
18. Eames C, Clark JM, Rousse G, Tarascon J-M, Islam MS (2014) Lithium migration pathways and van der Waals effects in the $LiFeSO_4OH$ battery material. Chem Mater 26:3672–3678
19. Anji Reddy M, Pralong V, Caignaert V, Varadaraju UV, Raveau B (2009) Monoclinic iron hydroxy sulphate: a new route to electrode materials. Electrochem Commun 11:1807–1810
20. Radha AV, Subban CV, Sun ML, Tarascon JM, Navrotsky A (2014) Possible correlation between enthalpies of formation and redox potentials in $LiMSO_4OH$ (M = Co, Fe, Mn), Li-ion polyanionic battery cathode materials. J Mater Chem A
21. Tiwari B, Bhattacharya I (2018) Layered P2-type novel $Na0.7Ni0.3Mn0.59Co0.1Cu0.01O_2$ cathode material for high-capacity & stable rechargeable sodium ion battery. Electrochim Acta 270:363–368
22. Su J, Zhou G, Pei Y, Yang Z, Wang X (2015) First-principles investigation on structural and electrochemical properties of $NaCoO_2$ for rechargeable Na-ion batteries. J Cent South Univ 22:2036–2042
23. Sun A, Beck FR, Haynes D, Poston JA, Narayanan SR, Kumta PN, Manivannan A (2012) Synthesis, characterization, and electrochemical studies of chemically synthesized $NaFePO_4$. Mater Sci Eng B 177:1729–1733
24. Nakayama M, Yamada S, Jalem R, Kasuga T (2016) Density functional studies of olivine-type $LiFePO_4$ and $NaFePO_4$ as positive electrode materials for rechargeable lithium and sodium ion batteries. Solid State Ion 286:40–44
25. Wang B, Han Y, Wang X, Bahlawane N, Pan H, Yan M, Jiang Y, Wang H, Wang Y, Yan N, Jiang M (2018) Prussian blue analogues for rechargeable batteries. ISCIENCE
26. Pan W, Guan W, Liu S, Bin Xu B, Liang C, Pan H, Yan M, Jiang Y (2019) $Na_2Fe(SO_4)_2$: an anhydrous 3.6 V, low-cost and good-safety cathode for a rechargeable sodium-ion battery. J Mater Chem A 7:13197–13204

27. Barpanda P, Lander L, Nishimura S, Yamada A (2018) Polyanionic insertion materials for sodium-ion batteries. Adv Energy Mater 8:1703055
28. Barpanda P, Oyama G, Nishimura S, Chung S-C, Yamada A (2014) A 3.8-V earth-abundant sodium battery electrode. Nat Commun 5:4358
29. Singh P, Shiva K, Celio H, Goodenough JB (2015) Eldfellite, $NaFe(SO_4)_2$: an intercalation cathode host for low-cost Na-ion batteries. Energy Environ Sci 8:3000–3005
30. Nasir NAM, Badrudin FW, Idrus A, Sazman FN, Taib MFM, Yahya MZA (2020) First-principles study on structural and electronic properties of Prussian blue cathode material for sodium-ion battery. Mol Cryst Liquid Cryst 693:115–122
31. Eftekhari A, Jian Z, Ji X (2017) Potassium secondary batteries. ACS Appl Mater Interfaces 9:4404–4419
32. Idrus A, Badrudin FW, Fariz M, Taib M, Hassan OH, Malik AB, Ali M, Zu M, Yahya A (2021) Ab Initio study on structural properties of $NaFeSO_4OH$ cathode material for rechargeable sodium ion battery. Solid State Phenom 317:400–405
33. Clark SJ, Segall MD, Pickard CJ, Hasnip PJ, Probert MIJ, Refson K, Payne MC (2005) First principles methods using CASTEP. Zeitschrift Für Krist. Cryst Mater 220:567–570
34. Langreth DC, Mehl MJ (1983) Beyond the local-density approximation in calculations of ground-state electronic properties. Phys Rev B 28:1809
35. Perdew JP, Burke K, Ernzerhof M (1996) Generalized gradient approximation made simple. Phys Rev Lett 77:3865–3868
36. Vanderbilt D (1990) Soft self-consistent pseudopotentials in a generalized eigenvalue formalism. Phys Rev B 41:7892–7895
37. Monkhorst HJ, Pack JD (1976) Special points for Brillouin-zone integrations. Phys Rev B 13:5188–5192
38. Melot BC, Rousse G, Chotard JN, Kemei MC, Rodríguez-Carvajal J, Tarascon JM (2012) Magnetic structure and properties of $NaFeSO_4F$ and $NaCoSO_4F$. Phys Rev B Condens Matter Mater Phys 85:094415
39. Singh S, Jha PK, Avdeev M, Zhang W, Jayanthi K, Navrotsky A, Alshareef HN, Barpanda P (2021) Marinite $Li_2Ni(SO_4)_2$ as a new member of the bisulfate family of high-voltage lithium battery cathodes. Chem Mater 33:6108–6119
40. Arcelus O, Nikolaev S, Carrasco J, Solovyev I (2018) Magnetism of $NaFePO_4$ and related polyanionic compounds. Phys Chem Chem Phys 20:13497–13507
41. Zeng Y, Chiu H-C, Ouyang B, Song J, Zaghib K, Demopoulos GP (2019) Unveiling the mechanism of improved capacity retention in Pmn21 Li_2FeSiO_4 cathode by cobalt substitution. J Mater Chem A 7:25399–25414
42. Tsevelmaa T, Odkhuu D, Kwon O, Cheol Hong S (2013) A first-principles study of magnetism of lithium fluorosulphate $LiFeSO_4F$. J Appl Phys 113:17B302
43. Zhou H, Upreti S, Chernova NA, Hautier G, Ceder G, Whittingham MS (2011) Iron and manganese pyrophosphates as cathodes for lithium-ion batteries. Chem Mater 23:293–300
44. Björkman T (2014) Testing several recent van der Waals density functionals for layered structures. J Chem Phys 141:074708
45. Aykol M, Kim S, Wolverton C (2015) van der Waals interactions in layered lithium cobalt oxides. J Phys Chem C
46. Das T, Tosoni S, Pacchioni G (2019) Structural and electronic properties of bulk and ultrathin layers of V_2O_5 and MoO_3. Comput Mater Sci 163:230–240
47. Stöhr M, Van Voorhis T, Tkatchenko A (2019) Theory and practice of modeling van der Waals interactions in electronic-structure calculations. Chem Soc Rev 48:4118–4154
48. Grimme S (2006) Semiempirical GGA-type density functional constructed with a long-range dispersion correction. J Comput Chem 27:1787–1799
49. Badrudin FW, Taib MFM, Hassan OH, Yahya MZA (2016) Effect of lithium intercalation on the structural and electronic properties of layered $LiFeSO_4OH$ and layered $FeSO_4OH$ using first-principle calculations. Comput Mater Sci 119:144–151
50. Abraham KM (2020) How comparable are sodium-ion batteries to lithium-ion counterparts? ACS Energy Lett 5:3544–3547

51. Moreau P, Guyomard D, Gaubicher J, Boucher F (2010) Structure and stability of sodium intercalated phases in olivine $FePO_4$. Chem Mater 22:4126–4128
52. Chakraborty S, Banerjee A, Watcharatharapong T, Araujo RB, Ahuja R (2018) Current computational trends in polyanionic cathode materials for Li and Na batteries. J Phys Condens Matter 30:283003
53. Kim M, Kim D, Lee W, Jang HM, Kang B (2018) New class of 3.7 V Fe-based positive electrode materials for Na-ion battery based on cation-disordered polyanion framework. Chem Mater 30:6346–6352
54. Reynaud M, Ati M, Boulineau S, Sougrati MT, Melot BC, Rousse G, Chotard J-N, Tarascon J-M (2013) Bimetallic sulfates $A2M(SO_4)_2{\cdot}nH_2O$ (A = Li, Na and M = transition metal): as new attractive electrode materials for Li- and Na-ion batteries. ECS Trans 50:11–19
55. Barpanda P, Oyama G, Ling CD, Yamada A (2014) Kröhnkite-type $Na_2Fe(SO_4)_2{\cdot}2H_2O$ as a Novel 3.25 V insertion compound for Na-ion batteries. Chem Mater 26:1297–1299
56. Watcharatharapong T, Thienprasert JT, Barpanda P, Ahuja R, Chakraborty S (2017) Mechanistic study of Na-ion diffusion and small polaron formation in Kröhnkite $Na_2Fe(SO_4)_2{\cdot}2H_2O$ based cathode materials. J Mater Chem A 5:21726–21739
57. Reynaud M (2013) Design of new sulfate-based positive electrode materials for Li-and Na-ion batteries. Université de Picardie Jules Verne
58. Banerjee A, Araujo RB, Ahuja R (2016) Unveiling the thermodynamic and kinetic properties of $Na_xFe(SO_4)_2$ (x = 0–2): toward a high-capacity and low-cost cathode material. J Mater Chem A 4:17960–17969
59. Gnanavel M, Lebedev OI, Bazin P, Raveau B, Pralong V (2015) Reversible transformation from amorphouse $Na_3Fe_3(SO_4)_2(OH)_6$ to crystallized $NaFe_3(SO_4)_2(OH)_6$ Jarosite-type hydroxysulfate. Solid State Ion 278:38–42
60. Tripathi R, Ramesh TN, Ellis BL, Nazar LF (2010) Scalable synthesis of tavorite $LiFeSO_4F$ and $NaFeSO_4F$ cathode materials. Angew Chem Int Ed Engl 49:8738–8742
61. Ati M, Melot BC, Chotard J-N, Rousse G, Reynaud M, Tarascon J-M (2011) Synthesis and electrochemical properties of pure $LiFeSO_4F$ in the triplite structure. Electrochem Commun 13:1280–1283
62. Hautier G (2016) Prediction of new battery materials based on ab initio computations. In: AIP conference proceedings, vol 1765, p 020009
63. Lyu Y, Liu Y, Yu ZE, Su N, Liu Y, Li W, Li Q, Guo B, Liu B (2019) Recent advances in high energy-density cathode materials for sodium-ion batteries. Sustain Mater Technol 21:e00098
64. Reynaud M, Rousse G, Abakumov AM, Sougrati MT, Van Tendeloo G, Chotard JN, Tarascon JM (2014) Design of new electrode materials for Li-ion and Na-ion batteries from the bloedite mineral $Na_2Mg(SO_4)_2{\cdot}4H_2O$. J Mater Chem A 2:2671–2680
65. Minkin VI (1999) Glossary of terms used in theoretical organic chemistry. Pure Appl Chem 71:1919–1981
66. Rousse G, Rodriguez-Carvajal J, Patoux S, Masquelier C (2003) Magnetic structures of the triphylite $LiFePO_4$ and of its delithiated form $FePO_4$. Chem Mater 15:4082–4090
67. Zaghib K, Mauger A, Goodenough JB, Gendron F, Julien CM (2007) Electronic, optical, and magnetic properties of $LiFePO_4$: small magnetic polaron effects. Chem Mater 19:3740–3747
68. Maxisch T, Zhou F, Ceder G (2006) Ab initio study of the migration of small polarons in olivine Li_xFePO_4 and their association with lithium ions and vacancies. Phys Rev B Condens Matter Mater Phys 73:1–6
69. Zhong G, Li Y, Yan P, Liu Z, Xie M, Lin H (2010) Structural, electronic, and electrochemical properties of cathode materials Li_2MSiO_4 (M = Mn, Fe, and Co): density functional calculations. J Phys Chem C 114:3693–3700
70. Badrudin FW, Taib MFM, Mustapha RIPR, Hassan OH, Yahya MZA (2017) Effects of vanadium substitution in the layered $LiFeSO_4OH$: a first principles investigation. Mater Today Proc 4:5108–5115
71. Lee J, Pennycook SJ, Pantelides ST (2012) Simultaneous enhancement of electronic and Li^+ ion conductivity in $LiFePO_4$. Appl Phys Lett 101

Corrosion Study of Carbon Steel in High-Acidic Environment: H_2S and CO_2–H_2S Comparison

Nuur Fahanis Che Lah, Puteri Sri Melor Megat Yusoff, Joel Raj Yogaraj, Mazli Mustapha, Elnizam Hafidi Delan, and Jessica Fong Fung Yee

Abstract In this study, corrosion rates of carbon steel in hydrogen sulphide (H_2S) and carbon dioxide (CO_2)–H_2S condition are reported from Electrochemical Impedance Spectroscopy (EIS) measurements and potentiodynamic polarization. The electrochemical measurement method has been utilized in monitoring the behaviour of carbon steel in 0.01 M H_2SO_4 solution containing different concentrations of H_2S. As the concentration of H_2S increases, the corrosion rate of the carbon steel also increases. This is due to the fact that the hydrogen evaluation reaction has been encouraged as the H_2S accelerated in proportional of H_2S concentration. On the other hand, under CO_2–H_2S condition it shows a different trend compared to the H_2S-only condition. As the concentration of H_2S increases, the corrosion rate decreased.

Keywords Corrosion · Hydrogen sulphide · Carbon steel · Carbon dioxide

---

N. F. C. Lah · P. S. M. M. Yusoff · M. Mustapha (✉)
Centre of Corrosion Research, Universiti Teknologi Petronas, Seri Iskandar, Perak, Malaysia
e-mail: mazli.mustapha@utp.edu.my

J. R. Yogaraj · E. H. Delan
Department of Mechanical Engineering, Universiti Teknologi Petronas, Seri Iskandar, Perak, Malaysia
e-mail: joel_17002321@utp.edu.my

E. H. Delan
e-mail: elnizam_17005406@utp.edu.my

J. F. F. Yee
Sarawak Energy Berhad, The Isthmus, Kuching, Sarawak, Malaysia
e-mail: jessica.fong@sarawakenergy.com

F. Ahmad et al. (eds.), *ICREEM 2022*, Lecture Notes in Mechanical Engineering,
https://doi.org/10.1007/978-981-99-5946-4_24

1 Introduction

Corrosion is a natural risk and process that converts a refined metal into a more chemically stable oxide. The materials are gradually destructed by chemical or electrochemical reaction under environment that causes the corrosion. Referring to American Society for Testing Materials (ASTM), corrosion can be identified as "the chemical or electrochemical reaction between a material, usually a metal, and its environment that produces a deterioration of the material and its properties" [1]. The corrosion of metal occurred when it reacts with other element/substance such as hydrogen, oxygen, an electrical current or even dirt and bacteria. Under high stress can also causes the corrosion which led to material cracking. Another substance that contributes to the metal corrosion is hydrogen sulphide (H_2S). H_2S was rich in the Proterozoic's old, anoxic oceans, where it gave energy to early forms of life while also influenced biochemical and physiological processes [2]. The high concentration of oxygen caused by proterozoic oxidation will lower H_2S in most environments. Organically sustained and high-level concentrations of H_2S are found in a few aquatic habitats due to limited production through geological and biological processes. Furthermore, H_2S production in aquatic environments can be facilitated by geological or biological factors, as well as the action of sulphate-reducing bacteria.

Sulphide corrosion is triggered by the production of H_2S, from the bacterial-assisted process, which subsequently convert to sulphuric acid (H_2SO_4) that attacks concrete and steel [3]. In this process, the H_2S gas is oxidized by biochemical reaction in the existence of moisture to form H_2SO_4. Widely known, H_2SO_4 can be destructive on steel surfaces exposed to harsh environments. The cost of maintenance is likely to rise as the aging infrastructure continues to fail. Corrosion may occur where stale dam generates and releases H_2S gas into an atmosphere containing oxygen gas and high relative humidity [4].

H_2S is a key contributor in the global sulphur cycle. The oxidation of H_2S to sulphur dioxide (SO_2) and transformed to sulphate (SO_4) through three unique chemical pathways. The H_2S reacts with the moisture and converts to H_2SO_4, which is corrosive to metals, and attributes to acidic deposition to soil and water. In the presence of oxygen, the H_2S will be oxidized to H_2SO_4, which works as a catalytic compound in the process of electrochemical corrosion, increasing corrosion rate significantly.

During the corrosion process, H_2S, which is mildly acidic when dissolved in water, goes through a series of chemical reactions. The chemical reactions are [5]:

$$\text{H2S dissolution}: H_2S(g) \Longleftrightarrow H_2S(aq) \tag{1}$$

$$\text{H2S dissociation}: H_2S(g) \Longleftrightarrow HS^-(aq) + H^+(aq) \tag{2}$$

$$HS^-\text{dissociation}: HS^-(aq) \Longleftrightarrow H^+(aq) + S^{2-}(aq) \tag{3}$$

$$\text{H2S reduction}: 2H_2S(g) + 2e^- \rightarrow H_2(g) + 2HS^-(aq) \quad (4)$$

$$\text{FeS formation by precipitation}: Fe(s) + S^{2-}(aq) \Longleftrightarrow FeS(aq) \quad (5)$$

pH plays a role in affecting the reactivity of H_2S in aqueous solution. The sulphide generated in low pH solution is molecular H_2S, which remains in this form until the pH value of roughly 6 is reached, at which point a large amount of bisulfide appears. More bisulfides will be produced as the pH value rises. At pH greater than 8, the bisulfide ion is the main H_2S species [6]. Typically, in an acidic solution, the higher the acidity, the higher the corrosion rate taking place on the steel due to the availability of molecular H_2S [7].

In an H_2S environment, several factors can influence the corrosion product between H_2S and steel. Corrosion products of iron or steel form iron-sulphur compounds with diverse phases in an aqueous H_2S environment, which play an essential role in the environment. Steel exposed to gaseous H_2S in the atmosphere may corrode more severely than steel in aqueous brine solutions [8]. Previous study has discovered that the corrosion product mostly comprises of pyrite, troilite, and mackinawite with a pH of less than 4 [9]. In this study, the effect of different concentrations of H_2S in high-acidic environment was tested and compared with CO_2 addition to the system, to investigate the role and synergistic relationship between CO_2 and H_2S in corrosion. The corrosion rate was evaluated based on polarization study and impedance spectroscopy. This preliminary study will contribute to further understanding of sulphur cycle in corrosion mechanism of high-acidic environment.

2 Experimental

2.1 *Three-Electrode System Setup*

Figure 1 shows the test setup at atmospheric pressure in a glass cell. A three-electrode setup was utilized with an Ag/AgCl (6.0733.100, Metrohm, Switzerland) as the reference electrode, a platinum wire as the counter electrode, and a carbon steel specimen as the working electrode. CO_2 was introduced to the system to create a CO2-saturated solution by continuously purging CO_2 gas into the solution to create the CO_2–H_2S environment.

2.2 *Materials and Preparation of Test Specimen*

The metal utilized for this study was carbon steel. The specimens with a bare surface of 10.0 mm^2 were cut from the carbon steel sheet and implanted in the epoxy resin

Fig. 1 Electrochemical cell setup with three type of electrode (i) Platinum wire Counter Electrode (CE) (ii) Ag/AgCl Reference Electrode (RE) and (iii) Working Electrode (WE)

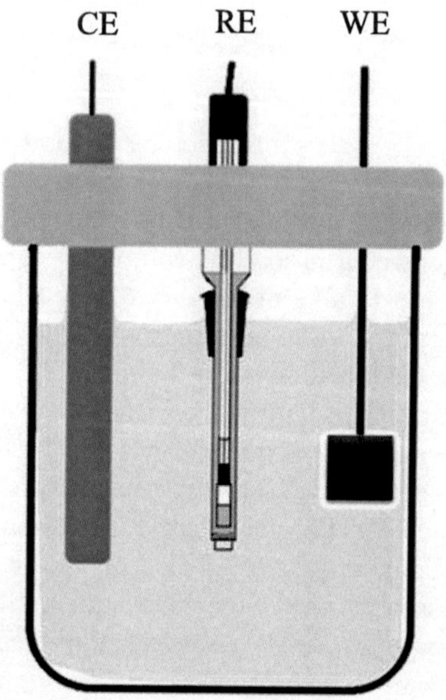

mould. The cross-section was in contact with the electrolyte. The electrode surface was polished with 600-grade wet carbide paper, degreased with acetone, rinsed with distilled water and dried by compressed hot air flow prior to each experiment.

2.3 *Preparation of Electrolyte Solution*

Analytical reagent grade Na_2S, H_2SO_4 and distilled water were employed to formulate the electrolyte solutions with various H_2S concentrations. The stock solutions were prepared; 2 M H_2SO_4 and 0.0025 M Na_2S. The electrolyte solutions that contain H_2S were formulated from the stock solution of Na_2S and H_2SO_4 in the following way: 1000 mL of distilled water solution was filled in the glass cell, and then the appropriate volume of 2 M H_2SO_4 stock solution was mixed into the distilled water to obtain 0.01 M of H_2SO_4 concentration, subsequently relevant amount of Na_2S solution was added to prepare acidic H_2S solution with concentrations of interest (1.25, 6.20 and 12.5 ppm). The testing was conducted with 1000 mL of testing solution. pH of each prepared solution was evaluated at the end of solution preparation.

2.4 *Electrochemical Measurement Setup*

The potentiodynamic polarization curve measurements were completed at a potential scan rate of 0.001 mV s^{-1}. All testing was set at a potential range from −0.10

to +0.10 V versus open-circuit potential (OCP) that evaluated with respect to the Ag/AgCl. The EIS measurements were performed at OCP over the frequency from 100 kHz to 10 MHz by using a 5 mV amplitude sinusoidal voltage. The experimental data were analysed by using the commercial software Nova 2.1.5. The data from EIS measurement were fitted using the Randles circuit without diffusion.

3 Result and Discussion

3.1 Microstructure

The surface morphology of the corroded steel is observed using optical microscope (OM) and is presented in Fig. 2 for immersion in 0.01 M of H_2SO_4 in H_2S and CO_2–H_2S condition. A white corrosion product with stone-like structure is observed to form on the surface of the carbon steel. The form of the corrosion based on the observation is localized and general. Black spot detected on the surface of carbon steel indicates the formation of iron sulphide consisting of mackinawite as one of the corrosion products [10]. The local supersaturation of FeS_x could be formed on the carbon steel surface via the following reaction, with the nucleation and growth of one or more of the iron sulphide, mackinawite [11]:

$$H_2S + Fe + H_2O \rightarrow FeS_X + H_2O + 2H \quad (6)$$

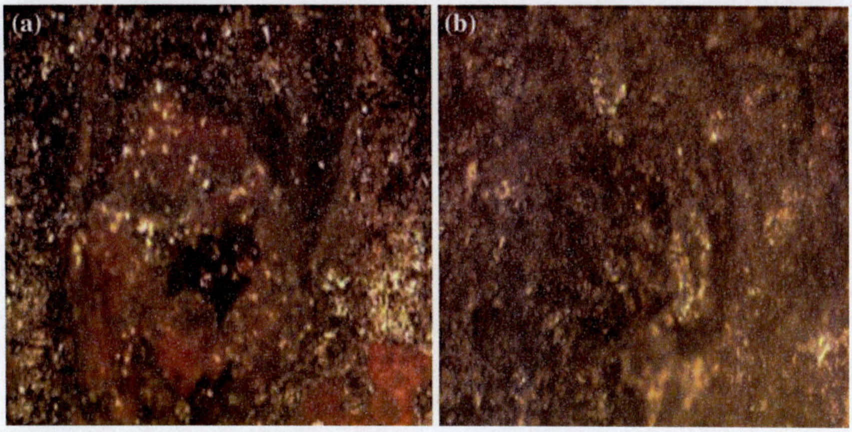

Fig. 2 Surface of corroded carbon steel after immersion for 30 days in **a** H_2S and **b** CO_2–H_2S condition at pH 2

3.2 Potentiodynamic Curves of Carbon Steel

Figure 3 shows the Tafel plot of both H_2S and CO_2–H_2S carbon steel corrosion test. Table 1 presents the corrosion rate of steel in H_2S and CO_2–H_2S solution from the Tafel plot. A different trend of corrosion rate was observed for H_2S and CO_2–H_2S where with increasing H_2S concentrations, corrosion rate is found increasing in the presence of oxygen with the lack of CO_2 purging while the opposite is observed under CO_2–H_2S condition. At low concentration of H_2S, higher corrosion rate is observed in the presence of CO_2 signifying the potential of CO_2 being the driving factor of corrosion on the specimen. This could be attributed to the absence of cathodic reaction involving reduction of oxygen since the purging procedure has created a CO2-saturated environment, removing dissolved oxygen in the tested system. Instead, the cathodic reaction dominating in the corrosion is suggested to be the hydrogen evolution reaction (HER). Both dissolved CO_2 and H_2S could form weak acids in the solution and participate in the HER in two main ways. One way is to provide H^+(aq) over dissociation or the direct reduction of H_2CO_3(aq) and H_2S(aq) [12]. Feng et al. (2017) suggested that while the corrosion behaviour was greatly dependent on the H_2S content, the corrosion potential was instead related to the CO_2 content and pH [13].

However, as the concentration of H_2S increased in the presence of CO_2, the corrosion rate decreased significantly. This may be due to the inhibition effect ascribed to the establishment of thin FeS film (tarnish) on the carbon steel surface that inhibited

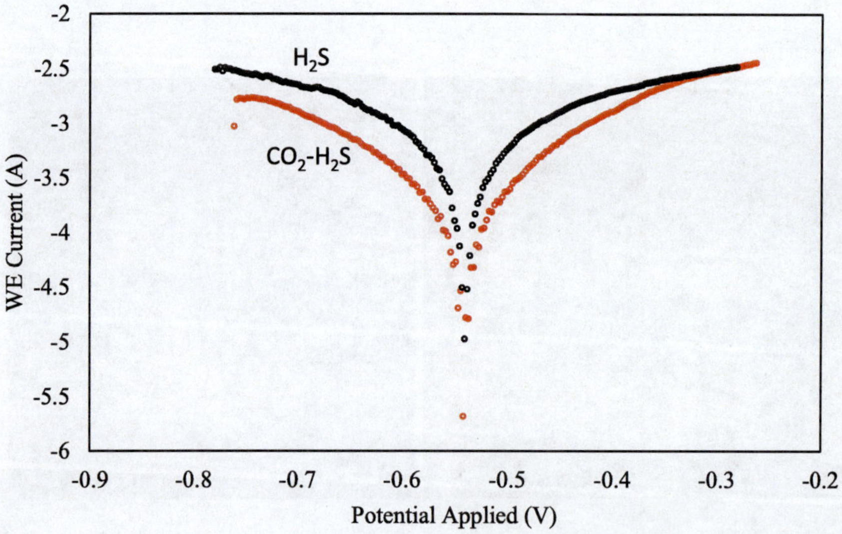

Fig. 3 Polarization curve for carbon steel under H_2S and CO_2–H_2S condition in 1.25 ppm of H_2S concentration at pH 2

Table 1 Corrosion rate from polarization study for carbon steel at pH 2

H_2S concentration (ppm)	Corrosion rate (mm/year)	
	H_2S	CO_2–H_2S
1.25	0.994 ± 0.129	3.367 ± 0.038
6.20	1.604 ± 0.199	2.665 ± 0.168
12.6	2.325 ± 0.143	0.561 ± 0.065

the anodic dissolution reaction. Choi et al. (2011) recommended that the precipitation of iron sulphide as well as iron carbonate film is feasible in the acidic solutions. This is attributed to the local supersaturation in regions directly on the surface of the carbon steel. These films offer corrosion inhibition in the acidic solutions.

3.3 *EIS of Carbon Steel in H_2S-containing Solutions*

EIS is one of the various electrochemical methods widely used for corrosion studies that is classified as a non-destructive approach due to the ability of EIS in differentiating range of electrochemical procedures happening at a corroding metallic interface. EIS data provide information on the kinetics and at the same time the mechanism of the corroding system. The EIS spectra were fitted with the electrochemical equivalent circuit using Nova 2.1.5 software. The equivalent circuit is by using the Randles circuit. Valcarce et al. (2006) believed that the electrodes that have gone through corrosion process may have numerous types of heterogeneities. This led to the consideration of a constant phase element (CPEs) usage in the circuit instead of a simple capacitor [14].

Fitting values from the EIS are shown in Tables 2 and 3. The process Q_p at higher frequencies signifies the capacitive behaviour of the passive film developed at the surface, paired with resistance R_p. As mentioned before, it was likely for all the examined metals to represent the oxide charging/discharging process as a simple capacitor. However, the double-layer formation at the electrolyte-solution interface revealed a time-constant distribution and had to be represented by a CPE. The polarization resistance, R_p is employed to evaluate the corrosion resistance. Polarization resistance is described as the resistance of the specimen to oxidation throughout the application of an external potential. The corrosion rate is proportionally related to the R_p and can be calculated from it.

From Table 3, it is evident that with the increase in concentration of H_2S under saturated CO_2 condition, the polarization resistance (R_p) increased. The increase in R_p values is due to the increased resistance to corrosion attack, possibly due to the FeS film layer formed on the metal surface in high acidic H_2S environment, creating protection to the material to a certain extent. This finding is aligned with the observation discussed in section B.

Table 2 Parameters obtained by fitting EIS data to the equivalent circuit using the Nova 2.1.5 for carbon steel (H_2S)

H_2S concentration (ppm)	Solution resistance, R_s (W)	R_p (W)	Q_p	n	χ^2
1.26	75.753	19.201	0.000293	0.91795	0.000123
6.20	76.387	8.9986	0.000352	0.87248	0.000137
12.6	74.906	6.2663	0.000381	0.82771	0.000662

Table 3 Parameters obtained by fitting EIS data to the equivalent circuit using the Nova 2.1.5 for carbon steel (CO_2–H_2S)

H_2S concentration (ppm)	Solution resistance, R_s (W)	R_p (W)	Q_p	n	χ^2
1.26	36.975	10.873	9 E–16	1.1	0.0556
6.20	61.426	30.878	9 E–16	1.1	0.0232
12.6	66.749	34.598	25.17 E–05	0.683	0.0186

4 Conclusion

In the increment of H_2S concentrations from 1.26 to 12.6 ppm, the corrosion rate of carbon steel increases from 0.994 to 2.995 mm per year. H_2S demonstrated a strong hastening effect on the cathodic hydrogen evolution of carbon steel, which cause the material to be severely corroded. The corrosion film developed in H_2S condition did not show any protective property. It was composed of mackinawite, which was a loose material in nature. Some destructive localized corrosion was observed on the carbon steel surfaces in H_2S-containing solutions. This is due to cementite stripped off from the grain boundary. However, different trends were observed in CO_2–H_2S condition. As the concentration increases, the corrosion rate decreased. The presence of H_2S with CO_2 would favour the formation of FeS(s) on the steel surface where the pH and Fe^{2+}(aq) concentration is higher. The formation of a thin sulphide film could be protective and reduce the corrosion rate.

Acknowledgements The authors acknowledge the financial support of Sarawak Energy Berhad in this research project (Cost centre: 015MD0-104).

References

1. NACE International/ASTM G193-21 (2022) Standard terminology and acronyms relating to corrosion
2. Kump L, Pavlov A, Arthur M (2005) Massive release of hydrogen sulfide to the surface ocean and atmosphere during intervals of oceanic anoxia. Geology 33. https://doi.org/10.1130/G21295.1

3. Nielsen J (2018) Why H_2S results in biogenic sulfide corrosion and how to eliminate it. ATS Innova Clean Water Solution
4. Woyciechowski P, Łukowski P, Szmigiera E, Adamczewski G, Chilmon K, Spodzieja S (2021) Concrete corrosion in a wastewater treatment plant—a comprehensive case study. Constr Build Mater 303:124388. https://doi.org/10.1016/j.conbuildmat.2021.124388
5. Suryaningsih S, Mindara JY (2016) Influence of wind and topography on distribution of H2S concentration at Cirata hydropower plant and its impact on corrosion rate. IOP Conf Ser Earth Environ Sci 29:012014. https://doi.org/10.1088/1755-1315/29/1/012014
6. Nafday OA (2004) Film formation and CO_2 corrosion in the presence of acetic acid
7. Sun W, Papavinasam S, Nesic S (2006) Kinetics of iron sulfide and mixed iron sulfide/carbonate scale precipitation in CO_2/H_2S corrosion. NACE Int Corros Conf Ser
8. Osarolube E, Owate I, Oforka NC (2008) Corrosion behaviour of mild and high carbon steels in various acidic media. Sci Res Essays 3:224–228
9. Li D, Zhang L, Yang J, Lu M, Ding J, Liu M (2014) Effect of H_2S concentration on the corrosion behavior of pipeline steel under the coexistence of H_2S and CO_2. Int J Miner, Metall Mater 2. https://doi.org/10.1007/s12613-014-0920-y
10. Choi Y-S, Nesic S, Ling S (2011) Effect of H_2S on the CO_2 corrosion of carbon steel in acidic solutions. Electrochimica Acta 56:1752–1760. https://doi.org/10.1016/j.electacta.2010.08.049
11. Wolthers M, Charlet L, Linde P, Rickard D, Weijden C (2005) Surface chemistry of disordered mackinawite (FeS). Geochimica et Cosmochimica Acta 69. https://doi.org/10.1016/j.gca.2005.01.027
12. Paul S, Pattanayak A, Guchhait SK (2014) Corrosion behavior of carbon steel in synthetically produced oil field seawater. Int J Metals 2014:628505. https://doi.org/10.1155/2014/628505
13. Feng R, Beck J, Hall D, Buyuksagis A, Ziomek-Moroz M, Lvov S (2017) Effects of CO_2 and H_2S on corrosion of martensitic steels in brines at low temperature. Corrosion 74. https://doi.org/10.5006/2406
14. Valcarce MB, De Sánchez SR, Vázquez M (2006) A comparative analysis of copper and brass surface films in contact with tap water. J Mater Sci 41:1999–2007. https://doi.org/10.1007/s10853-006-4499-1

Exploration of Possible Defects Originating in Al-5083 Laminates Synthesized Through Friction Stir Additive Manufacturing Technique

Adeel Hassan and Srinivasa Rao Pedapati

Abstract Green manufacturing is the concept of maximizing the productivity along with keeping healthy environment and strong economy by less material waste and preserving the resources. Friction stir additive manufacturing (FSAM) is the subdivision of additive manufacturing and lies in the category of sheet lamination (SL) according to the ASTM F2792. This additive technique is considered a competent candidate for sustainable manufacturing because of producing high-performance structural part with minimal material waste and utilizing less energy without generating fumes as compared to the existing fusion-based additive manufacturing techniques. In FSAM, mechanical properties and microstructure are based on the heat input and material mixing. These two inputs are commonly controlled with rotation speed (rpm), transverse speed (mm/min), plunge depth, tool geometry, and material to be processed. Therefore, in this study, an attempt is done to fabricate aluminum 5083 build of three plates at different rotation speeds (600, 800, 1000 rpm) and transverse speed of 20 mm/min, 30 rev/min. Defects originates, and microhardness of buildup plates were investigated in the build direction. Tunnel, groove, and incomplete stacking defects were noticed in the middle of joining track as compared to the beginning of the joining track. Maximum hardness of 90.9 HV found at 1000 rpm and 20 mm/min and overall, 1.83% hardness decreased in the process which may be improved by selecting optimized process parameters like rotation speed, tilt angle, transverse speed, etc. Further deep research is ongoing to explore the optimized set of process parameter window for preferable microstructure and mechanical properties.

Keywords Green manufacturing · Friction stir additive manufacturing (FSAM) · Solid-state additive manufacturing

A. Hassan
Department of Mechanical Engineering, University Teknologi PETRONAS, 32160 Seri Iskendar, Perak Darul Ridzuan, Malaysia
e-mail: adeel_21002773@utp.edu.my

S. R. Pedapati (✉)
College of Engineering, The University of Texas Permian Basin, 1105 West Highway 191Midland, Texas, 79705, USA
e-mail: pedapati_s@utpb.edu

F. Ahmad et al. (eds.), *ICREEM 2022*, Lecture Notes in Mechanical Engineering,
https://doi.org/10.1007/978-981-99-5946-4_25

1 Introduction

In today's global society, population and resource shortages are big problems. The environment is crucial, and any climatic shift causes a global imbalance. A new product quality management system and an environmental management system have both been proposed by the ISO. This era's main objective is to lessen the harm that industries do to the environment. New manufacturing techniques that are fit for a sustainable development strategy are required, such as Green Manufacturing [1]. A sustainable society encompasses a healthy environment, social equality, and a robust economy. Current manufacturing techniques and processes now place a higher emphasis on process sustainability than they do on social needs and technical hurdles [2]. An intensifying amount of energy and materials are used in the global manufacture of metal parts, which results in environmental degradation and resource overload due to the depletion of nonrenewable resources. For instance, the minimum energy need to refine 1 kg of aluminum ore is significantly greater about 21.6 MJ, which is the same as consuming 6.59 kg of coal [3]. Keeping in view the material waste, "Additive Manufacturing" (AM) is a technique of converting 3D data models into physical objects by layer-by-layer fashion with minimum material waste. Metal additive manufacturing (MAM) is a key component of additive manufacturing, which produces metal parts in layered manner and is adopted by critical engineering fields such as automobile, aerospace, and biomedical since the last three decades [4]. According to ASTM F2792 standard, additive manufacturing is formally categorized into seven routes such as: cold spray AM, material extrusion AM, electromechanical AM, powder bed fusion AM, material jetting AM, direct energy deposition AM, and sheet lamination AM. In cold spray AM, the metal powder is deposited on substrate at room temperature and atmospheric conditions, but less strength is achieved in deposited layers owing to unfused deposition. In material extrusion AM, feed material in the form of filament is deposited through nozzle on the substrate. Electromechanical AM is subtractive in nature with suitability of less complexity level parts production. Powder bed fusion AM deposited multiple layers of powder by getting melting through laser beam and produced dense layered parts. Another process is material jetting process, which is divided into three types: liquid metal jet, ink jet, and aerosol jetting. Next route of AM is direct energy deposition (DED), in this technique, substrate of multiple layered is obtained with excellent properties and it is used for repair work. The most advanced method used for AM is sheet metal lamination (SL), which is essentially the layer-by-layer addition of metal sheets to produce layered metal. To encapsulate, MAM is classified into two broader categories [5]:

1. Fusion-based additive manufacturing
2. Solid-state/fusion-free additive manufacturing

The fusion-based AM involves melting of feed material and subsequent solidification [6, 7]. A high-energy beam is required to melt the feed material, which creates problem like metal vaporization and excessive splashing. Because of this, highly textured columnar grains and non-uniform structures in the microstructure, residual stress, and deformation generated during the process are also affected the mechanical properties of the sample [8]. Similarly, conventional welding like gas metal arc welding and tungsten gas arc welding consumes 1.5–2 times more energy than solid-state manufacturing and also produces fumes causing disadvantageous effect of health and environment [9]. Solid-state additive manufacturing involves feed material in the layer forms and joined through plastic deformation below the melting point. If we consider the energy efficient and ecofriendly factor, the solid-state additive manufacturing is the better choice over existing fusion-based additive manufacturing.

Solid-state manufacturing is categorized as: ultrasonic additive manufacturing, friction stir additive manufacturing, and friction stir deposition. Friction stir additive manufacturing (FSAM) is a subdivision of sheet lamination (SL) [8], and it is similar to the friction stir lap welding (FSLW). It is basically layer by layer addition of metal sheets to produce layer metal parts having less complexity level [10]. Feed material in FSAM is a plate of desired dimensions. The plates are joined due to the severe plastic deformation owing to the heat produced by contact of non-consumable tools and process is done below the melting temperature of the feed material. Thus to avoid liquid state phase transition, FSAM technique was introduced by White [11] in 2002. The proposed technique is followed by Airbus and Boeing [12] in 2006 and witnessed the excellent interbonding of layers with more environmentally and greater production rate along with minimum material waste [5]. Palanivel et al. 2015a [7] and 2015b [13] successfully performed FSAM on magnesium-based WE43 alloy and AA 5083-O alloy, respectively, by stacking four sheets of each alloy. For Mg-based WE43 alloy, FSAM employed at two rotation speeds (800 and 1400 rev/min) with constant with constant transverse speed of 102 mm/min. A non-homogeneous trend observed in microhardness but average hardness of 115 HV achieved in the build, and it was further improved to 135 HV by aging at 160 °C for 60 h. Similarly, a build of four sheets was made of AA 5083-O alloy at a tool rotational rate of 500 rev min^{-1} and a tool traverse speed of 152 mm min^{-1}. Notable increment in the hardness (18%) was declared in build than base material. Later, in Yuqing et al. [14], a nine-layer stack of AA 7075-O is fabricated through FSAM with two process parameters such as tool rotation speed and transverse speed of 600 rpm and 60 mm min^{-1}, respectively. Microhardness increases from bottom to top and maximum hardness of 102 VH achieved in the top layer. FSAM of cast AA 2050 and T3 tempered alloys indicated a knockdown in tensile strength and improved ductility. Fracture analysis demonstrated a complex behavior whereas fracture resistance of welded zone was more than the base metal [15]. Aluminum lithium alloy 2195-T8 synthesized through FSAM up to five layers with 2 mm layer thickness of each plate and interfacial features were investigated. The parameter sets for the welding are 800, 900, 100 rpm with constant transverse speed of 100 mm/min. Hook defect perceived due to poor material mixing with cylindrical pin and conical pin having three flats. Overall maximum hardness reaches to 116.8 VH [16]. Zhang et al. [17, 18] studied AA6061-T6 numerically as well as experimentally. As build height going to increase, the peak temperature of newly added layer decreases but finer grain structure is produced, which leads to improved hardness and yield strength. Layer thickness is directly proportional to the hardness. However, after two to three layers, new added layers don't affect the mechanical properties of stacked layers. He et al. [19] joined 12 sheets of 7N01-T4 aluminum alloy by FSAM process. In build direction inhomogeneous microstructure formed and precipitate dissolution decreased from top to bottom which results decline in hardness. Li [20] improved the microhardness of aluminum 6XXX series in FSAM through precipitate hardening (controlling silicon content) and similarly Li [21] performed underwater FSAM and improved hardness and ultimate tensile strength at low-temperature aging treatment.

Al 5083 alloy is typically used in TV towers, drilling rigs, transportation equipment, armor plate, plate for ship building, rail and other vehicles, various structural applications that make use of the high strength and corrosion resistance. For this purpose, wrought aluminum alloy 5083 in the form of extruded rods, rolled plates, and sheets are used as a base material. The methodological framework of current metal additive manufacturing (MAM) is melting and depositing one thin layer at a time, and it is like microcasting or microwelding. Due to localized melting, large heat-affected zone (HAZ) is produced. Consequently, the part fabricated through this process shows anisotropic behavior and hence mechanical and microstructural properties of obtained structures are inferior to the base material. Therefore, key challenge is to maintain structural performance along with improved mechanical properties such as strength, hardness, and fatigue resistance. Thus, FSAM is the promising alternative technology of current AM processes to fabricate high structural part. In FSAM, the melting-based defects are not produced but other defects related to material plastic deformation are occurred due to the improper heat generation and material mixing, which advocates the deficiency of build's mechanical properties. Therefore, in this study, the suitability of the proposed process is explored to produce laminates of Al 5083 by investigating the possible defects occurred and microhardness achieved to promote clean green manufacturing. It is still a nascent technology and required deep research to come up commercially.

2 Material and Experimental Procedure

Aluminum 5083 plates of dimensions 100 mm × 150 mm × 2 mm were used in the current study as a research material. The additive manufacturing tool is having 20 mm concave shoulder and threaded conical pin of length 2.3 mm is used. Pin root and tip diameter were 3.5 mm and 3.2 mm, respectively. The experimental setup is shown in Fig. 1. Experiments were carried out for three set of rotation speed (600, 800, and 1000 rpm) and two set of transverse speed (20 and 30 mm/min) with zero tilt angle and constant dwell time of 10 s. A total of three plates were stacked together and total build height of 6 mm was achieved. The process parameters and their levels used in the current work are shown in Table 1.

Friction stir welding machine model FSW-F16 was used to perform additive manufacturing, and after executing FSAM, the samples were sectioned with power

Fig. 1 Experimental setup: FSW machine

Table 1 Parameters and their levels

Sample number	Number of plates stacked	Levels	Rotation speed (RPM)	Transverse speed (mm/min)	Depth (mm)	Dwell time (s)	Direction
A	3	1	600	20	3	10	CCW
		2	600	30	3	10	CCW
B	3	1	800	20	2.3	10	CCW
		2	800	30	2.3	10	CCW
C	3	1	1000	20	2.3	10	CCW
		2	1000	30	2.3	10	CCW

saw for confirmation of uniformity of material and hardness measurement. Standard metallographic procedure is used for microhardness sample preparation. The hardness of buildup plates along the building direction was measured with Vickers hardness tester (Leco LM 247AT, St. Joseph, MI, USA). Load of 500 g of force is applied for 10 s and hardness is measured at three different locations and then taken the average value.

3 Results and Discussions

3.1 Macro-defect Exploration

The stack of three aluminum plates was fabricated successfully and produced defects were investigated in two ways, first physically from top surface and secondly, defects in the cross-sectioned samples. Initially, sample A showed a constant digging of tool with "enable constant pressure" ON and this defect doesn't exhibit with "enable constant pressure OFF" for remaining samples B and C, it occurs due to excessive softness of material because of constant applied pressure and the working small material thickness. After turning it OFF, only the forging force applied in terms of applied plunge depth is illustrated in Fig. 2a. For the first lap, on upper plate, there was a smooth surface, no groove/line defect was observed for sample B (800 rpm and 20, 30 mm/min) but this was prominent at 1000 rpm, 20 mm/min. After adding third plate, a groove line defect throughout the welded zone observed for all samples is shown in Fig. 2c. In the beginning, there was no such defect seen, it occurs after traveling some transverse distance. Main causes of this defect (groove defect) are: high heat input and improper pressure in term of plunge depth, tilt angle, and incomplete contact of tool shoulder and workpiece. During the dwelling time at the beginning, enough heat is produced to plasticize and bond the plates while tool transverse, the plates are already heated because of small thickness, which produce excessive heat that leads to the groove generation. Poor clamping leads to detract the weld line, which would affect the joining path, exhibited in Fig. 2d. Clamping becomes loose because of huge vibrations produced; thus, it should be ensured before the next pass.

To reconnoiter the defects inside the final layered build, the build cross-sectioned from the start and from the middle. The cross-sectioned view is portrayed in Fig. 3. In the beginning of the friction stir welding, the material gets fully soften due to the application of 10 s dwell time and plates are joined together without any defect as illustrated in Fig. 3a. Similarly, tool moves transversely after completing dwell time, tunnel defect or worm hole originates at beneath of tool pin in first and second layer as indicated in Fig. 3b. This may be due to the zero-tilt angle, tool pin geometry, or more time in the form of rotational and transverse speed to plasticize the material with high heat input. At the top of first lap, when third layer is added, the material of third layer is joined with tool pin stirring effect and make the nugget zone. Material under the tool shoulder area in retreating and advancing side is plasticized but not joined with the top of first lap. There was a thin gap between the top of first lap and bottom of third layer. It was due to the inappropriate surface smoothness after removing the thin chips produced in the first pass of friction stir welding. So, it is recommended to prepare surface well after each pass before going to add up the next layer. In this study, the surface was smoothened by manual filling process, but the best way to remove the extra chips by performing milling operation instead of manual filling which yields full smooth top surface and consequently the next added layer will join accurately.

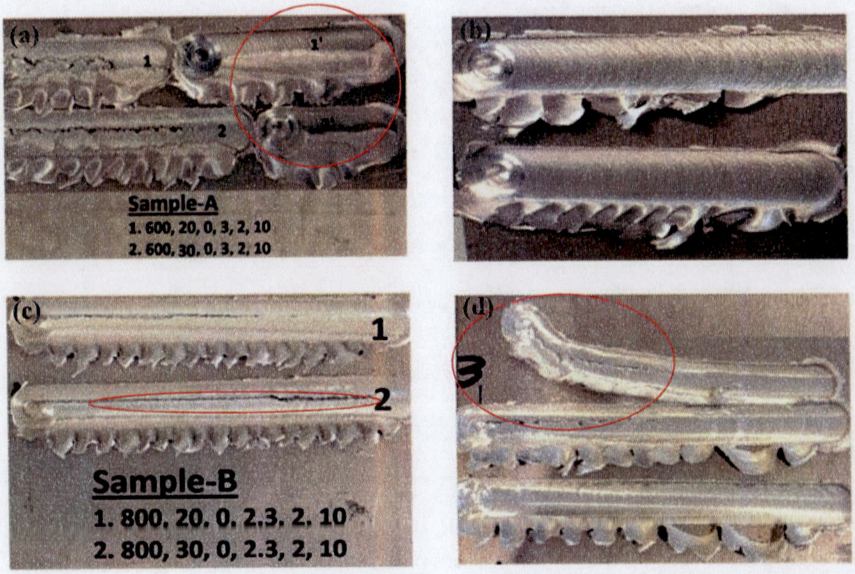

Fig. 2 Physical appearance of synthesized plates. **a** Constant tool digging **b** no groove defect in second layer **c** groove defect in third/top layer **d** poor clamming issue

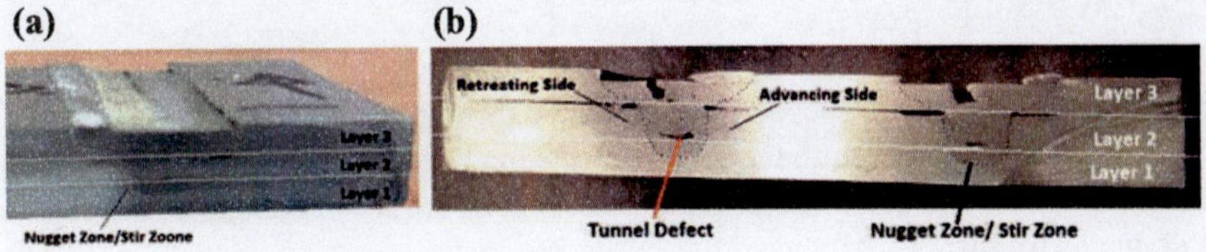

Fig. 3 Cross-section view of final build of three plates. **a** Cross-sectioned view from beginning. **b** Cross-section view from middle

3.2 Microhardness Investigation

Microhardness testing was employed on the nugget zone of buildup plates/samples, cross-sectioned from the defect-free region. Normally, the hardness value is found lesser in the second layer compared to the first layer at both transverse speed of 20 and 30 mm/min. The hardness value increases with increase of rotation speed in the second layer, it is because of more heat produced with increase of rotation speed, which gives better material mixing and further leads to improve hardness as shown below in the graph (Fig. 4). Maximum of 90.9 HV and minimum of 53.5 HV value were found in the top layer nugget zone at 1000/20 and 600/30, respectively. Recorded average hardness value was less than the hardness of base material along with inhomogeneous trend, which is due to the cyclic thermal variation, same results also reported by other researchers [13, 14]. The base material hardness was 92.6 HV

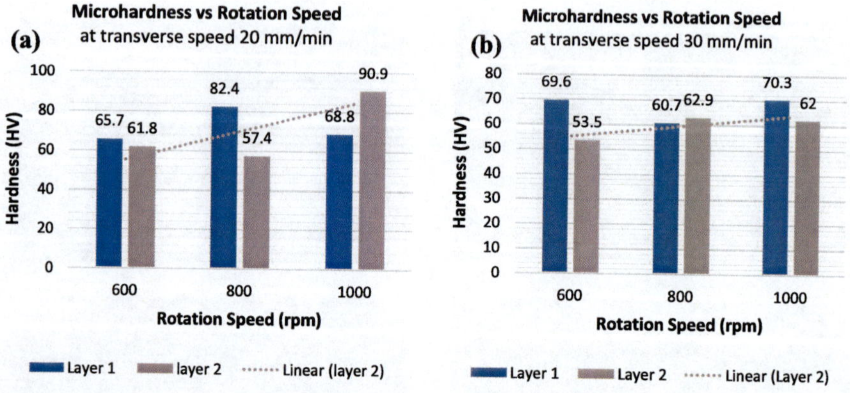

Fig. 4 Microhardness versus rotation speed: **a** at 20 mm/min transverse speed **b** at 30 mm/min transverse speed

and overall, 1.83% hardness decreased in the process which may be improved by selecting proper tool geometry and process parameters like rotation speed, tilt angle, welding environment, transverse speed, etc.

4 Conclusions

Laminates of Al 5083 are successfully produced through friction stir additive manufacturing process; however, the induced defects may be further investigated to make the process more feasible. The primary conclusions from the conducted study can be drawn as follow:

- Three plates were synthesized successfully; however, these two major parameters (rotation speed and feed rate) need to be optimized to produce sufficient heat produced for better material mixing.
- Before going to add new layer, the newly born lap surface should be milled instead of manual filling for better joining and ensure the tightness of material clamping to avoid the misalignment of joint.
- Maximum of 90.9 HV and minimum of 53.5 HV value were found in the top layer nugget zone at 1000/20 and 600/30, respectively.
- Overall, 1.83% hardness decreased in the process, which may be improved by selecting improve tool geometry and process parameters like welding environment, suitable post-aging, rotation speed, tilt angle, transverse speed, etc.
- Further study is underway to explore the exact range of parameters to produce defect-free build with sound mechanical properties.
- This novel approach with optimized process parameters would be beneficial on shop floor in transportation sector like aerospace, automobile etc. and will promote clean green manufacturing.

Acknowledgements This study was supported by Department of Mechanical Engineering, University Teknologi PETRONAS, Perak, Malaysia under the YUTP grant number 015LC0-185.

References

1. Paul ID, Bhole GP, Chaudhari JR (2014) A review on green manufacturing: it's important, methodology and its application. Procedia Mater Sci 6(Icmpc):1644–1649. https://doi.org/10.1016/j.mspro.2014.07.149
2. Akinlabi S, Akinlabi E (2018) Laser beam forming: a sustainable manufacturing process. Procedia Manuf 21:76–83. https://doi.org/10.1016/j.promfg.2018.02.097
3. Gottwald RB, Griffiths RJ, Petersen DT, Perry MEJ, Yu HZ (2021) Solid-state metal additive manufacturing for structural repair. Acc Mater Res 2(9):780–792. https://doi.org/10.1021/accountsmr.1c00098
4. Rathee S, Srivastava M, Pandey PM, Mahawar A, Shukla S (2021) Metal additive manufacturing using friction stir engineering: a review on microstructural evolution, tooling and design strategies. CIRP J Manuf Sci Technol 35:560–588. https://doi.org/10.1016/j.cirpj.2021.08.003
5. Derazkola HA, Khodabakhshi F, Simchi A (2020) Evaluation of a polymer-steel laminated sheet composite structure produced by friction stir additive manufacturing (FSAM) technology. Polym Test 90(June):106690. https://doi.org/10.1016/j.polymertesting.2020.106690
6. Palanivel S, Mishra RS (2017) Building without melting: a short review of friction-based additive manufacturing techniques. Int J Addit Subtr Mater Manuf 1(1):82–103
7. Palanivel S, Nelaturu P, Glass B, Mishra RS (2015) Friction stir additive manufacturing for high structural performance through microstructural control in an Mg based WE43 alloy. Mater Des 65:934–952
8. Gao H, Li H (2021) Friction additive manufacturing technology: a state-of-the-art survey. Adv Mech Eng 13(7):1–29. https://doi.org/10.1177/16878140211034431
9. Bevilacqua M, Ciarapica FE, Forcellese A, Simoncini M (2020) Comparison among the environmental impact of solid state and fusion welding processes in joining an aluminium alloy. Proc Inst Mech Eng Part B J Eng Manuf 234(1–2):140–156. https://doi.org/10.1177/0954405419845572
10. Srivastava AK, Kumar N, Dixit AR (2021) Friction stir additive manufacturing—an innovative tool to enhance mechanical and microstructural properties. Mater Sci Eng B 263:114832
11. White D (2002) Object consolidation employing friction joining. Google Patents, 01 Oct 2002
12. Lequeu P, Muzzolini R, Ehrstrom JC, Bron F, Maziarz R (2006) High-performance friction stir welded structures using advanced alloys
13. Palanivel S, Sidhar H, Mishra RS (2015) Friction stir additive manufacturing: route to high structural performance. JOM 67(3):616–621. https://doi.org/10.1007/s11837-014-1271-x
14. Yuqing M, Liming K, Chunping H, Fencheng L, Qiang L (2016) Formation characteristic, microstructure, and mechanical performances of aluminum-based components by friction stir additive manufacturing. Int J Adv Manuf Technol 83(9):1637–1647
15. Lu I (2019) Friction stir additive manufacturing (FSAM) of 2050 Al-Cu-Li Alloy. University of South Carolina
16. Zhao Z, Yang X, Li S, Li D (2019) Interfacial bonding features of friction stir additive manufactured build for 2195-T8 aluminum-lithium alloy. J Manuf Process 38(August 2018):396–410. https://doi.org/10.1016/j.jmapro.2019.01.042
17. Zhang Z, Tan ZJ, Li JY, Zu YF, Liu WW, Sha JJ (2019) Experimental and numerical studies of re-stirring and re-heating effects on mechanical properties in friction stir additive manufacturing. Int J Adv Manuf Technol 104(1–4):767–784. https://doi.org/10.1007/s00170-019-03917-6
18. Zhang Z, Tan ZJ, Li JY, Zu YF, Sha JJ (2020) Integrated modeling of process–microstructure–property relations in friction stir additive manufacturing. Acta Metall Sin Lett 33(1):75–87. https://doi.org/10.1007/s40195-019-00945-9

19. He C, Li Y, Zhang Z, Wei J, Zhao X (2020) Investigation on microstructural evolution and property variation along building direction in friction stir additive manufactured Al–Zn–Mg alloy. Mater Sci Eng A 777(November 2019):139035. https://doi.org/10.1016/j.msea.2020.139035
20. Li JY, Kong SN, Liu CK, Bin Wang B, Zhang Z (2022) Chemical composition effect on microstructures and mechanical properties in friction stir additive manufacturing. Acta Metall Sin Engl Lett (April). https://doi.org/10.1007/s40195-022-01406-6
21. Li Y et al (2022) Effect of post-fabricated aging on microstructure and mechanical properties in underwater friction stir additive manufacturing of Al–Zn–Mg–Cu Alloy. Materials (Basel) 15(9):3368. https://doi.org/10.3390/ma15093368

Classification of Pump Failure Using a Decision Tree Technique

Ruwaida Aliyu, Ainul Akmar Mokhtar, and Hilmi Hussin

Abstract The oil and gas sectors employ pumps extensively. The dependability and good operation of the pumps contribute significantly to the manufacturing line's efficiency. Nevertheless, breakdowns can occur for a variety of causes, and a single failure can result in severe production delays and economic losses. Therefore, it is of the highest importance to anticipate future errors and swiftly address their root causes. Traditional pump maintenance and fault detection methods are ineffective in detecting possible defects. However, with the rapid rise of industry 4.0, the growing use of sensors, and the use of artificial intelligence techniques, smart plants may automate their operations to greatly enhance their efficiency and quality of output. Given that, Prognostics and Health Management (PHM) is essential for optimal machine performance. Meanwhile, Predictive Maintenance (PdM) is an emerging topic within maintenance methodologies with the objective of predicting failure prior to its occurrence in order to schedule maintenance only when it is necessary. These solutions have benefited pump maintenance decision-making by addressing complexity issues, reducing downtime, enhancing overall reliability, and reducing pump operating costs. This study examines the use of the classification learner to predict pump failure. The vibration sensor data for the pump were employed for prediction in the MATLAB software using the classification learner machine learning method. According to the dataset, the variable of interest is the machine status, which is categorized as normal, broken, and recovering. On the basis of the confusion matrix and the evaluation of the true positive rate (TPR), the false negative rate (FNR), the positive predicted values (PPV), and the false discovery value, the actual result was anticipated (FDR). The experimental result reveals that the accuracy of the training model was 91.94%, whereas the accuracy of the testing model was 74.4%.

R. Aliyu (✉) · A. A. Mokhtar · H. Hussin
Mechanical Engineering Department, Universiti Teknologi PETRONAS, Bandar, 32610 Seri Iskandar, Perak Darul Ridzuan, Malaysia
e-mail: ruwaida_18001619@utp.edu.my

A. A. Mokhtar
e-mail: ainul_akmal@utp.edu.my

H. Hussin
e-mail: hilmi_hussin@utp.edu.my

F. Ahmad et al. (eds.), *ICREEM 2022*, Lecture Notes in Mechanical Engineering,
https://doi.org/10.1007/978-981-99-5946-4_26

Keywords Pumps · Failure prediction · Machine learning

1 Introduction

Pumps are typical hydraulic devices used in both commercial and domestic settings, and their main application in high-volume fluid handling is to regulate oil flow processes. The pumps have a wide range of applications and are vital to the expansion of the national economy. However, continued use of high-speed pumps without maintenance results in the mechanical components of the complete assembly being severely damaged. The lifespan data of the pumps are the main component of the standard reliability evaluation approach, and failure data may be difficult to get. According to surveys, maintenance costs often represent more than a third of total operating costs. Alraghbi [1] estimates that maintenance expenses are 15–70% of production cost. Shiels [2] states that temperature, oil level, motor RPM, cavity, liquid flow rate, and total pump duration are the primary variables that might shorten a centrifugal pump's anticipated life cycle.

Prognostics and health management (PHM) is an important aspect of production that monitors and analyzes equipment state to guarantee sufficient machine maintenance and operation. Reactive maintenance, which fixes problems after they happen, has made quick progress toward Predictive Maintenance (PvM), which conducts inspections at specified intervals. Predictive maintenance has gained popularity among manufacturing companies and academic research due to its many advantages over conventional maintenance practices, including reduced factory downtime, lower maintenance and manufacturing costs, higher levels of safety, production, and equipment life, and higher overall profits.

To determine when maintenance activities are necessary, the predictive maintenance technique uses predictive technology. It is founded on ongoing evaluation of a machine's or process's dependability, enabling maintenance to be carried out just when required. Additionally, it makes it possible to identify failures before they occur by employing prediction techniques based on past data. For instance, engineering approach, statistical inference approaches, machine learning techniques, and integrity aspects including visual characteristics, wear, and wearability Clustering algorithms, neural networks, and other technologies have a significant beneficial influence on the conventional fault monitoring techniques, making fault monitoring more accurate and intelligent [3]. This is due to the ongoing development of computer algorithms and artificial intelligence. As a result, our machine learning system may use these characteristics as features to provide the intended forecast, which is the probable downtime [4].

One of the most challenging aspects for a successful predictive maintenance plan is data availability. It's crucial to have the right kind and amount of data for training deep learning and machine learning algorithms. As a result, the data-driven approach is challenging to use when there is a lack of a significant amount of high-quality data.

Using a publicly available dataset with comparable attributes to temporarily train and test the models and methods was one way to overcome this problem.

The following outline constitutes the paper's organizational structure: The review of the relevant literature is covered in Sect. 2. The technique is broken forth in further detail in Sect. 3. The discussion of the results and the analysis may be found in Sect. 4. The final portion of the article is presented in Sect. 5.

2 Literature Review

2.1 Pump Operation and Failure Modes

2.1.1 Pump Classification

A pump's principal purpose is to transport liquids from low- to high-pressure locations. They are classified based on the manner in which they give extra energy to the fluid being pumped. There are two basic types of pumps: those that employ kinetic energy and those that use positive displacement [5]. Figure 1 is an illustration of a classification tree for pumps.

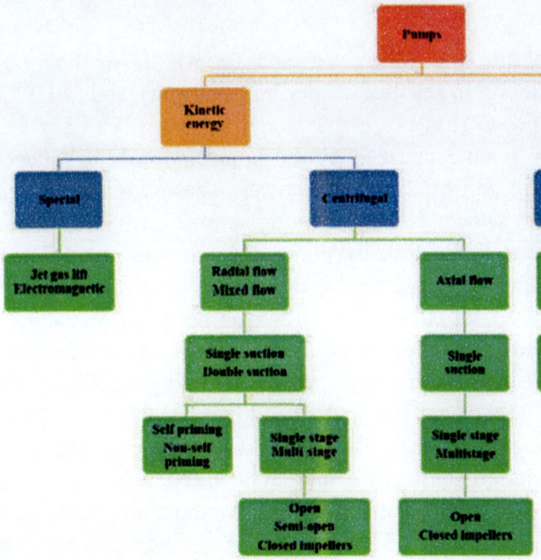

Fig. 1 Pumps classification tree [5]

2.1.2 Fundamental and Operating Principles of Pumps

Centrifugal pump's principal pumping units are the volute and impeller. The liquid reaches the eye of the impeller and becomes trapped between the impeller blades. The blades speed the liquid as it travels towards the tip of the blade. As the liquid accelerates, a low-pressure zone occurs in the impeller's eye (the Bernoulli principle, as velocity increases, pressure decreases). The liquid exits the impeller's exterior diameter at high speed and slams against the volute's inner casing. At this point, the liquid's centrifugal velocity equals zero, and the velocity is converted to pressure. Because the motor is spinning, the liquid has rotational velocity. The liquid is then routed around the internal volute housing in an ever-increasing escape path. The spinning velocity decreases as the conduit lengthens, and more pressure is supplied to the liquid, which leaves through the discharge nozzle ready to overcome system resistance. The usage of pumps is governed by a series of principles known collectively as affinity laws. These are a collection of recommendations that, when used collectively, may be used to determine the pump's operating characteristics in response to changing pump velocities. In Eq. (1), the following examples demonstrate these rules.

$$\begin{aligned} Q &\propto N \\ H &\propto N^2 \\ \text{BHP} &\propto N^3 \end{aligned} \tag{1}$$

where "Q" is the liquid volume in gallons per minute (gpm), "H" is the total head in feet, "N" is the shaft velocity in revolutions per minute (rpm), and "BHP" is the brake horsepower in horsepower (hp).

Performance of a pump is characterized by the interrelationships between the pump head, efficiency, flow, and braking horsepower. These correlations can be shown graphically on a chart known as the pump-characteristic curves. Four curves are displayed on a single graph to illustrate the pump-characteristic curve. The head-flow curve, also known as the H-Q curve, the efficiency curve, the energy curve, and the pump's minimum requirement curve, also known as net positive suction head required, NPSHr are included in these curves. Figure 2 displays a typical performance chart for centrifugal pumps. As flow increases, the total head gradually decreases. The braking horsepower (BHP) typically increases linearly with flow. The ideal efficiency point corresponds to the apex of the bell-shaped efficiency curve (BEP). Until the BEP zone, the required net positive suction head (NPSHr) stays constant. Until the curve hits the BEP threshold, the NPSHr grows exponentially. It is advised to operate a pump inside or near to the BEP zone. The operation of a pump to the extreme left or right of the BEP zone is unwanted and detrimental [6].

Fig. 2 Performance curve of a pump [7]

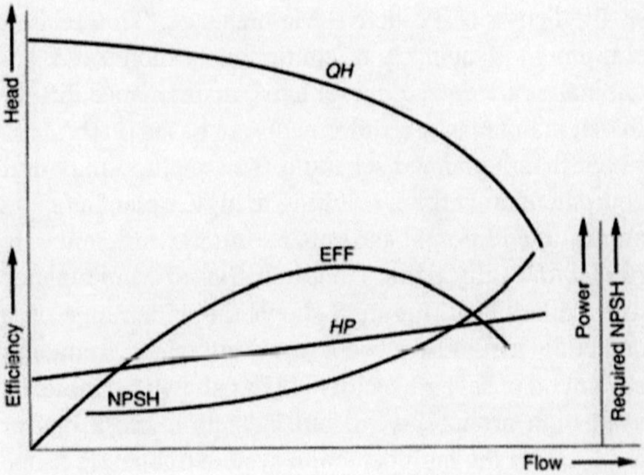

2.2 *Pump Failure*

Poor operating conditions or mechanical damage may contribute to pump failures. In no particular order, the following are a few of the most frequent unfavorable results of insufficient pump maintenance and operation: cavitation, very low or zero flow operation, dry running, bearing failures, impeller damage, and mechanical seal degradation.

2.2.1 Fault Detection Methods

In various industries, maintenance practices have evolved during the past decade. These methods are often categorized as:

Reactive Maintenance: Also referred to as "run until failure" strategy. The equipment is not maintained until it fails, and if it fails, it is either repaired or replaced, depending on the availability of cash. On the surface, it may appear that maintenance and labor costs are reduced, but in the long run, more money is spent on repairs and the purchase of new equipment. In anticipation of a breakdown, the equipment's lifespan actually decreases. Consequently, equipment must be changed more often. Unscheduled downtime of equipment, resulting in lost output, is one of the greatest challenges posed by this technique. Consequently, reactive maintenance causes equipment to operate inefficiently for extended durations, resulting in greater energy costs [8].

Preventive Maintenance: Sometimes known as time-based or planned maintenance (SMS). This technique to equipment maintenance is performed routinely according to a defined plan in time or process iterations in order to predict equipment failure. It is a normally effective strategy for avoiding failures. Although proactive maintenance is better to reactive maintenance, it cannot prevent unanticipated equipment downtime and necessitates unnecessary repair operations that may harm other components [9].

Predictive or Proactive Maintenance: This technique is based on the fact that equipment is regularly or continuously monitored, and maintenance is scheduled if anomalies are detected. Predictive maintenance differs from preventive maintenance in that maintenance requirements are based on the actual condition of the equipment, rather than a planned schedule. This method may dramatically reduce unscheduled equipment downtime, resulting in higher plant availability and improved plant operations. In addition, it can enhance energy efficiency by reducing the time equipment runs with faulty parts. Condition-Based Maintenance (CBM) is another name for this method [10]. Figure 3 shows the wide range of applications for various defect detecting methods. Recent study indicates frequency percentage of maintenance employed in several sectors [11] as shown in Table 1.

A high probability of fault detection and a low probability of false alarms are required for the fault detection system to operate properly. If the algorithm routinely generates false warnings due to an overly sensitive detection mechanism, operators may question its usefulness. Nonetheless, if the detection technique is too lax, it may eliminate unneeded false alarms while disregarding anomalies and mistakes that might lead to a system failure. Downtime resulting from significant breakdown owing to overlooked faults. A technique for fault detection must be developed to be both sensitive to genuine problems and resistant to false alarms [12].

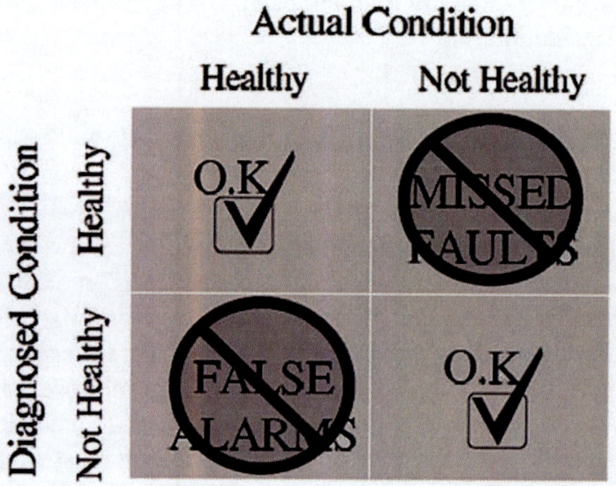

Fig. 3 Scenarios of fault detection [11]

Table 1 Maintenance application in industries [11]

Maintenance procedures	Percentage (%)
Reactive maintenance	55
Preventive maintenance	31
Predictive maintenance	12
Others	2

2.3 *Categories of Fault Detection Methods*

Improved production performance necessitates fault-detection processes that evaluate the operational condition of the equipment and identify defects early on. When classifying defect detection processes, model-based approaches, signal-based methods, and knowledge-based methods are the three basic categories employed. These categories are established by the methodology and a priori data used in the procedure [13].

2.3.1 Model-Based

Using mathematical and physical modeling, model-based processes imitate the behavior of the tested equipment. Model-based approaches demand expert understanding of the equipment as well as its physical, system, and application behavior. Figure 4 depicts how model-based fault diagnosis tools monitor the difference between the observed outputs of physical systems and the predicted outputs of the model.

A model-based defect detection system is based on the generation of residuals, which are defined as the disparities between measured and projected outputs. The system model might be a physics-based or empirical representation of the monitored system. The model establishes a connection between system outputs, system failures, system disturbances, and system inputs. The measured variables are the inputs and outputs of the system, whereas the predicted variables are the system model's outputs. Idealistically, these residuals should only be influenced by system failures and not by any changes in operating circumstances resulting from changes in system inputs and disturbances. Thus, produced residuals are exclusively sensitive to faults and insensitive to input or disturbance changes. If the system is in good condition, the residuals will resemble white noise. Any departures of the residuals from the behavior of white noise might be construed as a system error.

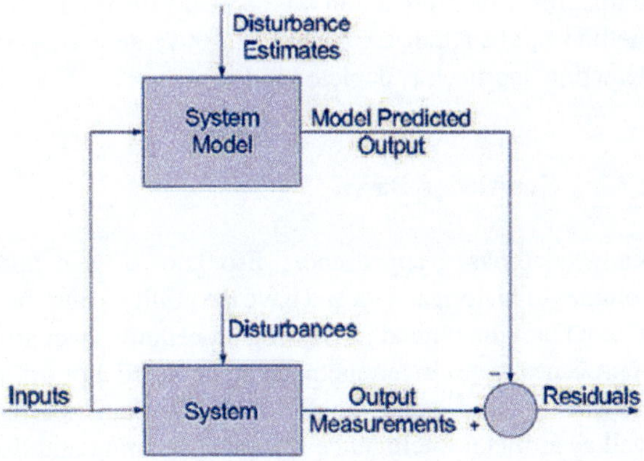

Fig. 4 Model-based fault detection [14]

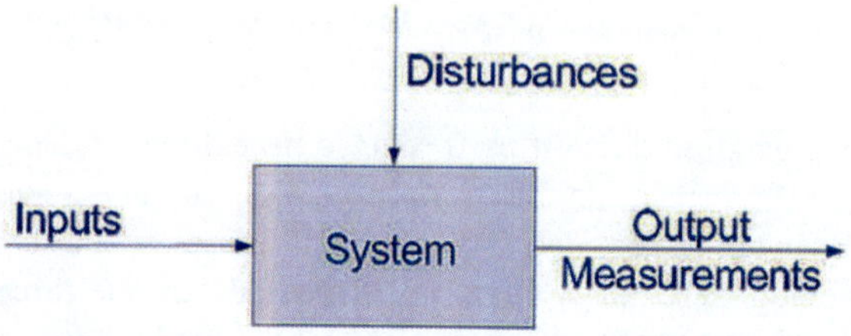

Fig. 5 Signal-based fault detection [14]

2.3.2 Signal-Based

Using signal processing techniques, signal-based procedures assess the measured operating state of the equipment, highlight, and extract features that would not otherwise be readily obvious, and deliver a diagnostic based on prior knowledge about healthy systems. These strategies nevertheless need specialist expertise in order to identify the indicators of erroneous information. Signal-based approaches are typically divided into time domains (Standard deviation, trends, slope, and magnitudes), frequency domains (motor current signature analysis), and time frequency domains (Fourier transform, wavelet transforms, Hilbert-Huang transform, instantaneous power FFT, high-resolution spectral analysis, wavelet analysis, bi spectrum, Park's vector approach, adaptive statistical time frequency method) based on the type of signal.

Harihara et al. [14] compare signal-based and model-based fault detection systems against an induction motor fault detector utilizing a coin toss. The study's findings can also be used for the identification of centrifugal pump failure. Receiver operating characteristic (ROC) curves are displayed for all three types of detection systems, and their false alarm and defect detection probabilities are compared. For the cases under examination, the defect detector based on a coin toss had a lower false alarm rate than the signal-based detection approach. The signal-based fault detection system was capable of 100% fault detection, but there was a significant probability of false alarms (about 50%). In contrast, the model-based fault detection approach had a false alarm rate of 0% and a fault detection capability of around 89%. If the false alarm probability restriction was decreased to around 10%, the model-based detection method might attain a capacity of 100% defect detection. The signal-based fault detection approach is depicted in Fig. 5.

2.3.3 Knowledge-Based

Knowledge-based approaches, also known as data-driven methods, utilize vast volumes of historical data and have the ability to automatically extract characteristics without previous signal processing procedures. In contrast to model and signal-based approaches, data-driven methods do not need a priori information and hence enable analysis without expert knowledge [12]. Statistical and probabilistic analyses, as well as artificial intelligence (machine learning and deep learning) approaches, are

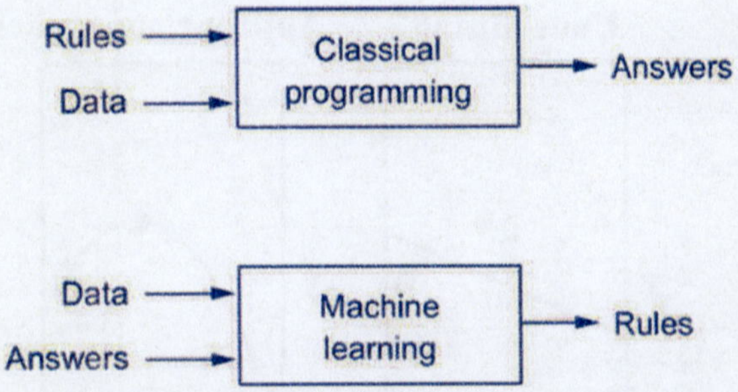

Fig. 6 Machine learning paradigm

frequently employed by data-driven methodologies. Academia is presently paying more attention to data-driven methodologies, with promising outcomes.

2.4 *Machine Learning*

Machine learning is a subset of artificial intelligence consisting of a set of techniques for automatically finding patterns in input data in order to create predictions about future data. In contrast to conventional programming, in which the designer specifies the rules (program) to be applied to the data in order to create a result, machine learning systems are trained to extract the rules from the data and the outcomes of previous iterations. Figure 6 illustrates ML concepts.

Some of the most popular problems that machine learning algorithms are used to perform include classification, regression, machine translation, and anomaly detection. These tasks are challenging for traditional programming languages to handle. Machine learning algorithms are widely divided into three categories: unsupervised learning, supervised learning, and semi-supervised learning depending on the type of data at hand.

2.5 *Supervised Learning*

On the other hand, utilizes labeled data to learn a mapping from inputs x to outputs y (also known as targets), such that $D = (x_i, y_i)N_i = 1$, where D represents the dataset and N represents the number of samples in the dataset [13]. The aim of supervised learning is to learn to predict y given unknown, unlabeled data inputs x in the future. Furthermore, supervised learning tasks may be divided into two categories: classification and regression. The goal of classification problems is to learn a mapping from inputs x to outputs y, where $y_1, \ldots, C$ is the number of discrete classes in a finite set. The issue is defined as a binary classification if the number of classes is equal to two, and as a multiclass classification if the number of classes

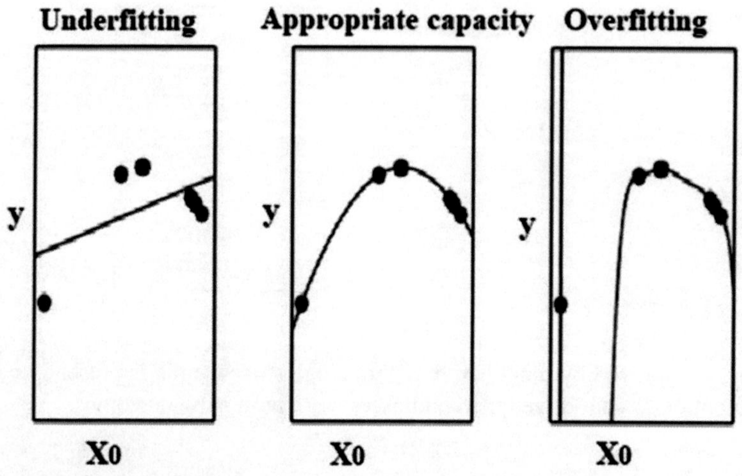

Fig. 7 Generalization of a model

is larger than two. Like classification problems, regression questions seek to learn a mapping from inputs x to outputs y; however, y is a continuous, real-valued output y_R.

2.5.1 Semi-supervised Model

May be employed instead of two distinct unsupervised and supervised models when just a small portion of the data is labeled. Semi-supervised learning predicts y from x using both unlabeled instances from $D = (x_i)N_i = 1$ and labeled examples from $D = (x_i, y_i)\ N_i = 1$. The ability of a machine learning model to generalize successfully onto new data, often known as the model's capacity, is the primary aim of machine learning models. Underfitting refers to the situation in which a model does not perform well on the data it was trained on and is also unable to generalize its results well. It is argued that a model is overfitting when it performs well on the data used for training but struggles when applied to new data. The optimal model is one that strikes a compromise between underfitting and overfitting, with the goal of maximizing both generalization and prediction performance. Figure 7 depicts the model generalization.

2.5.2 Unsupervised Learning

Employs unlabeled data with no explicit mapping from inputs x to outputs y, so that $D = (x_i)N_i = 1$. The purpose of unsupervised learning is to discover patterns in data by learning the probability distribution of the data. Unsupervised learning is frequently used to cluster data based on its attributes in order to find hidden elements inherent in the data.

To generalize and perform well on new, unobserved data, machine learning algorithms must undergo a model training phase. Training models iteratively requires comparing the predicted value y to the actual value y and calculating the prediction error. By reducing such errors, the models are able to learn and increase their predictive accuracy. A loss function assesses the performance of a model by calculating the prediction error. To measure such inaccuracy, there are a number of loss functions that take into consideration a number of elements. The optimization algorithm refers to the method through which the model is modified to reduce prediction error. The performance of the model is affected by the use of several loss functions and optimizers, each of which has its own internal parameters. The subsequent section describes the methodology utilized in this work.

3 Methodology

In this section, the methodology of the nonlinear model is clearly described. Consequently, the technique to prediction depicted in the figure was utilized in the study to classify pump failure. Figure 8 depicts the design for the proposed method for predicting pump failure. The method has been structured as shown in Fig. 8 to describe the steps used in this study.

Fig. 8 Process flow of prediction classification method

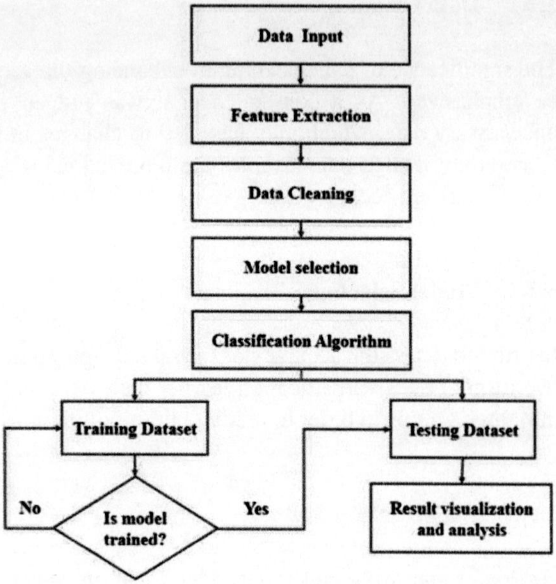

3.1 Data Input

This study's data are generated by the centrifugal pump, which is coupled to sensor readings. The data were acquired from the Kaggle dataset, which has a total of 52 sensors organized into three key categories. For forecasting centrifugal pump failure, they comprise timestamp data, sensor data (52 series), with all values being raw, and machine status.

3.2 Feature Extraction

In order to construct the model for the prediction of pump failure, the research study extracts significant elements from the collected data in order to fit the model. The data in this study contain thousands of features, and if the number of features increases or becomes similar, the number of observations retained in the dataset may cause the machine learning model to overfit the predicted data, which is one of the motivations for feature extraction in this work. Therefore, it is essential to employ regularization or dimensionality reduction strategies to avert the issue (Feature Extraction). Consequently, decreasing the regularization of the data reduces the risk of overfitting by increasing the chance that the model will be correct, reducing the possibility of overfitting, accelerating testing, and enhancing data presentation.

3.2.1 Data Cleaning

The significance of data cleaning in enhancing the accuracy of performance cannot be emphasized. As a consequence, it was judged fair to eliminate the pump's unnecessary data, which may have led to changes in normal operation conditions. Uncertainty-related data samples are removed using a filtration procedure in order to eliminate unnecessary data.

3.2.2 Model Selection

In order to assess the accuracy of the model's pump failure prediction and minimize overfitting issues, classification learner data were gathered. The data are separated into three groups in order to predict pump failure: normal, broken, and recovering.

3.2.3 Classification Method

For load monitoring and failure identification, supervised machine learning techniques are applied to the dataset generated from the publicly accessible pump data,

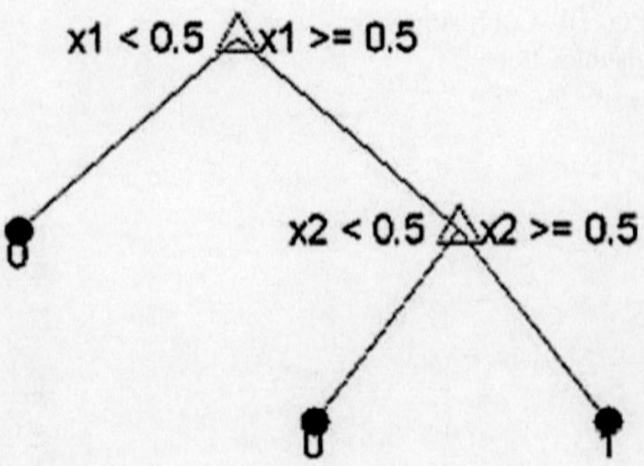

Fig. 9 Classification tree

which has a significant number of observations with 52 sensor observations. The fine decision tree technique was employed to train the classifier, as tree-based methods are prevalent in the business world and have been demonstrated to be effective at producing correct predictions. These methods are referred to be non-parametric since they make no assumptions about the distribution of the data and construct genuine models. These techniques need minimum data purification and are unaffected by anomalies or multicollinearity. Decision trees are more user-friendly than other supervised algorithms, can handle both categorical and numerical data, and are resistant to outliers. As a result, less data processing is required.

The classification trees are trained to predict responses to data. To predict a response, the decisions in the tree from the root (beginning) node down to a leaf node is followed. The leaf node contains the response. Statistics and Machine Learning Toolbox™ trees are binary. Each step in a prediction involves checking the value of one predictor (variable) as shown in the simple classification tree in Fig. 9. This tree predicts classifications based on two predictors, ×1 and ×2. To predict, start at the top node. At each decision, check the values of the predictors to decide which branch to follow. When the branches reach a leaf node, the data are classified either as type 0 or 1.

3.2.4 Dataset for Training

The dataset used to train the fine decision tree in the classification learner has 220,320 observations, 51 predictors, and machine status as the response, which is classified into three categories. For training, fivefold cross-validation was utilized.

3.2.5 Test Dataset

The trained model was tested using 20% of the dataset, which included 4395 observations, 51 predictors, and machine status as the response. The split criterion was Gini's, Diversity Index, and the maximum number of splits was 100.

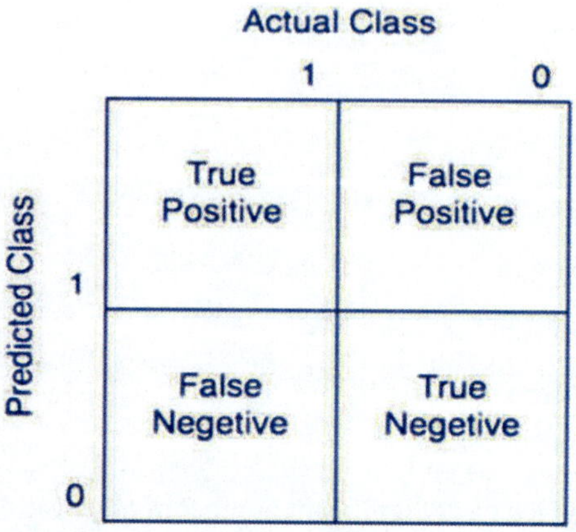

Fig. 10 Confusion matrix visualization

3.2.6 Classification Process

The confusion metrics also known as the error matrix as depicted in Fig. 10 was used to describe the performance of a classification model on a set of test data.

True positive (TP): An observation that is both anticipated and seen to be positive.

False positive (FP): An observation that was projected to be positive but turned out to be negative.

True negative (TN): An observation is anticipated to be negative but turns out to be negative.

False negative (FN): An observation is projected to be negative but turns out to be positive.

Accuracy gives the proportion of the total number of predictions that were correct and it is shown in Eq. (2) below:

$$\text{Accuracy} = \frac{\text{TP} + \text{TN}}{\text{TP} + \text{FP} + \text{TN} + \text{FN}} \tag{2}$$

Precision or the positive predictive value (PPV): is the fraction of positive values out of the total predicted positive instances. In other words, precision is the proportion of positive values that were correctly identified as shown in Eq. (3).

$$\text{Precision} = \frac{\text{TP}}{\text{TP} + \text{FP}} \tag{3}$$

The receiver operating characteristic (ROC) curve is a plot of sensitivity versus FP rate for different threshold levels. The area under the curve (AUC) is the area beneath this ROC curve, and it is used to assess the quality of a classification model. The greater the surface area, the higher the performance.

Sensitivity, recall, or the TP rate (TPR) as shown in Eq. (4), is the proportion of true positives out of all true positives (i.e., the proportion of true positives that are properly identified).

$$\text{Sensitivity} = \frac{\text{TP}}{\text{TP} + \text{FN}} \tag{4}$$

Specificity gives the fraction of negative values out of the total actual negative instances. In other words, it is the proportion of actual negative cases that are correctly identified as shown in Eq. (5). The FP rate is given by (1 − specificity).

$$\text{Specificity} = \frac{\text{TN}}{\text{TN} + \text{FP}} \tag{5}$$

4 Results and Discussion

The performance of the proposed technique was assessed using the MATLAB Classification learner app after the classifier was trained on the MATLAB 2021 version. The classifier architecture was built with a total of 220,320 training observations and a 20% validation split. The accuracy of the training dataset was 91.4%, whereas the accuracy of the validation dataset was 74.4%.

The three predicted classes in Fig. 11a, b are labeled "NORMAL," "BROKEN," and "RECOVERING," and each corresponds to a distinct condition of the pump for the training and tested models. The blue boxes reflect the observations that were successfully predicted, whereas the other boxes show the observations that were mistakenly predicted. When compared to the other classes in the training dataset, the RECOVERING class has the fewest observations. As a consequence, the RECOVERING class had the most wrong predictions.

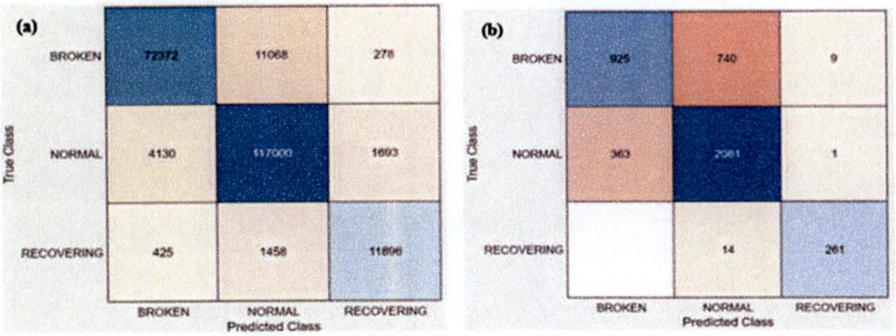

Fig. 11 Number of observation **a** trained and **b** tested model

Classification Learner was used to train the classifiers, and accuracy ratings were used to compare the models. Plotting class predictions helped to illustrate the findings, and the confusion matrix and ROC curve were used to assess performance. The measures for the third row from the top in Fig. 13 depict recovery with the genuine class. The blue cell in the TPR column represents the true positive rate for properly classified points in this class, which is 94.9%, and the columns display the projected classes recovering. 5.1% of the recovery was categorized erroneously in Fig. 12.

The false negative rate for incorrectly classified points in this class is 5.1%, shown in the orange cell in the FNR column. Figure 13 explains in more detail.

The PPV is the proportion of successfully classified data in each predicted class. The FDR counts the number of observations that are incorrectly assigned to each predicted class. When this decision has been taken, summary rows are now shown beneath the table in the confusion matrix. In Fig. 14, the false discovery rates for

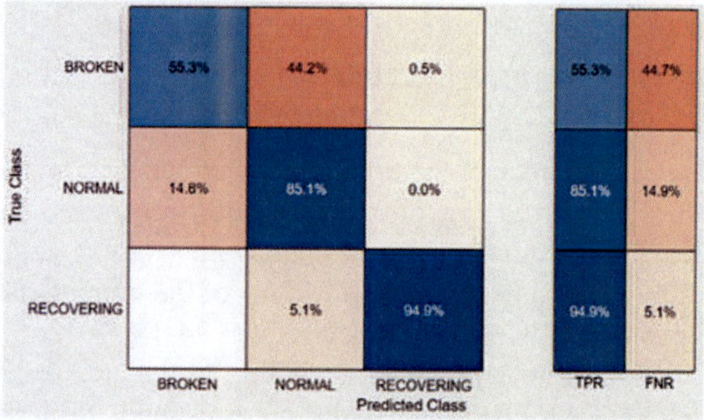

Fig. 12 Confusion matrix for true positive and false negative rates (TPR, FNR)

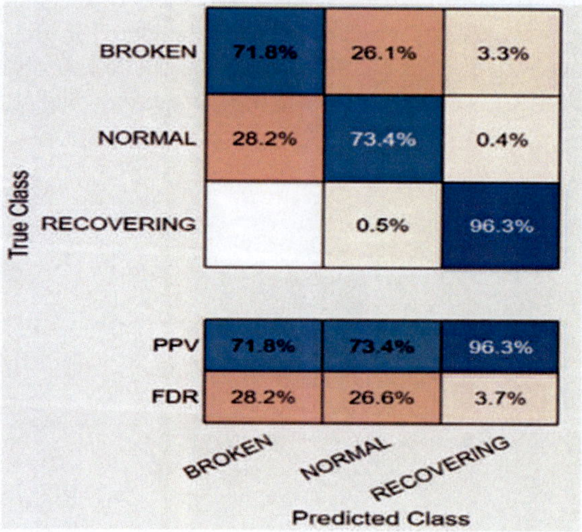

Fig. 13 Confusion matrix for true positive and false negative rates (PPV, FDR)

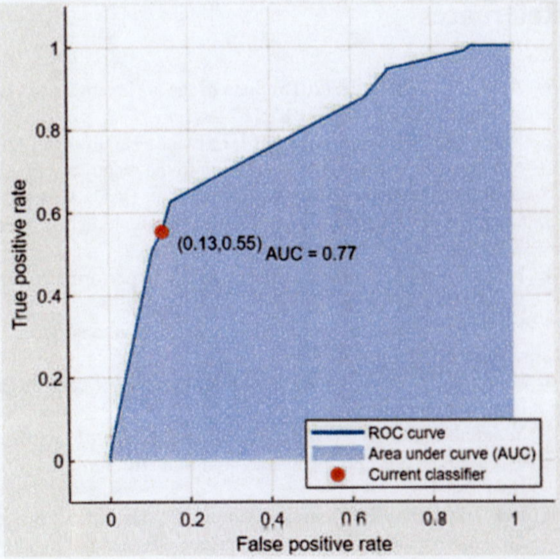

Fig. 14 The receiver operating characteristics (ROC) curve

incorrectly predicted points in each class are shown in orange, whereas positive predictive values for correctly predicted points in each class are shown in blue.

From Fig. 14, the ROC curve explains the sensitivity versus the FP rate for different threshold levels and it shows from the graph, the current classifier is at 0.13, 0.55. The area under the curve (AUC) is 0.77, and it is used to assess the quality of a classification model.

5 Conclusion and Future Work

A prediction of the acquired result was generated using the confusion matrix, true positive rate (TPR), false negative rate (FNR), positive predicted values (PPV), and false discovery value evaluations (FDV). According to the experimental results, the trained model's accuracy was found to be 91.94%; however, the tested model utilized for validation's accuracy was found to be 74.40%. The results obtained show a low percentage that has to be improved; hence, several algorithms will also be used in the future and compared in order to choose the best model for the research.

Acknowledgements The authors would like to thank the Centre of Graduate Studies (CGS) and Universiti Teknologi PETRONAS (UTP) for funding this research.

References

1. Alrabghi A, Tiwari A (2015) State of the art in simulation-based optimisation for maintenance systems. Comput Ind Eng 82:167–182
2. McKee KK, Forbes G, Mazhar I, Entwistle R, Howard I (2011) A review of major centrifugal pump failure modes with application to the water supply and sewerage industries. In: ICOMS asset management conference, Gold Coast, QLD, Australia, pp 1–12
3. Chen Y, Yuan J, Luo Y, Zhang W (2021) Fault prediction of centrifugal pump based on improved KNN. Shock Vib 2021
4. Kulkarni SJ (2017) Factors affecting pump performance-an insight into research and investigation. Int J Res Rev 4(8):39
5. Bachus L, Custodio A (2003) News product and services book review know and understand centrifugal pumps, p 2003
6. Harihara PP (2007) Sensorless fault diagnosis of centrifugal pumps [online]. https://oaktrust.library.tamu.edu/handle/1969.1/157904
7. Yedidiah S (1996) Performance characteristics of centrifugal pumps bt - centrifugal pump user's guidebook: problems and solutions. In: Yedidiah S (ed) Springer, Boston, MA, US, pp 13–17
8. Ruiz-Sarmiento JR, Monroy J, Moreno FA, Galindo C, Bonelo JM, Gonzalez-Jimenez J (2020) A predictive model for the maintenance of industrial machinery in the context of industry 4.0. Eng Appl Artif Intell 87:103289
9. Vilarinho S, Lopes I, Oliveira JA (2017) Preventive maintenance decisions through maintenance optimization models: a case study. Procedia Manuf 11:1170–1177
10. Çinar ZM, Nuhu AA, Zeeshan Q, Korhan O, Asmael M, Safaei B (2020) Machine learning in predictive maintenance towards sustainable smart manufacturing in industry 4.0. Sustainability 12(19)
11. Andrés D, Salinas M (2022) Predictive maintenance of induction motors using deep learning anomaly detection using an autoencoder neural network and fault classification using a convolutional neural network. Degree Proj Technol
12. Harihara PP, Parlos AG (2012) Fault diagnosis of centrifugal pumps using motor electrical signals. Centrif Pumps
13. Jardine AKS, Lin D, Banjevic D (2006) A review on machinery diagnostics and prognostics implementing condition-based maintenance. Mech Syst Signal Process 20(7):1483–1510
14. Kamal B, Sherif El-Gizawy HA, Supervisor T (2016) Condition-based monitoring system for diagnostics and prognostics of centrifugal pumps condition-based monitoring system for diagnostics and prognostics of centrifugal pumps

Shape Memory Polymers in Textile Applications—State of the Art and Future Trends

Nabihah Sallih, Nurul Hidayah Abdullah, Rosniza Hanim Abdul Rahim, Fadzliana Ahmad, M. Shahir Misnan, Leong Yin Liong, Tang Tong Boon, and Sum Wei Siang

Abstract Shape memory polymers (SMPs) have garnered widespread attention as tailorable, made-for-purpose smart reactive materials with the capability to maintain a temporary shape, and ability to return to their original shape when subjected to external stimuli. This article presents an overview on the landscape of SMPs based on the analysis on market survey, benchmarking activity, and records of intellectual property (IP). Market analysis for global SMPs was conducted by Mordor Intelligence for conjecture time frame of 2022–2030. The records on IP for SMPs were obtained from Derwent Innovation, while the data for commercially available SMP products were obtained based on published information. Market analysis revealed that SMP has about USD 1.6 billion market size in 2030 at compound annual growth rate (CAGR) of 17.4%, with healthcare dominates the SMP-based market at 40–50% market share, followed by electronics and aerospace at 17%, and textile at 6%. This finding is coherent to the outcomes of IP analysis where medical/healthcare-related segment has the highest IP records (52% of total IPs). A selected overview on SMP properties such as mechanical performance, shape fixity, and recovery when subjected to thermal stimuli is also presented in this paper. Challenges and prospects for SMPs are also outlined. This review could provide insights on designing and fabrication of SMP products such that they are in line with market trend, demands, and requirements.

Keywords Shape memory polymer · Market survey · Benchmarking

N. Sallih (✉) · N. H. Abdullah · R. H. A. Rahim · T. T. Boon
Institute of Health and Analytics, Universiti Teknologi PETRONAS, 32610 Seri Iskandar, Perak, Malaysia
e-mail: nabihah.sallih@utp.edu.my

F. Ahmad · M. S. Misnan · L. Y. Liong · S. W. Siang
Petronas Research Sdn. Bhd, Jln Ayer Hitam, Kawasan Institusi Bangi, 43000 Bandar Baru Bangi, Selangor, Malaysia
e-mail: fadzliana_a@petronas.com

F. Ahmad et al. (eds.), *ICREEM 2022*, Lecture Notes in Mechanical Engineering,
https://doi.org/10.1007/978-981-99-5946-4_27

1 Introduction

SMPs are renowned for their unique ability to undergo a substantial shape deformation and subsequently recover their original permanent shape when exposed to external stimuli such as heat, chemical, pH, electrical current/voltage, mechanical stress, gamma radiation, electromagnetic radiation (visible light, UV, IR, NIR, etc.), microwave radiation and magnetic field [1, 2]. They possess outstanding properties such as high recoverable strains (up to 400%), excellent molding ability, lightweight, tailorable properties, easy to be processed, low production cost, large recovery capacity, flexibility in selection of stimulating methods, biodegradability, and biocompatibility [2, 3]. SMPs have been reported to be applied in various applications such as biomedical engineering, aerospace, robotics, automotive industries, and smart textiles [4–9]. Figure 1 shows potential applications of SMPs in various sectors.

1.1 Landscape of Shape Memory Polymers Based on Market Survey, Analysis on Intellectual Property Records, and Published Literature

A market forecast analysis produced in partnership with Mordor Intelligence on the global SMPs for year 2022–2030 estimated that SMPs' market size to be USD 1.6 billion in 2030 at CAGR of 17.4% for aerospace (264.87 million), automotive (31.59 million), textile (87.61 million), electronics (275.03 million), healthcare (833.93

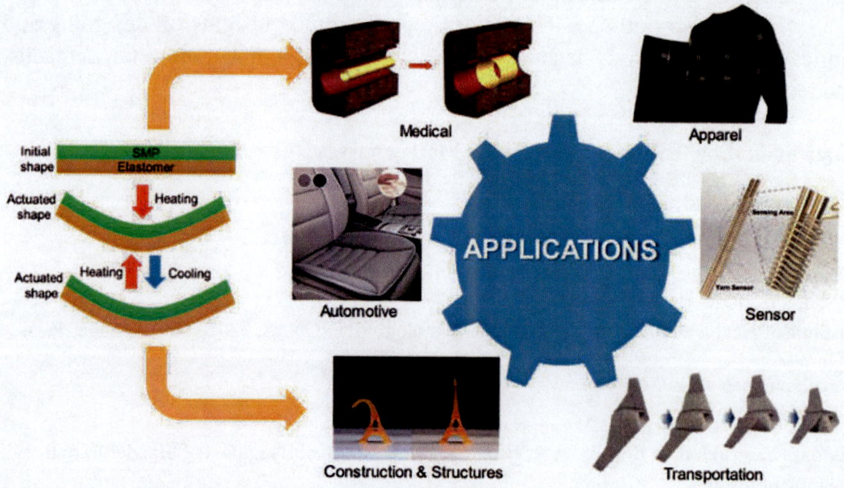

Fig. 1 Various potential applications of shape memory polymers

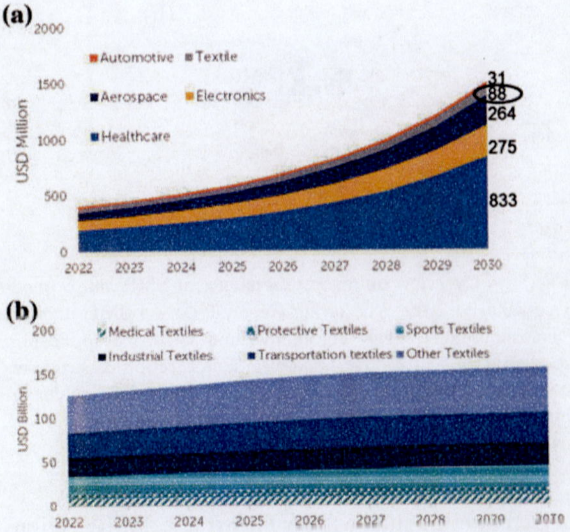

Fig. 2 A Revenue by end-user industry specifically for SMP and **b** technical textile global market size

million), and other end-user industries (106.82 million) (see Fig. 2). Although healthcare industry contributes the most to SMP market share, the textile industry has potential adaption in various applications. Healthcare dominates the SMP-based market at 52% market share, followed by electronics at 17%, transportation (aerospace and automotive) at 19%, and textile at 6%.

The orthopedic sector takes up the largest portion followed by implantable such as stents and catheters. minimally invasive surgical devices, drug delivery, bone repairs, vascular grafts, wound dressing, cell regulation, biodegradable sutures, actuators [4, 10–12]. Asia Pacific is expected to be the fastest growing region in the healthcare industry, as well as China and India have a significant position in the global medical market devices, fully supported by their respective government policies. Although many research and development projects on SMP have been focused on medical applications, their limited commercialization is due to limited suppliers of medical grade materials and precision of SMP. In addition, rigorous laws and standards that are needed to be passed complicate the adoption of SMP-based products. Besides medical/healthcare applications, the second and third largest industries for SMP adaption, with applications varying from the body of products, sensors, and actuators. Similar to healthcare industry, these two industries may also require rigorous standards and laws to be adhered to and they may take a longer time for commercialization purposes. At a CAGR of 12.7%, textile is a widely encompassing industry that can tap into various industries and thus holds a larger application range and market size. In addition, it is relatively an easier market to commercialize SMP-based products, and thus present itself as a promising lucrative sector to be ventured in.

Figure 3 shows the distribution of IP records on SMP extracted from Derwent Innovation. A total of 4150 IPs filed from 1968 till the end of July 2022 related to

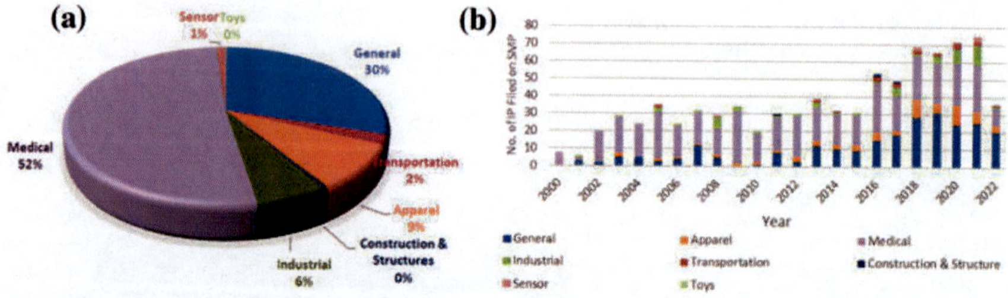

Fig. 3 A Overview on market segments for SMP, and **b** trending on applications of SMPs from year 2000 to 2022. The term "General" covers SMPs that are applicable to various segments, "Transportation" covers automotive and aerospace applications, "Apparels" includes garments, footwears, and protective apparels, "Construction & structures" includes building materials for construction and structural applications, "Industrial" includes repair patch, venting device, mold, hinges, soft robots, while "Medical" refers to any medical applications for internal and external uses

search phrases such as "SMP Textile", "SMP, Nylon", "SMP, Cellulose, Textile", "SMP, PLA", "SMP, Lignin", and "SMP, Alloy, Textile", were extracted. Out of 1270 IPs screened, only records that are relevant are further analyzed and reported in Fig. 3. Based on the IP records from 2000 to 2022, the top market segments are medical (52%), apparel (9%), and industrial applications (6%). The findings by IP analysis appear to be coherent to the outcomes of market survey. It is also noteworthy that SMPs developed for apparel and industrial applications are growing noticeably (Fig. 3(b)).

Specifically for smart textiles, the reported SMP products include thermo-regulation cloth [13], finishing agent [14], wearable garment and breathable outdoor/sportswear [15], color changing garment [15], bandages [15], hair styling agent [15], foot wears [15], and puncture resistance cloth [16]. Various fiber spinning methods, such as electrospinning [6], wet spinning [17], air-jet spinning [18], and centrifugal spinning [8], and yarn-processing methods, such as twisting [19] and core spun [20], have been used to manufacture such functional garments. In addition, textile processing methods such as knitting [13], weaving [21], braiding [22], and needle punching [23] were also employed.

1.2 Overview on Performance of SMP Based on Commercial Data and Published Literature

Table 1 shows the summary of the properties of selected SMP yarns, fibers, and filaments. The common base polymer used for SMP is polyurethane. The glass transition temperature is reported to be in the range of 20–65 °C. The tensile strength of SMP yarn can reach up to 45 MPa, and the strength of SMP filament is up to 11 MPa. The shape fixity of SMP has been reported to range from 68 to 97.5%. The shape recovery of SMP shows promising value, which is more than 85%.

Table 1 Properties of selected SMPs in the form of yarn, fiber, and filament

Form	Base polymer	Properties					Ref
		Stimuli	Glass transition temperature (°C)	Tensile strength (MPa)	Shape fixity (%)	Shape recovery (%)	
Fiber	Shape memory polyurethane	Thermal	20–38	-	68–74	90–94	[24]
Yarn	(Ring spinning core yarn)	Thermal	-	-	97.5	86	[20]
Yarn	(Friction spinning core yarn)	Thermal	-	-	94.5	87	[20]
Yarn	Shape memory polyurethane (Random electrospun yarn)	Thermal	65	26	97.5	-	[19]
Yarn	Shape memory polyurethane (Aligned electrospun yarn)	Thermal	65	45	97.0	-	[19]
Filament	PHA (poly(1,6-hxandiol adipate), MDI (4,4'-diphnylmethane diisocyanate) and BDO (1,4-butanediol)	Thermal	-	1.84	-	-	[25]
Filament	Thermoplastic polyurethane (Commercial SMP)	Thermal	55	11	-	-	[26]
Filament	Thermoplastic polyurethane (Commercial SMP)	Thermal	55	-	-	-	[27]

The development of SMPs has been mostly focused on material formulation/composition, but they lack adoption and translation into real applications that can be widely commercialized. The development of SMP materials with improvement of mechanical and thermal stability still need to be explored, and a better understanding on their physico-chemical and mechanical properties may lead to efficient, low cost and sustainable approaches to produce value-added products in industrial and engineering sectors. In addition, the know-how on efficient control mechanism and development of stimuli-triggering source are essential to accelerate the adoption of SMP materials into commercialized products.

2 Conclusions

This paper highlights the landscape of shape memory polymers from the perspective of market trending, records of intellectual property filed, and published data. In addition, the performance of commercially available SMPs and published data are presented to understand their thermal and mechanical properties. Potential areas for further research and development are also highlighted so that SMPs developed in the laboratory can be well adopted into commercialized products. The integration of SMPs in several market segments, including medical, industrial, construction and structures, apparel, transportation, and sensor may herald the dawn of a new era for smart materials with exceptional properties and remarkable potential. Innovative developments in fundamental research and industrial applications of smart materials have put forward the necessary steps for improving its mass and low-cost manufacturing. However, engineers and researchers working on shape memory technology should prioritize deep research on shape memory materials to utilize its potential in various areas.

References

1. Goda I, Zubair Z, L'Hostis G, Drean JY (2021) Design and characterization of 3D multilayer woven reinforcements shape memory polymer composites. J Compos Mater 55:653–673
2. Melly SK, Liu L, Liu Y, Leng J (2020) Active composites based on shape memory polymers: overview, fabrication methods, applications, and future prospects. J Mater Sci 55:10975–11051
3. Kim ML, Otal EH, Takizawa J, Sinatra NR, Dobson K, Kimura M (2022) All-organic electroactive shape-changing knitted textiles using thermoprogrammed shape-memory fibers spun by 3D printing. ACS Appl Polym Mater 4:2355–2364
4. Zare M, Davoodi P, Ramakrishna S (2021) Electrospun shape memory polymer micro-/nanofibers and tailoring their roles for biomedical applications. Nanomaterials 11:933
5. Kumar Patel K, Purohit R (2018) Future prospects of shape memory polymer nano-composite and epoxy based shape memory polymer- a review. Mater. Today Proc 5:20193–20200
6. Xiao YQ, Kan CW (2022) Review on development and application of 3D-Printing technology in textile and fashion design. Coatings 12:267
7. Chaudhary K, Kandasubramanian B (2022) Self-healing nanofibers for engineering applications. Ind Eng Chem Res 61:3789–3816
8. Barhoum A, Pal K, Rahier H, Uludag H, Kim IS, Bechelany M (2019) Nanofibers as new-generation materials: from spinning and nano-spinning fabrication techniques to emerging applications. Appl Mater Today 17:1–35
9. Chakraborty JN, Dhaka PK, Sethi AV, Arif M (2017) Technology and application of shape memory polymers in textiles. Res J Text Appar 21:86–100
10. Liu C, Qin H, Mather PT (2007) Review of progress in shape-memory polymers. J Mater Chem 17:1543–1558
11. Niyonzima I, Jiao Y, Fish J (2019) Modeling and simulation of nonlinear electro-thermo-mechanical continua with application to shape memory polymeric medical devices. Comput Methods Appl Mech Eng 350:511–534
12. Yeazel TR, Becker ML (2020) Advancing toward 3D printing of bioresorbable shape memory polymer stents. Biomacromol 21:3957–3965
13. Iqbal MI, Sun F, Fei B, Xia Q, Wang X, Hu J (2021) Knit architecture for water-actuating woolen knitwear and its personalized thermal management. ACS Appl Mater Interfaces 13:6298–6308

14. Korkmaz Memis N, Kaplan S (2020) Dual responsive wool fabric by cellulose nanowhisker reinforced shape memory polyurethane. J Appl Polym Sci 137:48674
15. Biswas MC, Chakraborty S, Bhattacharjee A, Mohammed Z (2021) 4D printing of shape memory materials for textiles: Mechanism, mathematical modeling, and challenges. Adv Funct Mater 31:2100257
16. Lin JY, Lin MC, Shiu BC, Lou CW, Lin JH, Chen YS (2022) Novel composite planks made of shape memory polyurethane foaming material with two-step foaming process. Polymers (Basel) 14:275
17. Ozipek B, Karakas H (2014) Wet spinning of synthetic polymer fibers. Woodhead Publishing Limited, pp 174–186
18. Tyagi GK (2010) Yarn structure and properties from different spinning techniques, in Advances in Yarn Spinning Technology, Woodhead Publishing Limited, pp 119–154
19. Shi Y, Chen H, Guan X (2021) High shape memory properties and high strength of shape memory polyurethane nanofiber-based yarn and coil. Polym Test 101:107277
20. Jing L, Hu J (2010) Study on the properties of core spun yarn and fabrics of shape memory polyurethane. Fibres Text East Eur 81:39–42
21. Grellmann H et al (2022) Fundamentals and working mechanisms of artificial muscles with textile application in the loop. Smart Mater Struct 31:023001
22. Holschuh B, Obropta E, Buechley L, Newman D (2012) Materials and textile architecture analyses for mechanical counter pressure space suits using active materials. AIAA Sp Conf Expo 2012:1–17
23. Mao N (2014) High performance textiles for protective clothing, in High Performance Textiles and Their Applications, Woodhead Publishing Limited, pp 91–143
24. Aslan S, Kaplan S (2018) Thermomechanical and shape memory performances of thermo-sensitive polyurethane fibers. Fibers Polym 19:272–280
25. Gupta P, Mohanty J, Garg H, Kumar B (2022) Memory behaviour of polyester knitted fabric integrated with temperature-responsive shape memory polymer filament. J Ind Text 51:5952–5972
26. Convena polymers. https://www.convena-polymers.de/
27. SMP Technologies Inc. http://www2.smptechno.com/en/tech/

Prescriptive Analytics for Dynamic Risk-Based Naval Vessel Maintenance Decision-Making

Mat Esa Mohd Adha, Muhammad Masdi, and Husin Hilmi

Abstract Prescriptive analytics is a novel maintenance approach by recommending the maintainer, the right maintenance action at the right time through prioritization of maintenance based on the dynamic risk assessment. It assists in improving the existing preventive maintenance approach by offering a dynamic maintenance plan based on risk or risk-based maintenance scheduling. To date, there is no clear definition of risk-based maintenance as well as its framework, and its application has not been applied extensively in naval domain, which is peculiar due to its function-system relationship compared to the plant maintenance. This paper applies risk-based maintenance decision-making framework in naval domain using prescriptive analytics for optimal maintenance scheduling to improve asset availability and reduce maintenance costs while meeting stakeholders' expectations. Objectives and considerations in the case study are also defined for peculiarities application of naval vessels. The novel dynamic risk-based maintenance methodology also aims to improve the effectiveness of the current preventive maintenance practice and the inconsistency issue of the existing maintenance decision-making model using the expert judgment in reliability-centered maintenance.

Keywords Prescriptive analytics · Risk-based maintenance scheduling · Dynamic risk assessment · Function-System hierarchical structure · Naval vessels maintenance

1 Introduction

Prescriptive analytics using risk-based maintenance (RBM) methodology is a novel maintenance decision-making framework using the combination of qualitative and quantitative approaches of the RBM [4]. This framework can be used to make decisions in selecting and scheduling of the right maintenance based on the dynamic

M. E. Mohd Adha (✉) · M. Masdi · H. Hilmi
Mechanical Engineering Department, Universiti Teknologi Petronas, Perak, Malaysia
e-mail: mohd_17005329@utp.edu.my

F. Ahmad et al. (eds.), *ICREEM 2022*, Lecture Notes in Mechanical Engineering,
https://doi.org/10.1007/978-981-99-5946-4_28

risk assessment of a system or equipment, using Condition Based Maintenance Assessment (CBMA). Normally, rotating machinery can be prognostically [6] operated in acceptable wear and tear conditions without causing its main component to deteriorate.

The elements of reliability, availability and maintainability play a crucial role in optimizing the operations of a naval vessel due to its specific issues [2]. The current operational condition needs to be continuously monitored to detect any symptom of deterioration in the early stage to avoid catastrophic event from occurring, which could damage the main components or even the whole naval vessel. All maintenance procedures have so far tried to compensate for the differences between the theoretical solutions offered by the maintenance strategies and the practical implementation onboard ships. Difficulties were encountered due to the complexity of the onboard equipment and systems on several levels: data collection and large volume of maintenance information, implementation of correct maintenance procedures, monitoring of operating parameters and maintenance, etc. [8].

The implementation of RBM will save the maintenance cost [2] as it requires maintenance to be carried out only when it is required based on the risk assessment. Further CBM assessment to diagnose a faulty condition is based on detected anomalies (deterioration symptom) in the early stage and the prognosis of its degradation path to predict the Remaining Useful Life (RUL) of a system or equipment [10]. It also identifies which part needs to be repaired/replaced. Its application in the naval domain is yet to be a significant area of research and its peculiar domain characteristic [2] contributes to the novel approach as the risk assessment for naval vessels will rank the criticality of onboard system or equipment based on the Failure Mode type and Float-Move-Fight function (Fig. 1) compared to the qualitative approach to seek a hierarchy process based on soliciting expert knowledge and opinions [4] in the RCM. However, it is not intended to replace conventional maintenance [5]. The effectiveness of conventional maintenance will be improved by optimizing maintenance schedule (planning the right maintenance at the right time or Just in Time (JIT) concept of maintenance) and at the same time minimizing Corrective Maintenance which is costly. The planning of the conventional maintenance with a fixed schedule based on calendar time or running hours will now become dynamic, based on the assessed risk (prioritization of maintenance tasks initiated by the system or equipment criticality) and the availability of the required resources (manpower, money and time). Thus, it will result in maintenance optimization, and it will save the operational cost.

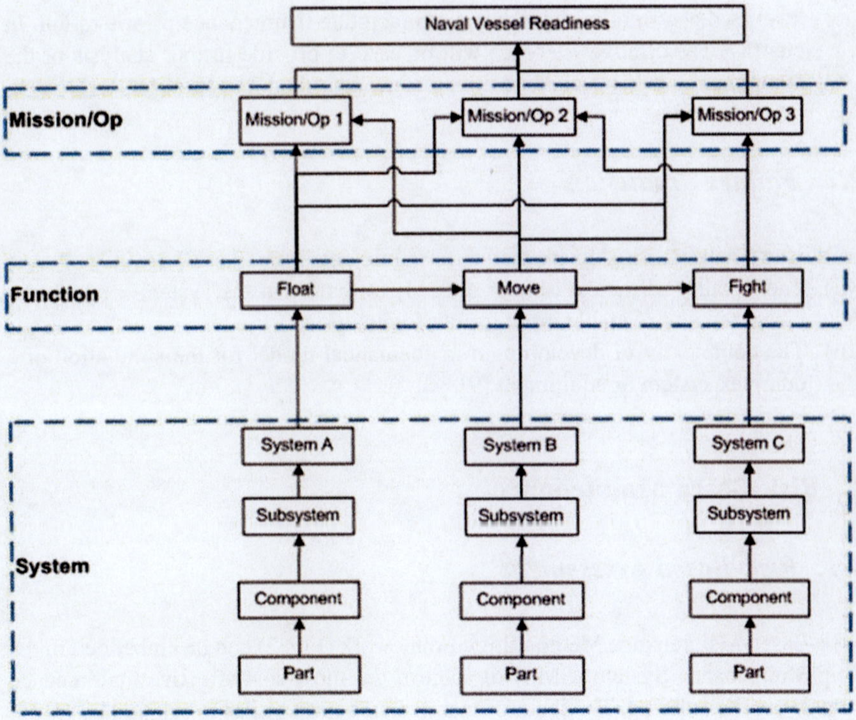

Fig. 1 Function-system hierarchy tree

2 Constraints and Issues

2.1 Risk Analysis

Several constraints and issues have been identified in the methodology of risk assessment as follows [2]:

(i) Conventional risk analysis using Failure Mode Effect and Effects Analysis (FMEA) is a one-off assessment, and it is conducted by experts. It is a tedious exercise and needs to be revised from time to time, as risks are dynamic.
(ii) Inconsistency/subjectivity issue resulted from dependence on expert judgment, whether caused by qualitative and quantitative analysis of risk assessment.
(iii) Unavailability/insufficient component's historical repair and failure data, e.g., running hours, type of defect, defect date, repair time (downtime), spare parts used and cost of repair. The data are required to determine the risk level of each component's failure modes in risk analysis in order to decide the maintenance strategy, as well as, to rank component criticality for maintenance task prioritization and scheduling. Thus, having considered the limitation of data availability, the repair cost data of the system can be used to represent the risk

level of the system in risk analysis to facilitate maintenance prioritization. In contrast, prescriptive analytics will be used to provide further analysis of the component levels to schedule maintenance based on its condition.

2.2 *Failure Prediction*

(i) Unavailability/insufficient historical failure data to predict RUL [7].
(ii) Poor quality of the csv sensor data [3] format from the system server to be converted into readable input format for analysis.
(iii) The complexity of developing a mathematical model for the simulation of a complex system or equipment [9].

3 Risk-Based Maintenance

3.1 *Risk-Based Assessment*

Risk-Based Maintenance Methodology Framework (Fig. 2) can be embedded in the Ship Maintenance System (SMS) to achieve the most cost-effective maintenance strategy and scheduling. The focus of maintenance strategy has to suit and be appropriate for the pre-selected ranked system or equipment by conducting a Risk-Based Assessment to the Function-System Hierarchy Tree, as indicated in Fig. 1, using the following criteria:

(i) Criticality: Based on the relation of Float-Move-Fight function to system, the subsystem (equipment), component defect or failure mode types:
   a. Urgent defect: The failure mode that affects the naval vessel's readiness and safety. It is categorized by system Category of Readiness (CAT) based on 1 to 4 and will be the first ranking process in risk analysis. A CAT 1 system is 100% operational whereby 4 is 0% (non-operational). Thus, a risk assessment will start with this failure-type ranking. CAT 4 has the highest risk score whereby for each CAT, the risk ranking will be based on the sequence of float, move and fight functions of its system/equipment.
   b. Normal defect: Failure mode that is minor and will not affect the naval vessel's readiness as the system/equipment can still be operational. It is not a complete breakdown and only requires minor/not urgent repair.

(ii) Repair Cost: The second-ranking process of risk analysis is by evaluating the repair cost of a normal defect as it already represents the combination of several influencing factors without the need for a multiple factors evaluation process using weightage [11]. Thus, it reduces the complexity of the analysis. The higher

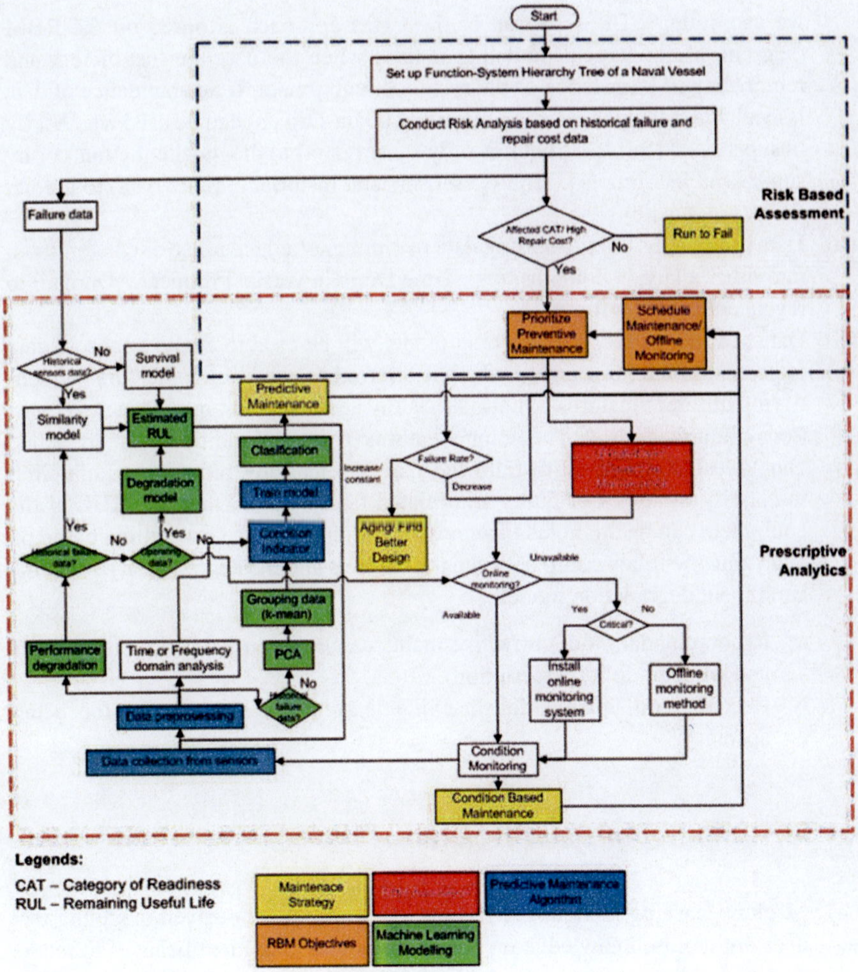

Fig. 2 Risk-based maintenance methodology framework

repair cost will reflect the complexity of the repair job, longer downtime/man-hour, etc. The system/equipment defect of each ranked function will be ranked based on the repair cost.

3.2 *Prescriptive Analytics*

The prescriptive analytics will be used in further CBM assessment as follows:

(i) Data Acquisition: Selection of the model approach for prescriptive analytics will depend on the online (sensor data) or offline (CBM Techniques) data that we can collect. The selection of the CBM approach is based on the RBM diagram. Mathematical modeling is used when the data are insufficient and require expert knowledge of the system or equipment. If an abundance of data is available, Machine learning modeling (Data-Driven) can be used whether by unsupervised learning (requires only sensor data) to diagnostic the fault or by supervised learning [1] (requires sensors and historical failure data) to predict RUL (prognostic).

(ii) Data processing: Preprocessing data to eliminate/reduce noise or convert them into other analysis domains, e.g., Time Domain versus Frequency Domain to reveal additional information.

(iii) Data analyzing: Machine learning model will be used to analyze sensors data to get additional information which is referred as condition indicators by identifying different features of normal/healthy and different types of faulty conditions (fault diagnosis). The unique features of faulty conditions that are called anomalies will be used to train the machine learning model, e.g., classifier to classify the types of faulty and related faulty parts. Later, the RUL of the equipment can be predicted (prognostic) for maintenance scheduling based on analyzing the relationship between the features and the degradation path of the equipment degradation model.

   a. Recommendation of corrective maintenance action for faulty components and when to do it (prescription).
   b. Records and updates for machine learning training model for better accuracy.

3.3 Prescriptive Analytics Techniques

(i) Machine learning modeling: Machine learning is a data-driven modeling that does not require knowledge on how a system or equipment behaves to model it. This is because its modeling techniques are purely based on data. Risk-based assessment can be made continuously and predicted in advance for early anticipation, However, having considered the limitation of data availability, repair cost data can be used to represent the risk level of the system in risk analysis for maintenance prioritization. In contrast, prescriptive analytics will be used for further analysis of the component's level to schedule maintenance based on its condition. Machine learning can improve the accuracy of the prediction by training the model using a large amount of dataset. It is suitable for modeling a complex system with many variables, and engine reading sheets can be used as the dataset.

(ii) Principle Component Analysis: Unsupervised machine learning using PCA technique will group data (k-mean) in determining its underlying structure that can give an interpretation or additional insight into the problems of the system or equipment [3] just based on sensor data.

(iii) Condition-Based Maintenance Assessment: CBMA feeds machine learning training input to differentiate the faulty type [2], for example, unique vibration signatures for different faulty types in frequency-domain analysis, especially for any rotating equipment. There are two types of CBM monitoring system for CBMA:

a. CBM Online/Permanent Monitoring: Fix the installation of sensors for CBM online monitoring together with the CBMA system, which will automatically analyze any anomalies detected and diagnose the problem. However, expensive budget allocation for sensor installation and CBMA system needs buy-in from the top management, and it depends on the success of the proof of concept.
b. CBM Offline/Routine Monitoring: Portable data logger and analyzer (single or multi-sensors) for vibration analysis are used to collect and store offline data. The measurement data are then uploaded into the PC for scanning dan automated diagnosis purposes. Other offline/routine monitoring includes ferrography and spectroscopy in oil analysis (tribology), infrared photography in thermographic analysis, motor current analysis and high frequency and ultrasonic analysis using the ultrasonic application (plat thickness) and high frequency (early-stage bearing defect). Thus, it still requires working station facilities with the CBMA software and knowledge on how to conduct the measurement and analyze the data (diagnostic and prognostic) using the software. The CBM tools also need calibration and limited capability as some CBM tools will be the constraint. For example, a portable vibration analyzer type single sensor takes more time to measure vibration for a complex/big system or equipment, e.g., propulsion system from various location points if compared with a two channeled vibration analyzer for the cause and effect analysis to diagnose a problem.

4 Case Study

Based on the Risk-Based Assessment to the Function-System Hierarchy Tree of a naval vessel (Fig. 3) using the defect and repair cost data, the Risk Analysis (Table 1) shows that a propulsion system (marine engine diesel) has the first rank of the most critical equipment of that is risky to move function, which is the second basic operational function in Float-Move-Fight function ranking after the float function to ensure the availability and operation of a naval vessel. The propulsion system consists of the marine diesel engines, which transmit power through gearboxes and shafting to the propeller to propel a naval vessel.

There are five engines to be studied in this case study. All engines are of the same type, but each engine starts with a different degree of initial wear and variations in the operating conditions, which are unknown. There are three optional settings that can be used to change the performance of each engine. Each engine has 12 sensors,

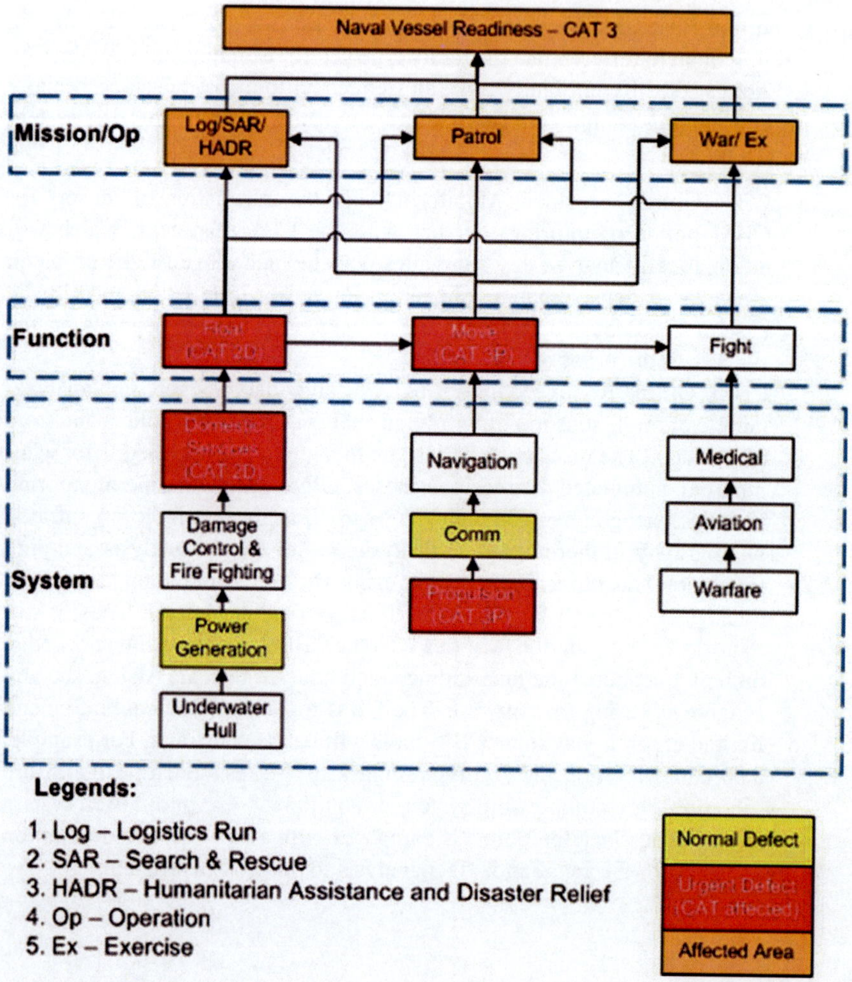

Fig. 3 Naval vessel function-system hierarchy tree

collecting different measurements related to the engine state at runtime. Over time, each engine develops a fault, which can be seen through sensor readings. The time series ends sometime before the failure. The data include the unit (engine) number, running hour, engine settings (RPM) and readings for 21 sensors. Table 2 illustrates the subset of the data.

Thus, sensor data are collected and analyzed using Predictive Maintenance Algorithm in prescriptive analytics to detect abnormalities and classify the potential type of faults as well as the kind of maintenance that is required and the parts that will be involved. By using the PCA technique, all 12 dimensions' datasets have been transformed into five principal components, which cover 86% variance. Figure 4 shows the scatter plot of the principal component that can distinguish the marine engine

Table 1 Risk analysis table

Readiness	Defect Type	Code	CAT	Function	Code	System	Code	Defect No	Repair Cost (RM)	Risk Score	Risk Ranking
Naval Vessel Readiness	Urgent Defect (URDEF)	U	4	Float	F	Underwater Hull	H			24	
						Power Generation	L			23	
						Damage Control & Fire Fighting	F			22	
						Domestic Services	D			21	
				Move	M	Propulsion System	P			20	
						Communication	C			19	
						Navigation	N			18	
			3	Float	F	Underwater Hull	H			17	
						Power Generation	L			16	
						Damage Control & Fire Fighting	F			15	
						Domestic Services	D			14	
				Move	M	Propulsion System	P	√ (1)		13	1
						Communication	C			12	
						Navigation	N			11	
			2	Float	F	Underwater Hull	H			10	
						Power Generation	L			9	
						Damage Control & Fire Fighting	F			8	
						Domestic Services	D	√ (1)		7	2
				Move	M	Propulsion System	P			6	
						Communication	C			5	
						Navigation	N			4	
				Fight	W	Warfare	W			3	
						Aviation	A			2	
						Medical	E			1	
	Normal Defect	N		Float	F	Underwater Hull	H				
						Power Generation	L	15	5,100.00		4
						Damage Control & Fire Fighting	F				
						Domestic Services	D	14	41,326.00		3
				Move	M	Propulsion System	P	25	4,482.00		5
						Communication	C	1	975.00		6
						Navigation	N				
				Fight	W	Warfare	W				
						Aviation	A				
						Medical	E				

Table 2 Marine engine sensors dataset.

	A	B	C	D	E	F	G	H	I	J	K	L	M	N	O	P
1	Engine	Classification	Hour	RPM	FRA	CWLSP	CWRSP	SWP	FP	CALSP	CARSP	LOP	PCOP	CWT	LOT	EGT
2	5	Finish W5	1	539	7.21	4	0.638	0.404	0.781	0.003	0.001	3.56	2.5	63.7	60.5	131
3	5	Finish W5	2	663	8.17	0.6	0.504	0.465	0.789	0.031	0.029	3.09	2.5	74.5	70.2	317
4	5	Finish W5	3	500	8.18	3.6	0.524	0.169	0.781	0.029	0.011	2.92	2.5	73.3	71.4	168

conditions between an engine that has just undergone 2nd W5 routine maintenance in blue dotted color and those due for the maintenance at 6000 running hours in the yellow dotted color by showing its condition grouping. It also shows the degradation trends as the grouping moves to the left bottom corner of the plot in tandem with the increase in the engine running hours. The results of several scoring methods in Table 3 provide five main sensor data (PCOP, LOT, FP, CWLSP and CWT) that are significant in distinguishing both engine degradation conditions that can be used in training the model.

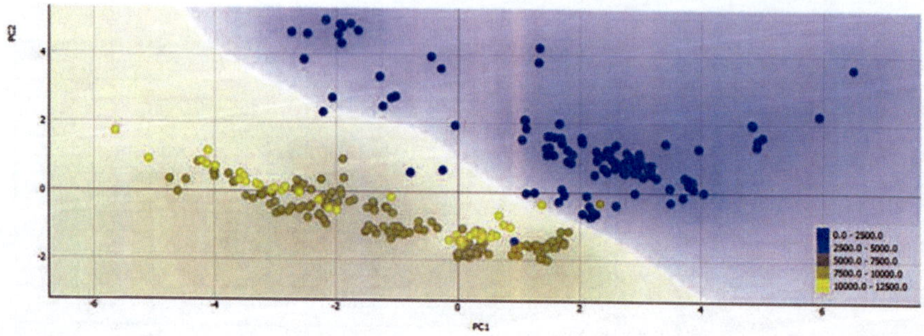

Fig. 4 Principal component analysis of engine dataset

Table 3 Scoring result

	#	Info. gain	Gain ratio	Gini	ANOVA	χ^2	ReliefF	FCBF
PCOP		0.694	0.353	0.380	293.172	189.933	0.250	0.910
LOT		0.675	0.338	0.337	483.433	182.732	0.139	0.844
FP		0.657	0.330	0.328	450.447	189.635	0.251	0.809
CWLSP		0.631	0.322	0.336	972.056	174.611	0.188	0.766
CWT		0.405	0.204	0.214	224.482	122.771	0.103	0.000
FRA		0.247	0.123	0.151	145.257	83.637	0.098	0.000
CARSP		0.228	0.114	0.142	10.721	51.107	0.024	0.000
CWRSP		0.120	0.061	0.079	18.766	25.984	0.012	0.000
SWP		0.119	0.060	0.075	4.126	37.917	0.033	0.000
CALSP		0.086	0.043	0.055	46.490	33.865	0.158	0.000
EGT		0.074	0.037	0.048	16.683	14.001	0.095	0.000
LOP		0.048	0.024	0.030	15.375	2.738	0.067	0.000
RPM		0.034	0.017	0.022	1.781	12.682	0.094	0.000

There are two models that have been used to classify the fault type of the engine based on sensor reading. The sensors data plot in Fig. 5 indicates the degradation of engine health over time. By comparing the later and earlier time series of the sensors' data in the plot, different condition indicators are shown in Fig. 6 between a marine engine, which has just undergone the 2nd W5 Maintenance Routine and engine as part of the maintenance routine (over 6000 running hours), which can give different condition indicators for the classification of the model training.

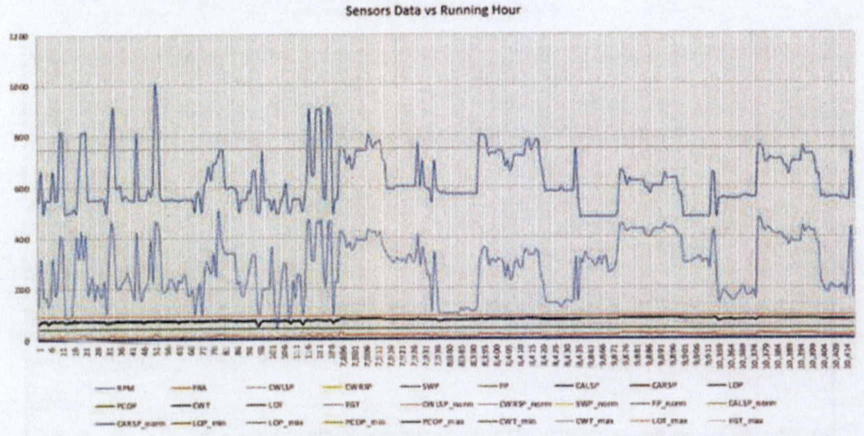

Fig. 5 Time domain analysis of sensor data

Having zoomed into the sensors data plot starting at 7,931 h in Fig. 7, an abnormal sensor data reading of EGT and CALSP from a marine engine has been identified as a faulty condition named unbalanced firing order has been used to train the Classification Tree, Neural Network and Logistic Regression models. The result in Table 4 shows that the Classification Tree model is highly accurate in predicting, with 97% accuracy. It is supported by the confusion matrix result, as shown in Fig. 8. The model only misclassified one predicted unbalance firing, which was supposed to be a normal condition. The model was tested using the testing dataset, and the comparison classification result of the three models is shown in Fig. 9.

The failure of the fuel injectors will be the root cause of the problem. Fuel injectors produce combustions at different rates and different combustion power at each chamber. Thus, it affects the unbalanced in the engine's firing, which will then cause lower exhaust gas temperature and higher charge air pressure. The maintenance action that needs to be scheduled for the pressure test of all the injectors and engine firing timing needs to be fine-tuned, especially at the affected engine speed.

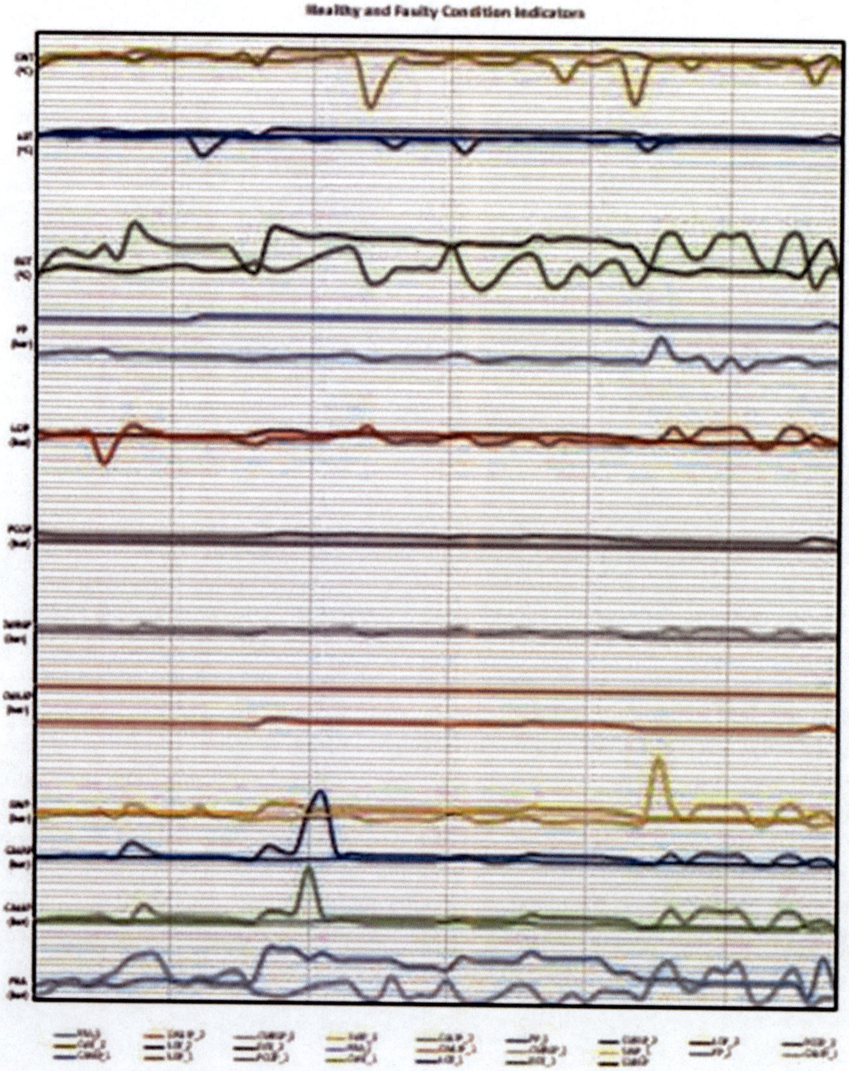

Fig. 6 Condition indicator of marine Engine (1) After 2nd W5 routine maintenance and (2) Marine engine due for 2nd W5 routine maintenance

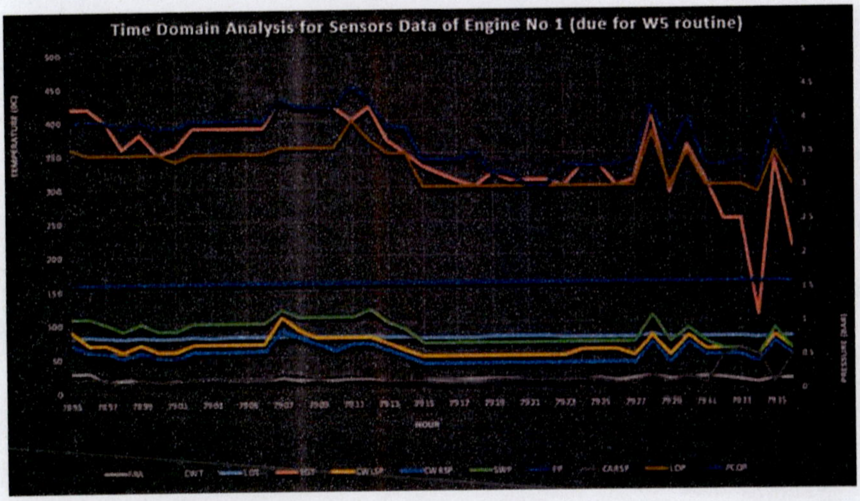

Fig. 7 Marine engine faulty condition

Actual \ Predicted	Normal	Unbalance firing	Σ
Normal	35	1	36
Unbalance firing	0	6	6
Σ	35	7	42

Fig. 8 Confusion matrix of classification tree model

	Tree	Logistic Regression	Neural Network	Hour	RPM	FRA	CWLSP	CWRSP	SWP	FP
1	1.00 : 0.00 → Normal	0.00 : 1.00 → Unbalance firing	0.71 : 0.29 → Normal	1.0	539.0	7.21	4.000	0.638	0.404	0.7810
2	1.00 : 0.00 → Normal	0.98 : 0.02 → Normal	1.00 : 0.00 → Normal	2.0	663.0	8.17	0.600	0.504	0.465	0.7890
3	1.00 : 0.00 → Normal	0.00 : 1.00 → Unbalance firing	0.66 : 0.34 → Normal	3.0	500.0	8.18	3.600	0.524	0.169	0.7810
4	1.00 : 0.00 → Normal	0.00 : 1.00 → Unbalance firing	1.00 : 0.00 → Normal	4.0	549.0	8.11	0.600	0.554	0.470	0.7810
5	1.00 : 0.00 → Normal	0.00 : 1.00 → Unbalance firing	0.70 : 0.30 → Normal	5.0	549.0	7.23	4.000	0.638	0.404	0.7810
6	1.00 : 0.00 → Normal	0.00 : 1.00 → Unbalance firing	1.00 : 0.00 → Normal	6.0	551.0	8.18	0.600	0.538	0.465	0.7820
7	1.00 : 0.00 → Normal	0.52 : 0.48 → Normal	1.00 : 0.00 → Normal	7.0	663.0	13.17	0.600	0.504	0.465	0.7810
8	1.00 : 0.00 → Normal	0.00 : 1.00 → Unbalance firing	1.00 : 0.00 → Normal	8.0	549.0	8.11	0.600	0.554	0.470	0.7810
9	1.00 : 0.00 → Normal	0.00 : 1.00 → Unbalance firing	1.00 : 0.00 → Normal	9.0	552.0	10.72	0.600	0.526	0.453	0.7810
10	1.00 : 0.00 → Normal	0.81 : 0.19 → Normal	0.96 : 0.04 → Normal	10.0	818.0	15.53	0.600	1.067	0.974	0.8210
11	1.00 : 0.00 → Normal	0.68 : 0.32 → Normal	0.96 : 0.04 → Normal	11.0	815.0	15.72	0.600	1.069	0.994	0.8210
12	1.00 : 0.00 → Normal	0.00 : 1.00 → Unbalance firing	0.10 : 0.90 → Unbalance firing	12.0	498.0	7.60	2.500	0.414	0.444	0.7400
13	1.00 : 0.00 → Normal	0.00 : 1.00 → Unbalance firing	0.05 : 0.95 → Unbalance firing	13.0	498.0	7.66	2.500	0.600	0.370	0.7330
14	1.00 : 0.00 → Normal	0.00 : 1.00 → Unbalance firing	0.09 : 0.91 → Unbalance firing	14.0	503.0	7.66	2.500	0.600	0.610	0.7330
15	1.00 : 0.00 → Normal	0.00 : 1.00 → Unbalance firing	0.04 : 0.96 → Unbalance firing	15.0	503.0	7.66	2.500	0.600	0.370	0.7330

Fig. 9 Prediction result of engine condition using testing dataset

Table 4 Cross-validation result of the three machine learning models

Model	AUC	CA	F1	Precision	Recall
Tree	0.986	0.976	0.977	0.980	0.976
Neural Network	0.991	0.952	0.952	0.952	0.952
Logistic Regression	0.755	0.929	0.926	0.925	0.929

5 Conclusion

The risk-based maintenance application framework is a novel maintenance approach that can be applied in the naval domain in the selection and scheduling of the correct maintenance. It is based on the dynamic risk assessment of a system or equipment using CBMA. It is novel due to it peculiar application in naval domain as the risk assessment will rank the criticality of onboard system or equipment based on failure mode types and Float-Move-Fight function to system relationship. This will improve the inconsistency issues of the conventional risk analysis using a risk matrix and expert judgement. The novelty is also featured when the methodology framework is coupled with prescriptive analytics using the machine learning modeling as a data processing engine for complex system or equipment onboard naval vessels whereby in this case study the marine engine. It is to be modeled mathematically and the availability of failure data. In this study, the unsupervised machine learning technique using PCA, together with time domain analysis, is used for prescriptive analytics to distinguish marine engine conditions and train the selected machine learning models in classifying the faulty conditions. The results have been compared between three selected models to establish the most accurate model for predicting the pre-selected faulty engine condition, which shows that the Classification Tree model has the best accuracy, 97% with only 1 being misclassified out of 42 predictions.

References

1. Cipollini F, Oneto L, Coraddu A, Murphy AJ, Anguita D (2018). Condition-Based maintenance of naval propulsion systems with supervised data analysis. https://doi.org/10.1016/j.oceaneng.2017.12.002
2. Cullum J, Binns J, Lonsdale M, Abbassi R, Garaniya V (2018) Risk-Based Maintenance Scheduling with application to naval vessels and ships. Ocean Eng 148:476–485. https://doi.org/10.1016/j.oceaneng.2017.11.044
3. Delhi N, Gugulothu N, Tv V, Malhotra P, Vig L, Agarwal P, Shro G (2017) Predicting remaining useful life using time series embeddings based on recurrent neural networks ∗. https://doi.org/10.1145/nnnnnnn.nnnnnnn
4. Kimera D, Nangolo FN (2020) Maintenance practices and parameters for marine mechanical systems: a review. J Qual Maint Eng 26(3):459–488. https://doi.org/10.1108/JQME-03-2019-0026
5. Lazakis I, Raptodimos Y, Varelas T (2017). Predicting ship machinery system condition through analytical reliability tools and artificial neural networks. https://doi.org/10.1016/j.oceaneng.2017.11.017
6. Muhammad MB, Sarwar U, Tahan M, Karim ZAA (2017) Intelligent fault diagnostic model for rotating machinery. 1858–1862
7. Sena Eruguz A, Tan T, van Houtum G-J (2017) A survey of maintenance and service logistics management: Classification and research agenda from a maritime sector perspective. Comput Oper Res 85:184–205. https://doi.org/10.1016/j.cor.2017.03.003
8. Simion D, Purcărea A, Cotorcea A, Nicolae F (2020) Maintenance onboard ships using computer maintenance management system. Sci Bull Nav Acad, 23(1), 134–141. https://doi.org/10.21279/1454-864X-20-I1-017

9. Tahan-bouria M, Muhammad M, Karim ZAA (2016). Adaptive Neuro-Fuzzy inference system for performance health monitoring of industrial gas turbines. 1365–1373
10. Tahan M, Tsoutsanis E, Muhammad M, Karim ZAA (2017) Performance-based health monitoring, diagnostics and prognostics for condition-based maintenance of gas turbines : A review. Appl Energy 198:122–144. https://doi.org/10.1016/j.apenergy.2017.04.048
11. Tang Y, Liu Q, Jing J, Yang Y, Zou Z (2017) A framework for identification of maintenance significant items in reliability centered maintenance. Energy 118:1295–1303. https://doi.org/10.1016/j.energy.2016.11.011